Swarms of Ions and Electrons in Gases

Edited by
W. Lindinger, T. D. Märk, and F. Howorka

Springer-Verlag Wien New York

Prof. Dr. Werner Lindinger
Prof. Dr. Tilmann D. Märk
Prof. Dr. Franz Howorka

Institut für Experimentalphysik,
Universität Innsbruck, Austria

Printed in Austria

With 82 Figures

Library of Congress Cataloging in Publication Data. Main entry under title: Swarms of ions and electrons in gases. Includes index. 1. Ions – Congresses. 2. Electrons – Congresses. 3. Gases – Congresses. 4. Collisions (Nuclear physics) – Congresses. I. Lindinger, W. (Werner), 1944–. II. Märk, T. D. (Tilmann D.), 1944–. III. Howorka, F. (Franz), 1937–. QC702.S93. 1984. 530.4'4. 84-10485.

ISBN 3-211-81823-5 Springer-Verlag Wien-New York
ISBN 0-387-81823-5 Springer-Verlag New York-Wien

Preface

Our understanding of elementary processes in plasmas has been increasing dramatically over the last few years. The development of various swarm techniques, such as the temperature variable selected ion flow tube or the selected ion flow drift tube, has provided the prerequisite for detailed investigations into ion molecule reactions both in binary and three body collisions, and the mechanisms of many reactions are now understood quite satisfactorily.

This information could not have been obtained without a detailed knowledge of the transport phenomena involved. Some of these, such as the internal-energy distribution of drifting ions, have only very recently been tackled both theoretically and experimentally; a consistent model is now being developed.

As the interactions between the various branches of swarm research have become more and more intense, the most obvious thing to do was putting together a review on the present state of this subject, which is the aim of this book.

Since about ten years we have been involved in swarm research ourselves. This was made possible through the generous funding of the Innsbruck-Forschungsschwerpunkt Plasmaphysik by the Austrian Science Foundation (Fonds zur Förderung der wissenschaftlichen Forschung), for which we would like to express our gratitude. In this context a cooperation has developed with outstanding scientists in the field, many of whom have contributed to this book. We also would like to thank our colleagues throughout the world for their continuous encouragement and support, which has been providing constant nourishment for swarm research at the Leopold Franzens Universität, Innsbruck.

Innsbruck, April 1984 **W. Lindinger, T. D. Märk,** and **F. Howorka**

Contents

Historical Development and Present State of Swarm Research

E. W. McDaniel

Georgia Institute of Technology, Atlanta, GA 30332, U.S.A.

This book addresses several areas of research in the field of atomic physics that may be investigated by swarm techniques. Most of the chapters assume that the reader is conversant with the topic, and the emphasis is on the latest results. However, in order to establish a broader perspective and a suitable background, this introduction provides some definitions, sketches some of the historical developments, and gives some general references that may be of interest to the reader.

Basic Considerations

We shall define an electron or ion swarm to be a large number of the particles in question moving about with a random velocity component in a gas of neutral molecules. The number density N of the charged particles may be large enough that space charge effects are important; in most of the situations covered in this book that is not the case. In the absence of an externally applied electric field, there is no net motion of the charged particles, and their velocity distribution is Maxwellian. If an external field $\vec{E}$ is present, the distribution is non-Maxwellian, and the swarm moves along the field lines. The average velocity of the charge carriers is called the drift velocity $\vec{v}_d$; it is constant if the gas is uniform, the field is constant and uniform, and steady-state conditions have been achieved. These conditions should be assumed to be fulfilled here unless there is an indication to the contrary.

Diffusion of the ions or electrons occur if there is a gradient in the relative concentration of the charge carriers. This diffusion is characterized by a scalar diffusion coefficient D in the limit $E/N \to 0$. When a field is present, diffusion occurs at different rates in directions parallel to $\vec{E}$ (longitudinal diffusion) and perpendicular to $\vec{E}$ (transverse diffusion). In this case, the diffusion coefficient is a tensor quantity

$$\overleftrightarrow{D} \equiv \begin{vmatrix} D_L & 0 & 0 \\ 0 & D_T & 0 \\ 0 & 0 & D_T \end{vmatrix} \qquad (1)$$

where D_L and D_T are the longitudinal and transverse diffusion coefficients, respectively. Measurements of steady-state drift velocities and diffusion coefficients are performed, for the most part, under conditions where space charge effects are negligible. An important exception occurs in certain afterglow experiments, where the diffusion is ambipolar. Another useful quantitity is the mobility of the charged particles, $K \equiv v_d/E$. We should note that K, v_d, D, D_L, and D_T are examples of "transport coefficients" because they are measures of the rate of transport of physical quantities through the gas.

Most measurements of ionic and electronic transport coefficients have been made with "drift tubes", and recently with "flow tubes" in the case of ions, by techniques described herein. These instruments can also be used to obtain reaction rate coefficients and sometimes cross sections for various kinds of atomic collision processes that have a direct bearing on drift and diffusion and that are of great relevance to all kinds of swarm phenomena. These processes include momentum transfer, ion-molecule interchange and association reactions, charge transfer, excitation and de-excitation of molecules, electron attachment and collisional detachment, photo- detachment of electrons from negative ions, and photodissociation of molecular ions. The energy dependence of these processes is obtained by varying the energy parameter E/N, which, along with the gas temperature T, determines the average energy of the charged particles in the steady state. The transport properties depend only on E/N, T, and the charge carrier-gas combination.

Great interest centers on these aspects of electron and ion swarm research. Additional interest is provided by applications of swarm data to the modeling of laboratory discharges, gas lasers, radiation detectors, lightning and other kinds of natural discharges, radio communications, flames, ionospheric phenomena, and planetary atmospheres. These applications will not be treated here, nor will the connection of swarm to plasma physics and plasma chemistry be systematically explored.

Some Historical Comments and General References

Recently a friend who is a Professor of English at the University of Strathocmulgee told me that he had come across the name of James Clerk Maxwell, read a biography of him, and been duly impressed. He then asked me if I though that Maxwell would be interested in the current research in my field were he to be brought back here to observe what is going on. My initial reaction was one of shock, during which I had the disturbing vision of the shade of Maxwell circulating around the various laboratories, kicking the tires, and demanding to know what all of us are up to. After recovering, I answered my friend in the affirmative, and I believe that my answer can be justified in two ways.

First, between about 1859 and 1866, Maxwell laid the foundations of much of the kinetic theory of gases, which is the main theoretical basis of all transport phenomena in gases. He also pointed out the unique significance of the $1/r^4$ potential in his collision integrals, thus foreshadowing the role to be played by the "polarization potential" decades later. Finally, he personally made measurements of the viscosity of gases, demonstrating his administrative ability by inducing his wife to serve as stoker during the runs at high temperature. At this time (30 or more years before the discovery of X-rays, radioactivity, the electron, and the nature of gaseous ions) experiments on ion and electron swarms were far in the future. However, it is hard to resist the temptation to believe that

Maxwell would have been one of the first researchers in our field had the requisite tools become available at the appropriate time.

The second reason for Maxwell's putative interest is provided by the contents of this book, which convey a sense of the breadth, vitality, and sophistication of the research now being performed. A description of the book and its place in the literature will be given in the next section. First, we shall enumerate a few general references in chronological order and gain some historical perspective while discussing them.

We shall confine ourselves to books published after World War II and start with the encyclopedic book by Loeb (1), which gives an exhaustive account of ion and electron swarm research from its beginnings immediately after the discovery of X-rays in 1895 up until the mid- 1950's. Among the earliest workers were Rutherford, Thomson, Townsend, Tizard, and of course Langevin, whose 1905 landmark paper contained a calculation of the zero-field diffusion coefficient and mobility of gaseous ions. (Were it not for the fear of being accused of gerrymandering, we would include Einstein in this hagiology. After all, we constantly use the "Einstein relation" between the mobility and diffusion coefficient of charge carriers in the limit $E/N \rightarrow 0$ that Einstein derived in 1905. It had been obtained earlier by Nernst (1888) and Townsend (1899) in different contexts). This calculation was based on Maxwell's momentum transfer method and the assumption of a polarization attraction potential to describe the ion-neutral interaction. Results were obtained with and without an additional hard-core repulsion. Langevin's contribution was truly remarkable--it had no rival until the introduction of the Chapman-Enskog method described below, and Langevin's polarization limit of the mobility still plays an important role today. The next "generation" of swarm researchers discussed by Loeb included Chapman, Enskog, Tyndall, Powell, Druyvesteyn, Penning, Huxley, Brandbury, Nielsen, Allis, and Morse. Their work extended from about the middle of World War I on past World War II. The Chapman-Enskog method (circa 1917) in the kinetic theory of gases has been of great general interest and utility. In our field, it was especially important in permitting an extension of Langevin's results for the $1/r^4$ potential to an arbitrary (but still spherically symmetric) potential. Unfortunately the Chapman-Enskog method is not well suited for extension of the transport calculations upward out of the region of very low E/N. The first major improvement along these lines came in the early 1950's when Wannier used the Boltzmann transport equation (published in 1872) to investigate the motion of gaseous ions at high E/N. Wannier (whose work Loeb discusses at length) was able to obtain detailed solutions in the high-field region only for two simple limiting models-- constant mean free time and constant mean free path. These models, which correspond to the polarization potential and the hard sphere potential, respectively, provided much needed insight into the high-field regime. Wannier's paper also contains results that are quite useful at low E/N. We are now at the end of our discussion of Loeb's book except to say that he not only treats ion and electron transport, but also electron velocity distributions, the theory and use of probes, the formation of negative ions, ionic recombination, electron avalanches, and the Townsend ionization coefficients. I find the book to be of considerable historical interest.

By the mid- 1950's, swarm research had grown to such proportions that it is impossible to outline the subsequent history of the entire field in this short introduction and even to give credit specifically to the many important contributors. The remaining reference (2-8) are useful in this connection, but they should be supplemented by books on electrical discharges and plasma physics.

Thus far we have emphasized the swarm transport phenomena themselves, making only peripheral mention of atomic collision phenomena. It is now appropriate to take a closer look at the latter, because so many of them enter into the formation of swarms, their detailed properties, and their ultimate destruction in the gas phase or on surfaces. First some definitions: By "atomic collisions" we mean collisions between electrons, photons, heavy particles and surfaces in the energy range where the phenomena are electronic and atomic,not nuclear. The maximum energy considered in each case is below that at which nuclear interactions become significant. By "heavy particles" we mean atoms, molecules and ions. A comprehensive reference and source of data on atomic collisions is the five-volume set of books by Massey, et al. (2), published between 1969 and 1974. The notes appended to the citations of these books in our list of references at the end of this introduction indicate the scope of the field of atomic collisions. Rough brackets on the impact energies of main interest are 0 - 1 MeV for electron collisions, 0 - 10 MeV for heavy particle impacts, and microwave to far ultra-violet for photons. In the books by Massey, et al. (2), attention is concentrated on two-body and three-body collisions, but there are useful chapters on electron and swarm phenomena (Vols. 2 and 3). Many kinds of single-beam and crossed-beam apparatus for cross section measurements are described, and various kinds of swarm techniques are discussed. Methods of calculating collision cross sections are also outlined. It is appropriate that this theoretical discussion appear in Massey's book. Massey's department at University College London, along with that of Bates at the Queen's University of Belfast, has been an outstanding center of theoretical collisions research for many years.

Now back to the swarm narrative, starting with the perspective provided by Vols. 2 and 3 of ref. (2). By the late 1960's, electron swarm measurements had become highly refined, due in large measure to the excellent work at the Australian National University in Canberra and at the Westinghouse Research Laboratories. The theory of electron drift and diffusion, based on the Boltzmann transport equation for the most part, was also in an advanced state. On the other hand, a general theory of ionic transport was still lacking - realistic calculations could be made only in the region of very low E/N, for spherically symmetric potentials, with only elastic collisions being considered. A major advance had been made, however, on the experimental side with the development of the drift tube mass spectrometer in the 1960's, good mobility data obtained with this instrument first being published in 1968. Previously (except in a few special cases such as alkali ions in the noble gases) ion-molecule reactions in the drift tube gas made the identities of the ions uncertain, the proper analysis of the ionic arrival time spectra impossible, and the mobility data dubious. Furthermore, no free (as opposed to ambipolar) diffusion coefficients had been directly measured. The development of the drift tube mass spectrometer depend on the development of techniques for constructing large ultra-high-vacuum systems, and on the availability of electron multiplier single-particle detectors, quadrupole mass spectrometers, and fast pulse circuitry (including multi-channel time analyzers).

The next book to be cited is that of McDaniel and Mason (3). It describes several kinds of drift tube mass spectrometers and their use for the masurement of ionic mobilities, diffusion coefficients, and ion-molecule reaction rate coefficients over a wide range of E/N. The state of ionic transport theory in about 1972 is treated at length, with considerable attention to the work of Wannier (1951-1953) and Kihara (1951-1953), whose contributions poved quite important in the great leap forward to be discussed in ref. (5).

Adhering to our chronological order, however, we next turn to the treatise of Huxley and Crompton (4) which deals with the diffusion and drift of electrons in gases. Their historical survey goes "back to the creation", and is especially interesting and authoritative because Huxley was a student of Townsend's and later Crompton picked up the cudgel from Huxley. Modern experimental techniques are examined in great detail, and the theory of electron transport is comprehensively discussed, first via the Boltzmann equation and then by mean free path techniques. A carefully selected set of experimental data is also provided in (4).

Proceeding now to our final reference on ion transport, we come to the book by Mason and Mc Daniel (5), to be published in 1985. Here the experimental portion of ref. (3) is updated by inclusion of new techniques involving drift tube mass spectrometers, flow tubes, and combinations of these two kinds of instruments with selected ion inputs. A key to the experimental ion transport data published by about mid-1983 is included.

Also a detailed discussion of the new theoretical work of Viehland and Mason is provided, starting with their initial paper in 1975 and going up to 1984. In several stages, Viehland and Mason succeeded in developing a kinetic theory of ionic drift and diffusion that applies at arbitrary E/N (the energy parameter) and for any ion-neutral interaction potential that is spherically symmetric. Only elastic collisions are considered. This theory has allowed interaction potentials to be obtained from transport data that extend to, and are consistent with, potentials derived from ion beam scattering data for many ion- gas combinations. These potentials are extremely important in the calculation of collision cross sections. Viehland and Mason, and others, are now studying the excitation and de-excitation of molecular ions in collisions, and the effects of such collisions and non-spherical potentials on transport phenomena. The goal is to lessen or remove the restrictions on the present form of the Viehland-Mason theory.

The reader may wonder at this point why progress toward a rigorous kinetic theory for ion transport at $E/N > 0$ was so much slower than that for electron transport. The answer lies mainly in the difference in the ratio of the charge carrier mass m to the gas molecule mass M for ions, the ratio is of the order unity; for electrons, m/M is of the order 10^{-4}. This enormous difference means that it is permissible to make approximations in the analysis of electronic motion that are not legitimate in the ionic case. Furthermore, the small mass of the electron causes its scattering to be nearly isotropic at low impact energies such as apply in drift tubes. Hence, if the electron distribution function is expanded in spherical harmonics, as is usually the case, all harmonics beyond the first or second can be neglected, with an attendant simplification of the calculation. This procedure does not work for ions. Viehland and Mason, however, did hit upon a "natural" set of basis functions for a description of ion transport--the "Burnett functions" at a suitably elevated temperature--that allows expansion in a rapidly converging series and per. mits a moment solution of the Boltzmann equation in a "two-temperature" theory. They then made an extension to a "three-temperature" theory that provides a means of calculating accurately not only K, but also D_L and D_T, at arbitrary E/N for an arbitrary spherically symmetric potential, when only elastic collisions are considered. By "three-temperature", we mean that the gas molecules have the temperature of the drift tube gas, while one effective temperature is associated with the longitudinal motion of the ions, and another with the transverse motion. The ionic velocity distribution is not obtained directly in this solution.

Finally, it is appropriate to mention several additional recent references dealing with aspects of atomic collisions that bear heavily on electron and ion swarms. The first of these is the treatise by Massey (6) which covers the formation, destruction, structure, and other properties of negative ions. The second, by Märk and Dunn (7),deals with the ionization of atoms, molecules, and ions by electron impact. Both books contain experimental and theoretical material. Also, we have the five-volume set of books by Massey, et al. (8) that is concerned with applications of atomic collision processes in astrophysics, planetary atmospheres, controlled fusion research, gas lasers, and many other areas. These five volumes contain much valuable data on cross sections and reaction rates and on applications involving swarms.

This Book

The present volume does not purport to develop the basic theory of ion or electron transport or to discuss the measurement of transport properties systematically. Instead, it covers a variety of individual experimental and theoretical topics that are currently the subject of active research. These status reports provide valuable insights into the microscopic processes basic to swarm phenomena observed on a macroscopic scale.

There are five chapters on ion transport. One deals with nonequilibrium ion velocity distributions, a subject on which little progress had been made either theoretically or experimentally until recent years. Two other chapters also treat topics on which spectacular progress has recently been made: internal energy distributions of molecular ions in drift tubes, and the use of mobility data to obtain ion-neutral interaction potentials. (The Viehland-Mason theory played an essential role here). An evaluation of techniques for measuring transverse diffusion coefficients and a discussion of current research comprise another chapter. The fifth chapter is devoted to a phenomenon discovered only in 1979, runaway ion mobilities.

Many kinds of collisions are included in the six chapters on ion reactions, and both positive and negative ions are considered. Flannery discusses various aspects of the theory of elastic scattering, heavy particle interchange, charge transfer, and rotational, vibrational, and electronic excitation in ion-molecule collisions. Ferguson considers vibrational excitation and de-excitation and charge transfer of molecular ions in drift tubes. Thereafter, the emphasis is on ion-molecule reactions, including clustering.

The final section of the book deals with electron swarms in dense gases and in inert gases, and with experimental studies of plasma reaction processes using a flowing afterglow/ Langmuir probe instrument. Electron attachment in dense gases receives detailed attention, as does electron-ion recombination.

The reader should find much that is new and interesting in this book. Don't stop here-- the best is yet to come !

References

1. Loeb, L.B. (1955). "Basic Processes of Gaseous Electronics", Univ. of Cal. Press, Berkeley. Second Ed. (1960)-1028 pp.
2. Massey, H.S.W. and Burhop, E.H.S. (1969). "Electronic and Ionic Impact Phenomena". Vol. 1, 664 pp. Oxford Univ. Press (Clarendon), London and New York. This volume deals with elastic and inelastic collisions of electrons with atoms. The experimental sections of each volume of this

treatise are full of detail and contain many graphs and tables of data.

Massey, H.S.W. (1969). "Electronic and Ionic Impact Phenomena", Vol. 2, 630 pp. Oxford Univ. Press (Clarendon), London and New York. Volume 2 covers electron collisions with molecules, photoionization, photodetachment of electrons from negative ions, radiative recombination, and bremsstrahlung. All kinds of electron collisions are discussed: elastic scattering; rotational, vibrational, and electronic excitation; dissociation; ionization; spin exchange; negative ion formation; and electron ion recombination. Also included are electron transport phenomena.

Massey, H.S.W. (1971). "Electronic and Ionic Impact Phenomena". Vol. 3, 819 pp. Oxford Univ. Press (Clarendon), London and New York. Volume 3 is concerned with thermal energy collisions involving neutral and ionized atoms and molecules, diffusion, and ionic mobilities.

Massey, H.S.W., and Gilbody, H.B. (1974). "Electronic and Ionic Impact Phenomena". Vol. 4, 1045 pp. Oxford Univ. Press (Clarendon), London and New York. Volume 4 covers higher energy collisions involving neutral and ionized atoms and molecules, electron-ion recombination, and ion-ion recombination. The higher energy phenomena that are treated include elastic scattering, excitation, ionization, charge transfer, dissociation, and ion-atom interchange at impact energies up into the MeV range.

Massey. H.S.W., Burhop, E.H.S., and Gilbody, H.B. (1974). "Electronic and Ionic Impact Phenomena", Vol. 5, 567 pp. Oxford Univ. Press (Clarendon), London and New York. This volume deals with positrons, positronium, muons, muonium, and mesic atoms, and it also contains extensive notes on recent advances in the entire field of atomic collisions.

3. Mc Daniel, E.W., and Mason, E.A. (1973). "The Mobility and Diffusion of Ion in Gases". Wiley, New York.
4. Huxley, L.G.H., and Crompton, R.W. (1974)." The Diffusion and Drift of Electrons in Gases". Wiley, New York.
5. Mason, E.A., and Mc Daniel, E.W., "Transport Properties of Ions in Gases", Wiley, New York. To be published in 1985.

6. Massey, H.S.W. (1976). "Negative Ions". Cambridge Univ. Press, London and New York
7. Märk, T.D., and Dunn, G.H., Eds. (1984), "Electron-Impact Ionization", Springer-Verlag, Wien
8. Massey, H.S.W., Mc Daniel, E.W., and Bederson, B., (1982-4), "Applied Atomic Collision Physics", 5 vols., Academic, New York.

Velocity Distribution Functions of Atomic Ions in Drift Tubes

H. R. Skullerud[1] and S. Kuhn[2]

[1] Electron and Ion Physics Research Group, Physics Department, Norwegian Institute of Technology, N-7034 Trondheim, Norway

[2] Institute for Theoretical Physics, University of Innsbruck, A-6020 Innsbruck, Austria

1. Introduction

Drift tube experiments constitute a class of low energy ion-molecule collision experiments, where the kinetic energies of the ions are varied typically in the range 0.01 to 10 eV by varying the ratio E/n_o between an applied electrostatic field E and the number density n_o of the gas molecules.

The ions in the drift tube undergo multiple collisions with the gas molecules, and thereby acquire some distribution function $f(\vec{r},\vec{v},t)$ in 6-dimensional phase space $(\vec{r},\vec{v})$. This distribution function is determined by E , n_o , the gas temperature T , the cross sections for the various ion-molecule collision processes, and initial and boundary conditions.

For molecular ions, the distribution function $f(\vec{r},\vec{v},t)$ can be subdivided into separate distribution functions for the different vibrational/rotational internal states, as discussed by Viehland /1/ in the subsequent article. In the present article, a one-state description of the ions will be used, and the methods surveyed therefore apply strictly only to atomic ions.

The drift tube experiments are usually arranged to measure one or several of the transport coefficients α (reaction rate), $\vec{v}_{dr}$ (drift velocity) and $\vec{\vec{D}}$ (diffusion tensor). These coefficients govern the long time evolution of *spatial moments* of $f(\vec{r},\vec{v},t)$, for an ion swarm moving in a homogeneous and unbounded gas:

$$\alpha = \lim_{t}\{-(d/dt)\ln N\} \tag{1}$$

$$v_{dr} = \lim_{t}\{(d/dt)\langle\vec{r}\rangle\} \tag{2}$$

$$\vec{\vec{D}} = \lim_{t}\{(d/dt)(\tfrac{1}{2}\langle\vec{r}^{*}\vec{r}^{*}\rangle)\} \qquad (\vec{r}^{*} = \vec{r} - \langle\vec{r}\rangle) \tag{3}$$

where

$$N(t) = \int f(\vec{r},\vec{v},t)\,d\vec{r}d\vec{v} \tag{4}$$

$$\langle\vec{r}\rangle = N^{-1}\int \vec{r} f(\vec{r},\vec{v},t)\,d\vec{r}d\vec{v} \tag{5}$$

$$\langle\vec{r}^{*}\vec{r}^{*}\rangle = N^{-1}\int(\vec{r} - \langle\vec{r}\rangle)(\vec{r} - \langle\vec{r}\rangle) f(\vec{r},\vec{v},t)\,d\vec{r}d\vec{v}. \tag{6}$$

A kinetic theory analysis of the experiments hence does not usually demand the calculation of the full 6-dimensional distribution function $f(\vec{r},\vec{v},t)$, and it is in most cases sufficient to know the 3-dimensional *velocity* distribution functions associated with the spatial moments

$$f^{(o)}(\vec{v},t) = N^{-1}\int f(\vec{r},\vec{v},t)\,d\vec{r} \tag{7}$$

$$\underset{\sim}{f}^{(1)}(\vec{v},t) = N^{-1}\int \vec{r}^{*} f(\vec{r},\vec{v},t)\,d\vec{r} \tag{8}$$

$$\underset{\sim}{f}^{(2)}(\vec{v},t) = N^{-1}\int\{\tfrac{1}{2}(\vec{r}^{*}\vec{r}^{*} - \langle\vec{r}^{*}\vec{r}^{*}\rangle)\} f(\vec{r},\vec{v},t)\,d\vec{r}. \tag{9}$$

The equations governing these velocity distributions functions will be established in Section 2.

The main concern of the present article is the solution of the very simplest problem; the calculation of the steady state and spatially averaged velocity distribution function for ions in an unbounded gas,

$$f_s(\vec{v}) = \lim_{t} f^{(o)}(\vec{v},t), \tag{10}$$

although the methods to be sketched can also be applied to other cases.

The distribution function $f_s(\vec{v})$ determines the reaction rate α uniquely, and when reactive collisions are absent or infrequent it also determines the drift velocity.

For sufficiently soft ion-molecule interactions, there is a finite probability of the ions being accelerated ad infinitum without being stopped by collisions, and a steady state distribution function then does not exist. This *runaway effect* is discussed by Howorka /2/ elsewhere in this book, and will in this article implicitly be assumed not to occur.

A number of computational methods may in principle be used to calculate ion velocity distribution functions.

Conceptually simplest is the method of computer simulation. Single ions are followed through a large number of free paths, and their trajectories are calculated exactly - with the path lengths, collision partner velocities and center-of-mass scattering angles chosen stochastically in accordance with some given probability distribution. For a good description of the method, we refer to Lin and Bardsley /3/.

Related to the computer simulation method is the path integral method of successive collisions, based upon an integral form of the Boltzmann equation. This method is surveyed in Section 3.

Velocity moment methods have been used since the days of Maxwell /4/ for the calculation of transport coefficients, but have only to a small extent - and sometimes incorrectly - been used to determine velocity distribution functions. We discuss the use of velocity moment methods for such purposes in Section 4.

In addition to the methods listed above, cubic spline methods /5/ and various finite difference schemes /6/, /7/ have been successfully applied to *electron* velocity distribution calculations, but not yet to problems where a small massratio approximation can not be used. We will not discuss these "electron methods" in this article.

2. Spatial Moment Decomposition of the Boltzmann Equation

The Boltzmann equation for the ion distribution function $f(\vec{r},\vec{v},t)$ may be written

$$\{\partial_t + \vec{v}\cdot\nabla + L\}f(\vec{r},\vec{v},t) = 0 \qquad (11)$$

where L is a linear operator in velocity space:

$$L = \vec{a}\cdot\nabla_{\underline{v}} + J. \qquad (12)$$

$\vec{a} = q\vec{E}/m$ is here the acceleration of an ion due to the electrostatic field $\vec{E}$, and J is the collision operator.

The collision operator may be split into an elastic part, an inelastic part and a reactive part:

$$J = J_{eL} + J_{ineL} + J_R. \qquad (13)$$

Explicit forms for the different terms can be found in the review paper of Kumar & al /8/.

Assume now boundaries to be absent, and the swarm to be confined within a finite region of space. One can then perform the operations

$$\int \vec{r}^{\,n}\{\text{Boltzmann equation}\}d\vec{r} \qquad (14)$$

$$\int \vec{r}^{\,n}\{\text{Boltzmann equation}\}d\vec{r}d\vec{v} \qquad (15)$$

and thereby obtain spatial moment equations, which to order $n \leq 2$ are

$$(d/dt)\ln N(t) \equiv -\alpha(t) = \int J_R f^{(o)}(\vec{v},t)d\vec{v} \qquad (16)$$

$$\{\partial_t + (L - \alpha(t))\, f^{(o)}(\vec{v},t) = 0 \qquad (17)$$

$$(d/dt)\langle\vec{r}\rangle \equiv \vec{v}_d(t) = \int \vec{v}f^{(o)}(\vec{v},t)d\vec{v} - J_R\underset{\sim}{f}^{(1)}(\vec{v},t)d\vec{v} \qquad (18)$$

$$\{\partial_t + (L - \alpha(t))\}\underset{\sim}{f}^{(1)}(\vec{v},t) = (\vec{v} - \vec{v}_d(t))f^{(o)}(\vec{v},t) \qquad (19)$$

$$(d/dt)\left(\tfrac{1}{2}\langle\vec{r}^{\,*}\vec{r}^{\,*}\rangle\right) \equiv \vec{\mathcal{D}}(t) = \int \vec{v}f^{(1)}(\vec{v},t)d\vec{v} - \int J_R\underset{\sim}{f}^{(2)}(\vec{v},t)d\vec{v} \qquad (20)$$

$$\{\partial_t + (L - \alpha(t))\}\underset{\sim}{f}^{(2)}(\vec{v},t) = (\vec{v} - \vec{v}_d(t))\underset{\sim}{f}^{(1)}(\vec{v},t) - \vec{\mathcal{D}}(t)f^{(o)}(\vec{v},t). \qquad (21)$$

Due to the cylindrical symmetry of the problem around the $\hat{a}$-axis, these equations are effectively only two-dimensional in velocity space.

With boundaries present in the form of parallel electrodes at $z = 0$ and $z = d$ $(\vec{a} = a\hat{z})$, similar spatial moment equations can be formed by integration over x and y alone. These equations are suitable e.g. for investigating the steady stream of ions in a Townsend - Huxley type lateral diffusion experiment.

3. Path Integral Methods

Theoretical calculations of ion swarm properties have most often been based upon the integro-differential form of the Boltzmann equation. The integral equation forms are however more suitable for a study of mathematical properties /9/, and also allow in a more natural way for the inclusion of boundary and/or initial conditions /10/.

To convert the Boltzmann equation (11) into a "path integral equation", one first writes the streaming operator $\partial_t + \vec{v}\cdot\nabla + \vec{a}\cdot\nabla_v$ as a total time derivative along a free path, and secondly writes the collision term Jf as the difference between a loss term $\bar{\nu}(v)f(\vec{r},\vec{v},t)$ and a gain term $G(\vec{r},\vec{v},t)$. Thereby one obtains a first-order inhomogeneous differential equation

$$\{(d/dt)_{path} + \bar{\nu}(v)\}f(\vec{r},\vec{v},t) = G(\vec{r},\vec{v},t) \tag{22}$$

which when formally solved gives the integral equation

$$f(\vec{r},\vec{v},t) = f(\vec{r}(o),\vec{v}(o),\ t=o)\exp\{-\int_o^t \bar{\nu}(v(t'))dt'\} + \int_o^t G(\vec{r}(t'),\vec{v}(t'),t')\exp\{-\int_{t'}^t \bar{\nu}(v(t''))dt''\}dt' \tag{23}$$

where

$$G(\vec{r},\vec{v},t) = \int K(\vec{v}\leftarrow\vec{v}_i)f(\vec{r},\vec{v}_i,t)d\vec{v}_i \tag{24}$$

$$\vec{r}(t') = \vec{r} - (t - t')\vec{v} + \tfrac{1}{2}\vec{a}(t - t')^2 \tag{25}$$

$$\vec{v}(t') = \vec{v} - (t - t')\vec{a}\ . \tag{26}$$

Expressions for the scattering kernel $K(\vec{v} \leftarrow \vec{v}_i)$ and the collision rate $\bar{\nu}(v)$ were given by Pidduck /11/ for the case of elastic collisions, and a thorough discussion of these and similar expressions has been given recently by Braglia /12/. The expressions are complicated, but not intractable for modern computers.

Eq.(23) can conveniently be rewritten in the form

$$f(\vec{r},\vec{v},t) = f_o(\vec{r},\vec{v},t) + SG(\vec{r},\vec{v},t) = f_o(\vec{r},\vec{v},t) + SKf(\vec{r},\vec{v},t) \qquad (27)$$

where S is a "source stretching operator", K the scattering operator, and f_o the distribution function for ions that have not yet undergone any collision.

At a given time, the swarm may formally be subdivided into groups having undergone zero, one, two .... collisions, and one may write /10,13/

$$f(\vec{r},\vec{v},t) = \sum_{n=o}^{\infty} f_n(\vec{r},\vec{v},t) \qquad (28)$$

$$f_n = SKf_{n-1} = (SK)^n f_o \; . \qquad (29)$$

The "partial distributions" f_n and thereby the full solution f can thus - in principle - be found by repeated application of the SK-operation.

This "method of successive collisions" is readily implemented for the calculation of the steady state and spatially averaged velocity distribution function $f_s(\vec{v})$ (eq.(10)), when no reactions occur /14/, /15/. The space and time dependence of the partial distributions can then be omitted, and $f_s(\vec{v})$ will necessarily - apart from a normalization factor - be identical with the partial distribution for ions that have undergone a large number of collisions,

$$f_s(\vec{v}) = \lim_n \{(SK)^n f_o(\vec{v}) / \int d\vec{v} (SK)^n f_o(\vec{v})\}. \qquad (30)$$

The choice of "starting distribution" $f_o(\vec{v})$ influences the rate of convergence of the iteration scheme, but not the shape of the limiting velocity distribution function $f_s(\vec{v})$.

The rate of convergence of the iteration scheme is illustrated in Fig.1, for a hard sphere, cold gas and unit mass ratio model. For other mass ratios, and for softer interactions, the convergence is alas always slower than this - but not unduly so, except for electrons and very heavy ions. The practical implementation of the iteration procedure is described in more detail in /15/.

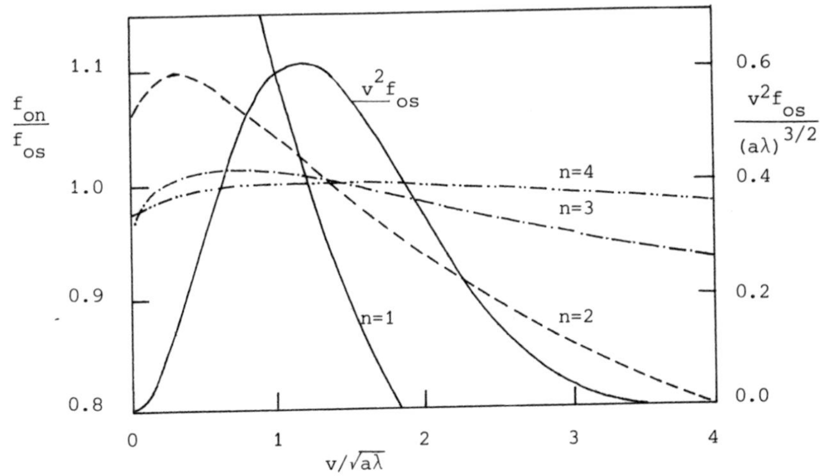

Fig.1. Speed distribution functions $f_{on}(v) = \int f_n(\vec{v})d\hat{v}/4\pi$, relative to the steady state speed distribution function $f_{os}(v)$, for a hard sphere, cold gas and unit mass ratio model. λ is the mean free path, and the ions are released with velocity $\vec{v}(t=0) = 0$.

4. Velocity Moment Methods

The Boltzmann equation can be converted into a set of velocity moment equations by performing the operation

$$\int \psi_j(v)\{\text{Boltzmann equation}\}d\vec{v} \tag{31}$$

where $\psi_j(\vec{v})$ belongs to some set of polynomials in $\vec{v}$. This operation may be applied to the equation (11) for the 6-dimensional distribution function $f(\vec{r},\vec{v},t)$, or to the spatial moment equations (17), (19) and (21). The lowest order *spatial* moment equation (17), which is our main concern, is thereby transformed to the form

$$(d/dt)\langle\psi_j\rangle + \langle\tilde{L}\psi_j\rangle = \alpha(t)\langle\psi_j\rangle \tag{32}$$

where $\langle F(\vec{v})\rangle \equiv \int F(\vec{v}) f^{(o)}(\vec{v},t)d\vec{v}$, $\alpha(t) = \langle J_R\rangle$, and $\tilde{L}$ is the operator adjoint to L , defined by the relation

$$\int F(\vec{v}) L G(\vec{v}) d\vec{v} = \int G(\vec{v}) \tilde{L} F(\vec{v}) d\vec{v} \tag{33}$$

i.e. $\tilde{L} = \vec{a}\cdot\tilde{\nabla}_{\underline{v}} + \tilde{J} = -\vec{a}\cdot\nabla_{\underline{v}} + \tilde{J}$. (34)

A main reason for transforming the Boltzmann equation to a set of *velocity* moment equations, as eq.(32), is that this simplifies the calculation of the collision integral tremendously. The operator $\tilde{J}$ works on *known and well-behaved polynomials* $\psi_j(\vec{v})$, and differential forms of $\tilde{J}$ can be used *without any approximation*. Construction rules for such forms may be found in /8/.

The velocity moment equations (32) are transformed to systems of linear differential equations by inserting for the velocity distribution function some basis set expansion

$$f^{(o)}(\vec{v},t) = \sum_{i=o}^{i_o} x_i(t)\phi_i(\vec{v}) \tag{35}$$

to get equations for the expansion coefficients $x_i(t)$ of form

$$\sum_{i=o}^{i_o} \{\langle\psi_j\rangle_i (d/dt) - \vec{a}\cdot\langle\nabla_{\underline{v}}\psi_j\rangle_i + \langle\tilde{J}\psi_j\rangle_i\} x_i = \alpha(t)\sum_{i=o}^{i_o} x_i\langle\psi_j\rangle_i \tag{36}$$

where $\langle F(\vec{v})\rangle_i \equiv \int F(\vec{v})\phi_i(\vec{v})d\vec{v}$ and $\alpha(t) = \Sigma x_k\langle J_R\rangle_k$. These equations must be supplemented with the normalization condition

$$\sum_{i=o}^{i_o} x_i\langle 1\rangle_i = 1 . \tag{37}$$

From eqs.(36) and (37) "the i_o'th approximation" to $f^{(o)}(\vec{v},t)$ and hence to the velocity moments $\langle\psi_j(\vec{v})\rangle$ can be found. It is essential that the basis size i_o be varied, and the calculation - of some quantity - is then said to have converged when the calculated value does not change with further increase in basis size. One should be aware that converged velocity moments do not ensure coverged $f^{(o)}(\vec{v},t)$-values.

Mathematical convergence (which is not always necessary for numerical convergence) of the expansion (35) will usually demand the basis set $\{\phi_i\}$ to be "complete" - in a Hilbert space encompassing the expanded function. To illustrate the meaning of this requirement, consider basis sets formed as products of a weight function $\omega(\vec{v})$ and polynomials $\Pi_i(\vec{v})$ orthonormal with respect to this weight function:

$$\phi_i(\vec{v}) = \omega(\vec{v})\Pi_i(\vec{v}) \tag{38}$$

$$\int\omega(\vec{v})\Pi_i(\vec{v})\Pi_j(\vec{v})d\vec{v} \equiv \int\omega^{-1}\phi_i(\vec{v})\phi_j(\vec{v}) = \delta_{ij}\,. \tag{39}$$

Such basis sets will - at best - be complete with respect to a Hilbert space of functions square integrable with a weight $1/\omega(\vec{v})$. The expansion (35) will hence only be expected to converge if $f^{(o)}(\vec{v},t)$ fulfills the condition

$$\int\omega^{-1}|f^{(o)}(\vec{v},t)|^2 d\vec{v} \neq \infty\,. \tag{40}$$

The most commonly used basis sets in transport theory employ gaussian type weight functions $\omega(\vec{v})$. With finite electric fields, non-extreme mass ratios and realistic ion-molecule interactions, the steady state velocity distribution $f_s(\vec{v}) = \lim_t f^{(o)}(\vec{v},t)$ will however always fall off asymptotically slower than any gaussian, and the corresponding integral in eq.(40) will diverge if $\omega(\vec{v})$ is a gaussian. For the calculation of steady state velocity distribution functions, gaussian weight polynomial expansions should therefore *not* be attempted.

A non-gaussian weight basis set, of form

$$\phi_i(\vec{v}) \rightarrow \phi_{\ell m}^{(r)}(\vec{v}) = e^{-x^2}x^{\ell+2r}Y_{\ell m}(\vec{v}) \tag{41}$$

$$x^2(v) = \int_o^v d(\tfrac{1}{2}m\tilde{v}^2)/\Theta(\tilde{v}) \tag{42}$$

$$\Theta(v) = kT + mva/\nu_1(v) \tag{43}$$

has recently been applied successfully to the calculation of $f_s(\vec{v})$, for elastic collision models and ion-molecule mass ratios $m/m_o \lesssim 1$ /16/. $Y_{\ell m}(\hat{v})$ is here a spherical harmonic, and $\nu_1(v)$ is the collision frequency for momentum transfer. The choice of "temperature function" $\Theta(v)$ ensures that the integral in eq.(41), with $\omega(\vec{v})\rightarrow\exp(-x^2)$ and $f^{(o)}(\vec{v},t)\rightarrow f_s(\vec{v})$, converges, and that the weight function transforms into a Maxwellian at vanishing fields $\vec{a} \to 0$.

Some illustrative results, calculated for simple cold gas models with cross sections of form

$$\sigma(v) = \sigma_o (v/v_o)^{\gamma-1} \tag{44}$$

are shown in Figs.2-4. For details of the computational schemes we refer to the original paper /16/.

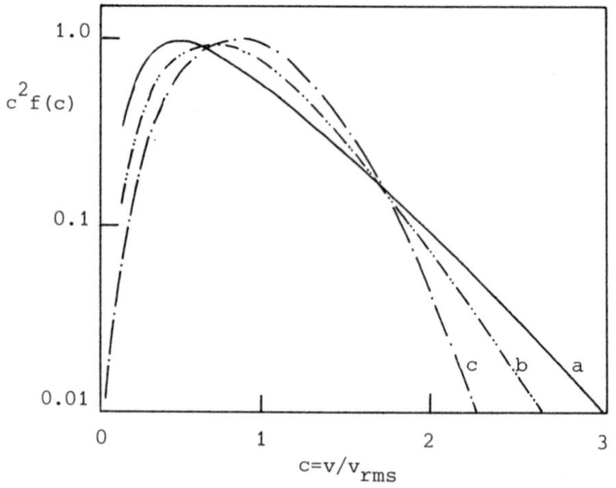

Fig.2. Speed distribution functions for constant collision frequency, isotropic scattering models.
a. $m/m_o = 0.5$
b. $m/m_o = 0.1$
c. $m/m_o << 1$

Fig.2 shows the effect of varying the mass ratio m/m_o . The speed distribution function (the velocity distribution function averaged over all direction $\hat{v}$) becomes narrower as m/m_o diminishes. but is far from the limiting "Davydov distribution function even at $m/m_o = 0.1$.

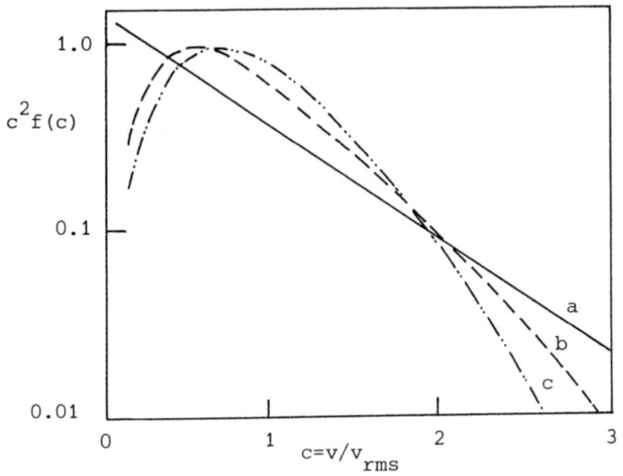

Fig.3. Speed distribution functions for constant collision frequency, unit mass ratio models.

a. Charge exchange ($\chi=\pi$)

b. Isotropic scattering

c. Repulsive r^{-4}-potential scattering

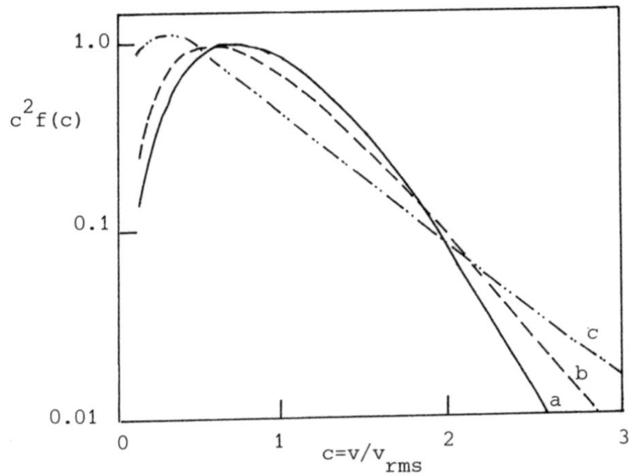

Fig.4. Speed distribution functions for mass ratio $m/m_o = 0.2$ and isotropic scattering models.

a. $\gamma = 0.5$

b. $\gamma = 0$

c. $\gamma = -0.5$

Fig.3 shows the effect of varying the angular distribution in the scattering. The collision frequency for momentum transfer, ν_1 , is the same for the three models used. It is clearly wrong to assume the speed distribution function to be determined essentially by ν_1 alone, as has been done in some ion-molecule reaction studies /17/.

Fig.4 shows the effect of varying the cross section parameter γ - and the effect is much as one would expect. One feature is however *not* apparent from the figure : $\gamma = -0.5$ was the smallest γ-value for which reasonably converged numerical results could be obtained. This has nothing to do with the runaway effect, but may possibly be connected with the fact that distribution functions falling off asymptotically slower than $\exp(-\beta v^{1/2})$ - where β is some constant - are not uniquely determined by their moments.

5. Conclusions

A survey has been given of two different methods that may be used for the numerical calculation of ion velocity distribution functions: The successive collision path integral method and the velocity moment method. In addition, methods for coping with spatially dependent problems by using spatial moment expansions have been sketched.

The path integral method described is in spirit close to computer simulation methods, but enjoys a wider flexibility in the choice of quadrature rules than the latter. The simulation methods may be regarded as path integral methods where all integrals are evaluated by Monte Carlo quadrature. The flexibility in the choice of quadrature rules allows the path integral method to be designed especially to investigate "difficult regions" in velocity space, e.g. regions where the velocity distribution function is very small or shows singularities. Path integral calculations are however computationally quite heavy, and will therefore mostly only be used when other methods fail.

The velocity moment method is best suited for calculating the bulk part of the velocity distribution function. The "difficult regions" in velocity space are not easily probed, when using reasonably sized basis sets. These regions are however difficult *because* they contain few ions, and are therefore most often unimportant.

A drawback with path integral and computer simulation methods is that they demand a total cross section to be defined, and this requires some kind of cut-off to be introduced in the ion-molecule interaction. In the velocity moment equations, on the other hand, only transport cross sections $\sigma_\ell(v)$ occur, and these are always finite quantities for ion-neutral interactions.

At last, we note that velocity moment methods always fail to converge when the ion-molecule interaction is very soft, regardless of what basis set is used. Runaway phenomena can therefore not be investigated using such methods.

Acknowledgement. One of the authors (S.K.) acknowledges partial support from the Austrian Research Funds under contract no. S-18/03.

References

1. Viehland, L.A. (1984) *in this volume*
2. Howorka, F. (1984) *in this volume*
3. Lin, S.L. and Bardsley, J.N. (1977) J. Chem. Phys. 66 435
4. Maxwell, J.C. (1866) Collected Papers No.2, pp.40-3
5. Pitchford, L.C., ONeil, S.V. and Rumble, J.R. (1981) Phys. Rev. A 23 294
6. Kitamori, K., Tagashira, H. and Sakai, Y. (1980) J. Phys. D 13 535
7. Segur, P., Bordage, M-C., Balaguer, J-P. and Yousfi, M. (1983) J. Comp. Phys. 50 116
8. Kumar, K., Skullerud, H.R. and Robson, R.E. (1980) Aust. J. Phys. 33 343
9. Molinet, F.A. (1977) J. Math. Phys. 18 984
10. Kuhn, S. (1983) Memorandum UCB/ERL M83/57, U.C. Berkeley,

CA 94720, *to be published*

11. Pidduck, F.B. (1915) Proc. London Math. Soc. 15 89
12. Braglia, G.L. (1980) Beitr. Plasmaphys. 20 147
13. Kuhn, S. (1980) Phys. Rev. A 22 2460
14. Kleban, P. and Davis, H.T. (1978) J. Chem. Phys. 68 2999
15. Skullerud, H.R. and Kuhn, S. (1983) J. Phys. D 16 1225
16. Skullerud, H.R. (1984) J. Phys. B, *in press*
17. Rebentrost, F. (1972) Chem. Phys. Lett. 17 486, 489
    Rebentrost, F. (1973) Int. J. Mass Spec. Ion Phys. 11 475

Internal-Energy Distribution of Molecular Ions in Drift Tubes

L. A. Viehland

Parks College of Saint Louis University, Cahokia, IL 62206, U.S.A.

The nonreactive motion of trace amounts of a single ion species through a dilute gas in a drift tube is influenced by the gas temperature, T, by the ratio, E/N, of the electric field strength to the gas number density, and by the details of the ion-neutral collisions. On the macroscopic level, this motion is described in terms of the gaseous ion transport coefficients such as the standard mobility, K_o, and the diffusion coefficients, $D_{\parallel}$ and $D_{\perp}$, parallel and perpendicular to the direction of the electric field. On the microscopic level, this motion is described in terms of the position, velocity and internal-energy state of each ion and neutral as a function of time, since quantum-mechanical effects are completely negligible except for electrons and for very light ions at extremely low values of T and E/N. The connection between these levels of description is through the distribution function $f_i(\underline{r},\underline{v},t)$ for ions in internal state i at position $\underline{r}$ with velocity $\underline{v}$ at time t, and through the similar distribution functions for each neutral species.

Since the neutral molecules are essentially in equilibrium under the conditions normally used in drift-tube experiments, their distribution functions must have the equilibrium form:

$$F_{j,k}(\underline{r},\underline{V}_j,t) = N_j \, (M_j/2\pi k_B T)^{3/2} \, Z_j^{-1} \exp[-M_j V_j^2/2k_B T - \epsilon_{j,k}/k_B T] \quad (1)$$

$$Z_j = \sum_k \exp[-\epsilon_{j,k}/k_B T] \quad (2)$$

Here N_j, M_j, and $\underline{V}_j$ are the number density, mass, and velocity, respectively, of neutral species j, $\epsilon_{j,k}$ is the internal energy of a j molecule in internal state k, and k_B is Boltzmann's constant.

At all but the smallest values of E/N used in a drift tube, the ion distribution function is decidedly non-equilibrium. In principle, it can be calculated from a knowledge of T, E/N, and the details of the ion-neutral collisions by solving the Wang Chang-Uhlenbeck-de Boer (WUB) equation [1] or some other "Boltzmann-like" kinetic equation appropriate to the situation. From this solution the macroscopic properties (such as K_o) can be calculated by integrating the product of the distribution function and the appropriate quantity (such as $\underline{v}$). In practice, this procedure cannot be carried out, and additional simplifications must be sought.

Special efforts are ordinarily made in drift-tube experiments to ensure that the "end effects" have been factored out of the experimental results and that only the steady-state motion of the ions through the gas is measured. When this is done, ion properties other than the number density are independent of position in the apparatus, the ion density gradients are small, the time scale for the variations of the ion density is much longer than for all other ion properties, and the distribution function can be written as:

$$f_i(\underline{r},\underline{v},t) = n(\underline{r},t) f_i^{(0)}(\underline{v}) + n(\underline{r},t) f_i^{(z)}(\underline{v}) \frac{\partial}{\partial z} n(\underline{r},t) + n(\underline{r},t) \, [f_i^{(x)}(\underline{v}) \frac{\partial}{\partial x} + f_i^{(y)}(\underline{v}) \frac{\partial}{\partial y}] n(\underline{r},t) \quad (3)$$

where $n(\underline{r},t)$ is the ion number density and where

$$\sum_i \int f_i^{(0)}(\underline{v})\, d\underline{v} = 1 \qquad (4)$$

$$\sum_i \int f_i^{(x,y,z)}(\underline{v})\, d\underline{v} = 0 \qquad (5)$$

Here the homogeneous electric field is assumed to define the z axis of a cartesian reference frame and to set up an x-y symmetry in the drift tube. The terms $f_i^{(x,y,z)}(\underline{v})$ are needed in order to properly describe diffusion, but in the remainder of this article we shall focus on the internal-translational energy distribution $f_i^{(0)}(\underline{v})$, which is of interest for mobility studies and for kinetic studies in which such a small amount of a reactive neutral is added to the gas that $f_i^{(0)}(\underline{v})$ is unchanged.

Inserting eq. (3) into the WUB equation and separating terms according to powers of the ion density gradient leads to an equation for $f_i^{(0)}(\underline{v})$ that is simpler than the WUB equation but is still too complicated to solve in general. The well-known Chapman-Enskog solution procedure [2] cannot be used because large values of E/N will result in large deviations from equilibrium. Furthermore, perturbation treatments based on a small ion-neutral mass ratio are useful only for drift-tube studies of electron motion in gases [3]. Recourse must therefore be made [4] to model studies, to purely numerical methods, or to moment methods in which $f_i^{(0)}(\underline{v})$ is expanded in terms of some complete set of basis functions orthogonal with respect to some weighting function $g_i^{(0)}(\underline{v})$. An alternative way to view a moment method is that $g_i^{(0)}(\underline{v})$ is chosen on physical grounds so as to represent some zeroeth-order approximation to $f_i^{(0)}(\underline{v})$, the basis functions orthogonal with respect to $g_i^{(0)}(\underline{v})$ are constructed by appropriate analytical or numerical techniques, $f_i^{(0)}(\underline{v})$ is expanded in terms of these functions, and solutions are sought for the first few expansion coefficients from which the experimentally-measured transport coefficients can be calculated. Note that the determination of $f_i^{(0)}(\underline{v})$ itself is not usually the goal of a moment method, since such a determination would entail solving for a very large number of expansion coefficients.

The translational energy distributions have been determined in some numerical studies of electron-molecule systems by Monte Carlo techniques [5]. However, most of the model calculations and numerical determinations of energy distributions in drift tubes have been limited to atomic ion-atom systems, as discussed in the preceeding article [6]. These studies show that there are differences between the true ion energy distributions and the zeroeth-order approximations used in the Gaussian moment methods that have been tried so far. It is important to note, however, that these differences are less remarkable than is the general overall accuracy of the Gaussian moment ideas. We therefore boldly assume--especially since we have little other alternative at present--that the $g_i^{(0)}(\underline{v})$ used in the Gaussian moment theory of polyatomic ion-neutral systems [7] gives a good representation of the true $f_i^{(0)}(\underline{v})$ in drift-tube studies

In the past decade it has been established that a successful moment theory of drift-tube experiments must, at the very least, make explicit use of the fact that at high E/N the trace ions can have an average kinetic energy appreciably greater than the thermal energy of the gas. A "two-temperature" theory of drift-tube experiments with atomic ions and neutrals uses a zeroeth-order ion energy distribution that is Gaussian:

$$g^{(0)}(\underline{v}) = (m/2\pi k_B T_k)^{3/2} \exp(-mv^2/2k_B T_k) \qquad (6)$$

Here m is the ion mass, T_k is the "ion temperature", and the subscript i has been dropped since there are no internal states. It seems plausible that T_k, which at this stage is simply a parameter for solving the moment-theory equations for the transport coefficients, should be set equal to the kinetic temperature of the ions in the laboratory frame. In first approximation this leads to the famous Wannier equation

$$\frac{3}{2}k_B T_k = \frac{3}{2}k_B T + \frac{1}{2}mv_d^2 + \frac{1}{2}Mv_d^2 \tag{7}$$

where $v_d = N_o K_o(E/N)$ is the ion drift velocity. The accuracy of the Wannier equation is well established, and the two-temperature theories provide an excellent description of mobilities and reaction rates for atomic ions in atomic gases. However, these theories are essentially useless for diffusion coefficients when the ion-neutral mass ratio is greater than about four.

The inability of a two-temperature theory to describe gaseous ion diffusion adequately appears to stem from two facts. First, compared with the mobility, the diffusion coefficients are more intimately connected with the anisotropic nature of the ion motion that becomes increasingly manifest as E/N and the ion-neutral mass ratio increase. Second, the two-temperature theories make use from the start of a single ion temperature to characterize the energy distribution, thereby assuming essentially isotropic conditions. These ideas led to the development of a "three-temperature" moment theory based on the zeroeth-order ion energy distribution:

$$g^{(0)}(\underline{v}) = (m/2\pi k_B T_\perp)(m/2\pi k_B T_\parallel)^{1/2} \exp\left[-m(v_x^2+v_y^2)/2k_B T_\perp - m(v_z - v_{dis})^2/2k_B T_\parallel\right] \tag{8}$$

The parameters $T_\perp$, $T_\parallel$ and v_{dis} may be chosen in any manner that facilitates the solution of the moment equations for the transport coefficients, but rapid convergence is obtained when these are defined, in first approximation, to be the temperatures characterizing the ion energies perpendicular and parallel to the electric field and the ion drift velocity, respectively.

For polyatomic ion-neutral systems a successful moment theory must, at the very least, contain one additional ion temperature, T_i, that characterizes the energy held in the internal degrees of freedom while the ion moves through the drift tube in steady state. A theory that is analogous to the two-temperature theory for atomic systems can be based on the use of spherical basis functions that are orthogonal with respect to the zeroeth-order distribution function:

$$g_i^{(0)}(\underline{v}) = g^{(0)}(\underline{v}) \exp\,[-\epsilon_i/k_B T_i]/\sum_i \exp[-\epsilon_i/k_B T_i] \qquad (9)$$

where $g^{(0)}(\underline{v})$ is given by eq. (6) and ϵ_i is the internal energy of an ion in state i. Similarly, a theory analogous to the three-temperature theory can be based on eq. (9) if $g^{(0)}(\underline{v})$ is given by eq. (8) and cartesian basis functions are used. Implicit in the existing theories [7] is the additional assumption that inelastic collisions do not have much of an effect on the momentum-balance equation but do greatly affect the energy-balance equation for the system. Although comprehensive tests of these approaches must still be performed, preliminary indications are that no special difficulties are likely to arise just because inelastic collisions and internal degrees of freedom have been introduced into the moment method. The spherical basis functions should give good results for those quantities that are not sensitive to the anisotropy of the ion distribution function at high E/N: ion mobility, total ion energy, and ion-neutral rate coefficients. For greater accuracy, and in order to obtain good results for ion diffusion, the cartesian basis functions will need to be used.

Using spherical basis functions, the moment method for molecular systems leads to a generalization of the Wannier expression, eq. (7). It is convenient to compare these expressions in terms of the relative kinetic energy between the ions and the molecules of the single-component neutral gas. While the Wannier equation gives

$$E_{rel} \approx \tfrac{3}{2}k_BT + \tfrac{1}{2}Mv_d^2 = \tfrac{3}{2}k_BT_{eff} \quad , \tag{10}$$

the moment method gives

$$E_{rel} \approx [1+M\xi/m]^{-1}[\tfrac{3}{2}k_BT+\tfrac{1}{2}Mv_d^2] = \tfrac{3}{2}k_BT'_{eff} \quad . \tag{11}$$

Here the prime has been added to the effective temperature so that in the discussion below it will be clear whether eq. (10)or (11) is being used. The quantity ξ is a dimensionless ratio of the collision integral for inelastic energy loss to that for momentum transfer. In combination with the factor of M/m in eq. (11), we can expect the presence of anisotropic potentials and inelastic collisions in molecular systems to have the largest impact on ion mobility and ion-molecule reaction rate coefficients in the case of light ions in heavy neutral gases. We will return to this point again below.

The internal ion temperature T_i is the temperature that results when the difference between the pre- and post-collision ion internal energies averages to zero. In general, this average energy balance will depend on the cross-sections for inelastic collisions as compared with the elastic collisions. A special case that is of great experimental importance is that of molecular ions in an atomic neutral gas, where $T_i = T'_{eff}$ in steady state. A simple physical explanation can be given for this. Energy is fed into the internal degrees of freedom of the ions by collisions with the structureless neutrals, so the source of the ion internal energy is its translational motion. Energy leaks out of both the internal and translational degrees of freedom of the ions only through the translational motion of the neutrals. Since the leak is the same for both forms of energy, and since the internal energy is fed by the translational, it is not surprising that T_i must equal T'_{eff} in steady state.

The Wannier theory [2] predicts that the transport coefficients for an atomic ion in an atomic gas should be the same at any combination of T and E/N that leads to the same value of T_{eff}. Although it was developed in the early 1950's, little experimental use of this prediction was made until the middle 1970's because the theory seemed to be specific to a special model (the Maxwell model) of the ion-neutral interaction, because other predictions of Wannier's theory are only semi-quantitative, and because there did not seem to be any systematic way of improving the accuracy of results obtained with this theory. In the last decade a number of theoretical and experimental tests have shown that it is accurate within about 10%, and methods for achieving even higher accuracy have been developed from moment theories. Ironically, we seem to have come full circle, and eq. (10) is now used in many situations where it is clearly not applicable. Figure 1 shows, for example, that it does not apply to molecular ion-neutral systems.

The moment theory [7] predicts that the transport coefficients for a polyatomic ion-neutral system should be the same at any

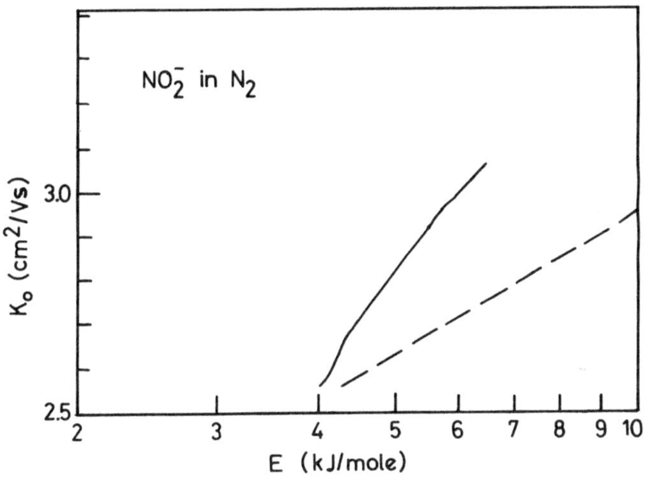

Figure 1: Test of the prediction that the standard mobility of NO_2^- ions in N_2 gas should be the same at different combinations of T and E/N that give the same values of E_{rel} according to eq. (10). The experimental values [8] at low E/N and variable T are indicated by the solid curve, while the values [9] at variable E/N and low T are indicated by the dashed curve.

combination of T and E/N that leads to the same value of T'_{eff}. The presence of ξ in eq. (11) makes it difficult to test this prediction, since no way of relating ξ to other measured properties is known and since the theoretical calculation of ξ is fraught with difficulties. However, it is possible to work in the other direction and extract values of ξ from two sets of experimental mobilities (such as shown in fig. 1) in order to use these values in other studies, for example in drift-tube studies of polyatomic ion-neutral reactions. The extraction of ξ is possible because at low E/N the principle of detailed balance shows that $\xi \rightarrow 0$. Consequently, if we combine eqs. (10) and (11), we find that

$$\xi = (m/M)\ (T_{eff}/T'_{eff} - 1) \quad , \tag{12}$$

where we interpret T'_{eff} in this equation as the actual temperature measured in a drift-tube experiment at low E/N that gives a particular value of the mobility and we interpret T_{eff} as the collision temperature calculated from eq. (10) based on the measured values of T and E/N in a low-temperature experiment giving the same value of the mobility. Values of ξ have been obtained in this manner [9] for NO_3^-, NO_2^-, NO^+ and Cl^- ions in N_2 gas.

What are the implications of the above theory for drift-tube studies of ion-molecule reaction rate coefficients? To answer this question, let us briefly review the usual procedures used to analyze the data obtained when a small amount of a reactive neutral is added to a much larger amount of a neutral gas "buffer". At the particular values of T and E/N used in the measurement of the rate coefficient, the drift velocity may be used to compute the corresponding average relative kinetic energy between the ions and the reactive neutrals, since from eq. (7):

$$E_{rel}^{(R)} \approx [M_R(\tfrac{3}{2}k_B T_k)+m(\tfrac{3}{2}k_B T)]/(m+M_R) =$$

$$\tfrac{3}{2}k_B T + \tfrac{1}{2}[(m+M_b)/(m+M_R)]M_R v_d^2 = \tfrac{3}{2}k_B T_{eff}^{(R)} \qquad (13)$$

Here the symbols b and R indicate the buffer and reactive neutral, respectively, and v_d is approximately the same as the drift velocity of the ions through the pure buffer gas at the same T and E/N. By changing E/N, the measured rate coefficients may be studied as a function of $T_{eff}^{(R)}$. This does not represent a study of the thermal rate coefficient as a function of temperature, because the internal degrees of freedom of the reactive neutral are always characterized by T rather than $T_{eff}^{(R)}$. However, by repeating these measurements at different T, the effects of translational energy and the internal energy state of the reactive neutral may be separately studied. The data obtained can have important implications (for example, for models of the upper atmosphere) and a large number of ion-neutral reactions have been studied in this way [10].

The two-temperature moment theory for atomic ion-atom systems provides a justification for the above procedure and indicates that it should be accurate within about 10%. Experimental justification for the procedure has been obtained by studying atomic ion-molecule reactions in several inert gas buffers. In most cases the rate coefficients are found to be the same at the same values of $T_{eff}^{(R)}$, no matter which buffer gas is used. Exceptions occur when the rate coefficients are strongly influenced by the internal energy state of the reactive neutral or when they change rapidly with $T_{eff}^{(R)}$ and therefore are sensitive to small differences in the tails of the energy distributions. These exceptions are reasonably-well understood, and techniques are available for treating them [11]. However, when molecular ions are used the exceptions seem to be more common and they are generally more substantial. The physical explanation for this effect seems clear: the internal energy distributions of the ions must be quite different in the different buffer gases. In order to show explicitly how this explanation applies, the moment theory presented above must be considered further.

The moment theory of polyatomic ion-neutral systems [7] leads to a generalization of eq. (13), namely:

$$E_{rel}^{(R)} \approx \frac{3}{2}k_B T[1+M_b\xi/m]^{-1}[1+(M_b-M_R)\xi/(m+M_R)] + \frac{1}{2}[1+M_b\xi/m]^{-1}[(m+M_b)/(m+M_R)]M_R v_d^2 = \frac{3}{2}k_B T_{eff}^{(R)\prime} \qquad (14)$$

Since the thermal energy term in eq. (14) is important only at low E/N where ξ is small, it should be reasonable assumption to replace eq. (14) by the equation:

$$E_{rel}^{(R)} \approx \frac{3}{2}k_B T + \frac{1}{2}[1+M_b\xi/m]^{-1}[(m+M_b)/(m+M_R)]M_R v_d^2 = \frac{3}{2}k_B T_{eff}^{(R)''} \qquad (15)$$

The primes have been added to the effective temperatures in order that eqs. (13)-(15) may be distinguished in the discussion below. We note that the moment theory of polyatomic species indicates that the measured rate coefficients should be approximately the same in different buffer gases, when they are compared at the same value of $T_{eff}^{(R)\prime}$ or $T_{eff}^{(R)''}$ but not at the same value of $T_{eff}^{(R)}$.

The difference between eqs. (13) and (15) is the factor of $[1+M_b\xi/m]^{-1}$, which is also the difference between eqs. (10) and (11). Since the definition [7] of ξ means that it is always positive, the effect of this factor is always to make $E_{rel}^{(R)}$ lower than would be calculated from eq. (13). This is consistent with the fact that at particular values of T and E/N the ions may acquire, on the average, only a certain amount of energy from the electric field during the time between collisions with the neutral molecules, and only a fraction of this energy can remain in the translation degrees of freedom when there are internal states present.

The factor $[1+M_b\xi/m]^{-1}$ also explains the large differences that are frequently observed when molecular ion-neutral reaction rate coefficients are measured in different buffer gases and analyzed in terms of $T_{eff}^{(R)}$. As a specific example, let us consider the reaction of O_2^+ ions with CH_4 neutrals, a reaction which has been studied [12] using both He and Ar as the buffer gas in a flow-drift tube. For this reaction the factor becomes $[1+0.125\xi]^{-1}$ in He buffer but $[1+1.25\xi]^{-1}$ in Ar buffer. The very large difference between the coefficients of the ξ terms certainly outweighs any errors made in assuming that the two ξ values are approximately equal. The error made in analyzing the experimental results in terms of $T_{eff}^{(R)}$ rather than $T_{eff}^{(R)\prime}$ or $T_{eff}^{(R)\prime\prime}$, and thereby ignoring this factor, is seen to be largest in Ar, while the He buffer data comes closer to showing no effect due to inelastic collisions. This is in accord with the previous observation that the presence of anisotropic potentials and inelastic collisions in molecular systems has the largest impact in the case of light ions in heavy neutral gases. It also shows that the difference between the results in the two buffer gases is due to the disparate ion-neutral mass ratios, and is not attributable to different ion-neutral potential energy surfaces, to different rates of inelastic energy transfer, etc. This justifies the experimental "rule" that molecular ions seldom show effects attributable to internal excitation when the drift tube used is filled with helium.

In order to quantitatively probe the effects of internal excitation of ions drifting in heavy inert gases upon the rate coefficients for reaction with small amounts of molecular gases in a drift tube, it is clear that ξ must be determined. The complicated nature of molecular scattering theory means that the theoretical calculation of ξ will probably be limited in the near future to model situations involving rather extreme approximations. The experimental determination of ξ is possible up to T'_{eff} values of 600 to 1000K by comparing two types of mobility data, such as shown in fig. 1; the temperature limitation arises because it is difficult to make mobility measurements at low E/N using drift tubes heated above about 900K. It is likely that a judicious combination of theoretical model calculations and experimental work using drift tubes and other experimental techniques will allow ξ values to be obtained with reasonable accuracy over wider ranges.

Once ξ values become available, additional research will be needed to verify the accuracy of the first approximation results for polyatomic species. In particular, there have been as yet no theoretical or experimental tests of eqs. (11) and (14), although it is hoped that they are about as accurate ($\pm$10%) as the similar eqs. (10) and (13) for atomic systems. Future theoretical work will probably focus on comparing first and higher approximations of the moment theory using reasonable models. Experimental work will probably entail using the values of ξ to check the accuracy of the prediction that the reaction rate coefficients in different buffer gases should agree when compared at the same values of $T_{eff}^{(R)'}$ or $T_{eff}^{(R)''}$.

Whether the research proposed above proves successful or not, it must be remembered that little information will be provided by such studies about the true ion energy distribution $f_i^{(0)}(\underline{v})$ in drift tubes. If the first approximation results of the moment theory for polyatomic species prove to be reasonably accurate, it is possible that the zeroeth-order distribution $g_i^{(0)}(v)$ given by eqs. (6) and (9) may be about as accurate a representation of $f_i^{(0)}(\underline{v})$ as $g^{(0)}(\underline{v})$ is a representation of $f^{(0)}(\underline{v})$ for monatomic ions in monatonic gases. This is certainly not guaranteed, however, and it must be remembered that there are differences between $g^{(0)}(\underline{v})$ and $f^{(0)}(\underline{v})$, especially in the tails of the distributions [6], that can become important in special situations, i.e., when the rate coefficients change rapidly with relative kinetic energy. Theoretical treatment of such differences is difficult, and requires the use of higher order approximations in the moment theory.

If the first approximation results of the moment theory for polyatomic species prove not to be accurate, then the theory must be modified to begin with more reasonable zeroeth-order approximations for $f_i^{(0)}(\underline{v})$. One obvious line of endeavor would be to use the distribution given by eqs. (8) and (9) to obtain more accurate expressions than eqs. (10) and (13). Another would be to introduce more than one internal ion temperature, in particular, one temperature for the rotational degrees of freedom and another for the vibrational.

In summary, it must be conceeded that we still have very little information about the true energy distributions of molecular ions in drift tubes. However, progress has been made on both the experimental and theoretical sides in the last few years, not in the direct determination of the energy distributions but in the use of reasonable assumptions about them to understand how they affect the transport and reactive properties of gaseous ions.

References

1. C. S. Wang Chang, G. E. Uhlenbeck and J. de Boer, in "Studies in Statistical Mechanics", Vol. 2, eds. J. de Boer and G. E. Uhlenbeck, North-Holland, Amsterdam, 1964.

2. E. W. McDaniel and E. A. Mason, "The Mobility and Diffusion of Ions in Gases", Wiley-Interscience, New York, 1973.

3. L. G. H. Huxley and R. W. Crompton, "The Diffusion and Drift of Electrons in Gases", Wiley-Interscience, New York, 1974.

4. K. Kumar, H. R. Skullerud and R. E. Robson, Aust. J. Phys. 33, 343, 1980.

5. S. R. Hunter, Aust. J. Phys. 30, 83, 1977.

6. S. Kuhn and H. R. Skullerud, preceeding article.

7. L. A. Viehland, S. L. Lin and E. A Mason, Chem. Phys. 54, 341, 1981.

8. F. L. Eisele, M. D. Perkins and E. W. McDaniel, J. Chem. Phys. 73, 2517, 1980.

9. L. A. Viehland and D. W. Fahey, J. Chem. Phys. 78, 435, 1983.

10. D. L. Albritton, At. Data Nucl. Data Tables 22, 1, 1978.

11. D. L. Albritton, in "Kinetics of Ion-Molecule Reactions", ed. P. Ausloos, Plenum, New York, 1979.

12. I. Dotan, F. C. Fehsenfeld and D. L. Albritton, J. Chem. Phys. 68, 5665, 1978.

Determination of Ion-Atom Potentials from Mobility Experiments

I. R. Gatland

School of Physics, Georgia Institute of Technology,
Atlanta, GA 30332, U.S.A.

Introduction

Drift tube experiments have been used to investigate the behavior of low density ion swarms, with applied electric fields, since the 1930's. Starting with the work of Tyndall and of Bradbury and Nielsen it was recognized that the evolution of the swarm consisted of a steady drift of the swarm center and diffusion out from this center. Thus the experimental data may be analyzed in terms of a drift velocity, $\vec{v}$, and a diffusion coefficient, D, such that the ion density at position $\vec{r}$ at time t is

$$n(\vec{r},t) = C[4\pi Dt]^{-3/2} \exp[-(\vec{r}-\vec{v}t)^2/4Dt] \tag{1}$$

assuming a concentration of C ions at the origin at time zero. The direction of $\vec{v}$ is, of course, that of the external electric field, $\vec{E}$, and their ratio defines the mobility

$$K = v/E \tag{2}$$

For low electric field strengths K is nearly constant but varies considerably at higher fields. Also at higher fields the diffusion coefficients in the longitudinal and transverse directions are not the same.

A major theoretical study by Wannier[1] in the late 40's and early 50's investigated the relationships between the macroscopic quantities and the microscopic ion motion. Using a Boltzmann Equation analysis and some model potentials with known cross sections Wannier derived expressions for the drift velocity and diffusion coefficients. Extrapolating from these model cases he also proposed a relation between the mean ion energy, ε, the neutral gas temperature, T, and the drift velocity

$$\varepsilon = \frac{3}{2} kT + \frac{1}{2} (m+M) v^2 \tag{3}$$

where k is Boltzmann's constant, m the ion mass, and M the gas atom or molecule mass. However general results of this type were beyond the scope of his analysis. One rigorous result of Wannier's work relates to the reduced mobility

$$K_o = KN/N_o \tag{4}$$

$$= K\ (P/760\ \text{Torr})(273\ \text{K}/T) \tag{5}$$

where N is the gas number density, N_o is Loschmidt's number, and P is the gas pressure. At a given gas temperature the reduced mobility is a function of the ratio E/N, not of E and N separately. The standard unit for E/N is the Townsend: 1 Td = 10^{-17} V cm^2.

The early experiments suffered from several uncertainties, in particular the identification of the ions involved. Accurate, unambiguous data were not generally available until the early 1960's when experiments were devised which included a mass spectrometer to identify the ions as well as techniques for removing end effects in the drift chamber. The drift tube-mass spectrometer of McDaniel and coworkers(2) was one of the first to meet all these requirements. Thereafter accurate measurements of drift velocities and longitudinal diffusion coefficients (and to a lesser extent transverse diffusion coefficients) were obtained for a variety of ion-gas combinations. Many of the measurements are of direct value in various tenuous plasma situations including atmospheric, laser, and engineering applications. Others were obtained as an initial step in the study of ion-neutral reactions in the drift chamber. However until the 1970's no adequate theory was available which related mobilities and diffusion coefficients to ion-atom or ion-molecule interaction potentials, except at extremely low values of E/N. An extensive survey of the experiments and analysis at that time has been given by McDaniel and Mason(2).

The first effective techniques for evaluating mobilities, given a potential, were Monte Carlo simulations. Skullerud investigated the general features of this method and applied it to 4-6-n type potentials(3). In particular he developed a potential based on observed K^+ - Ar mobilities(4). Lin and Bardsley also carried out Monte Carlo calculations for Li^+ - He, K^+ - He, and K^+ - Ar as well as some oxygen ion cases(5). As usual the primary drawback of Monte Carlo studies is the amount of computer time involved.

Indeed, without the use of the null collision concept, they would have been impractical[6].

In the mid 1970's Viehland and Mason developed a theory of ion motion in a gas with arbitrarily large external fields[7][8]. They assumed that the dominant feature was an increase in the width of the ion velocity distribution rather than a displacement of the center without appreciable broadening. This requires that the ion velocity distribution be characterized by a temperature, T_i, which is higher than that of the neutral gas. The effective temperature, T_f, for ion-neutral collisions is then intermediate between T and T_i. Given a set of basis functions which are orthogonal with respect to the gaussian weight function (with temperature T_i) the form of the velocity distribution is specified by the moments associated with these basis functions. The Viehland-Mason theory uses the Boltzmann Equation to obtain relations between these moments, as will be discussed in detail later. Further developments of the theory introduce different ion temperatures for the longitudinal and transverse velocity components so that diffusion as well as drift may be handled properly.

Once the ion-neutral interaction potential is specified, so that details of the Boltzmann collision term may be calculated, the ion mobility may be obtained for arbitrary T and E/N. In practice the infinite set of equations for the moments has to be truncated but third order results are usually accurate to better than 1%. The first order results are particularly instructive:

$$\frac{3}{2} kT_f = \frac{3}{2} kT + \frac{1}{2} Mv^2 \tag{6}$$

and

$$v = (3eE/8N)(\pi/2kT_f\mu)^{1/2}/\Omega''(T_f) \tag{7}$$

where e is the ion charge, $\mu = mM/(m+M)$ is the reduced mass in the collision, and

$$\Omega''(T_f) = \frac{1}{2} \int \exp(-x^2)\, Q_1(xkT_f)\, x^2 dx \tag{8}$$

with

$$Q_1(\varepsilon) = \int (1 - \cos\theta)\, \sigma(\varepsilon,\theta)\, d\cos\theta\, d\phi \tag{9}$$

Here $x = \varepsilon/kT_f$, ε is the collision energy, θ is the scattering angle, ϕ is the azimuthal angle, σ is the differential cross section, Q_1 is the momentum transfer cross section, and Ω'' is the first omega integral. Except at very low temperatures (< 10K) the scattering may be calculated classically so, in terms of the impact parameter b,

$$Q_1(\varepsilon) = \int [1 - \cos\theta(\varepsilon,b)]\, 2\pi b\, db \tag{10}$$

where the scattering angle, θ, is obtained by direct integration of the equations of motion or by using the scattering integral form. It is interesting to note that equation (6) is equivalent to the Wannier result; equation (3).

If the problem were to find the mobilities given the potential the analysis would be complete. However the inverse problem is much more pertinent. Since mobility data are available for a wide range of E/N for many ion-atom combinations (with centrally symmetric potentials) it may well be possible to determine the potential from the experimental results. In particular the range of collision energies is from about .05 eV to 1 eV or above; an energy range typical of the well depths associated with such ion-atom interactions. It may be expected that the mobilities are not too sensitive to the hard core of the potential which is associated with higher energy scattering. Also the long range part of the potential is already known to be the polarization potential, $-\alpha/2r^4$, where α is the polarizability. Hopefully this controls the low E/N mobilities. Unfortunately these expectations are not always realized in practice so that additional information from beam scattering experiments or low temperature mobilities is required.

Mason and Viehland proposed[9] that the direct inversion of the mobilities to yield potentials be carried out by a method similar to that of Smith for neutral-neutral interactions[10]. This assumes that the mobility is most sensitive to the potential at an inter-nuclear separation distance, r, given by

$$\pi r^2 = \Omega'' \tag{11}$$

Suppose one has both a set of experimental mobilities and a trial potential for which theoretical mobilities are calculated. Then both theoretical and experimental values of r may be determined for any chosen value of T_f by using the first order results, equations (6) and (7), and equation (11). The new trial potential at the experimental value of r is then given by the trial potential at the theoretical value of r. If the original trial potential is reasonably close to the true potential iterating this process should converge to the true potential. However the process works by warping the separation distances but cannot change the potential well depth. A separate technique must be included to change the well depth and an appropriate well depth found so that convergence is obtained.

This program has been applied to almost all alkali ion-rare gas combinations and to several negative halogen ion-rare gas combinations as well as other cases.

Problems with this program have been discovered in only two cases so far.

The first involved H^+ and D^+ ions in helium where ion runaway occurred[11][12]. The second involved Li^+ - Ar and is more problematic: the mobility calculations do not converge as the order of truncation is increased for E/N values where the mobility is increasing rapidly. One conjecture is that an instability, similar to that producing runaway, is involved[13]. On the other hand Skullerud attributes the problem to an inadequacy in the basis functions used, in particular the assumption of a dominant gaussian form of the velocity distribution. He has calculated the Li^+ - Ar mobilities using a modified weight function and basis set[14]. A direct comparison with experiment has not yet been published but his mobility curve does not show the unexpected structure given by the original third order calculations using the Viehland-Mason theory[15]. For all other cases investigated the Viehland-Mason theory appears to give good results.

Theory

The ion velocity distribution, $f(\vec{u})$, where $\vec{u}$ is the ion velocity, is expanded in the form[7][8]

$$f(\vec{u}) = \sum_{r\ell m} C_{r\ell m} \pi^{-3/2} e^{-w^2} \psi^r_{\ell m}(\vec{w}) \tag{12}$$

where

$$\vec{w} = \vec{u}/(2kT_f/m)^{1/2} \tag{13}$$

is the normalized velocity and

$$\psi^r_{\ell m}(\vec{w}) = L_r^{\ell+1/2}(w^2)\, w^\ell P^m_\ell(\cos\theta)\, e^{im\phi} \tag{14}$$

are the Burnett functions with θ and ϕ the polar and azimuthal angles of $\vec{w}$, respectively. Here L are the generalized Laguerre polynomials and P the associated Legendre functions[16]. For the present application the velocity distribution is axially symmetric so m = 0 always applies and the m index is often dropped. Since the Burnett functions are complete and orthogonal with respect to the weight function exp $(-w^2)$ the coefficients, C, are given in terms of the moments

$$\langle\psi^r_{\ell m}\rangle = n^{-1} \int f(\vec{u})\, \psi^r_{\ell m}(\vec{w})\, d^3u \tag{15}$$

The normalization constant, n, is the local number density and

$$\langle\psi^0_{00}\rangle = \langle 1\rangle = 1 \tag{16}$$

Also T_i is chosen so that

$$\langle\frac{1}{2} mu^2\rangle = \frac{3}{2} kT_i \tag{17}$$

i.e.,

$$\langle\psi^1_{00}\rangle = \langle\frac{3}{2} - w^2\rangle = 0 \tag{18}$$

The drift velocity is

$$\langle u_z \rangle = \langle \psi_{10}^{o} \rangle \, (2kT_i/m)^{1/2} \tag{19}$$

assuming that the electric field is in the z direction.

The controlling equation for f is the Boltzmann Equation

$$(eE/m)\ \partial f/\partial u_z = (\partial f/\partial t)_{coll} \tag{20}$$

assuming that the local velocity distribution is almost independent of time and position. The resulting equation for the moments is

$$(eE/m)\ \langle \partial\psi_\ell^r/\partial u_z \rangle = N\langle J\psi_\ell^r \rangle \tag{21}$$

where

$$J\psi_\ell^r = N^{-1} \int F(\vec{V})\ [\psi_\ell^r(\vec{w}) - \psi_\ell^r(\vec{w}\,')]\ |\vec{u}-\vec{V}|\sigma d\Omega d^3V \tag{22}$$

is the collision integral and

$$F(\vec{V}) = N(M/2\pi kT)^{3/2} \exp[-\tfrac{1}{2} MV^2/kT] \tag{23}$$

is the neutral gas velocity distribution. Here $\vec{V}$ is the neutral gas atom or molecule velocity and $\vec{w}\,'$ is the ion velocity after a collision which results in a center of mass scattering into the solid angle $d\Omega$. Since the ψ's are complete and the angular momentum is conserved

$$J\psi_\ell^r = \sum_s a_{rs}(\ell)\ \psi_\ell^s \tag{24}$$

and, using the orthogonality of the ψ's, the matrix elements of the collision operator, J, are given by

$$\begin{aligned} a_{rs}(\ell) = {} & [(2\ell+1)s!\Gamma(3/2)/\Gamma(\ell+s+3/2)\pi^3](m/2kT_i)^{3/2} \\ & \times (M/2kT)^{3/2} \int\int\int \exp(-mu^2/2kT_i - MV^2/2kT)\ [\psi_\ell^s(\vec{w})]^* \\ & \times [\psi_\ell^r(\vec{w}) - \psi_\ell^r(\vec{w}\,')]|\vec{u}-\vec{V}|\sigma d\Omega d^3V d^3u \end{aligned} \tag{25}$$

which may be calculated once the differential cross section

$$\sigma = \sigma(\tfrac{1}{2}\mu[\vec{u}-\vec{V}]^2, \theta) \tag{26}$$

has been specified. Since the cross section only involves two variables six of the eight integrations may be completed and the other two result in omega integrals. In particular

$$a_{oo}(1) = (8\mu/3m)(2kT_f/\pi\mu)^{1/2}\ \Omega''(T_f) \tag{27}$$

with

$$T_f = \mu(T/M + T_i/m) \tag{28}$$

the effective temperature for ion-neutral collisions. All of the other matrix elements also involve omega integrals at the temperature T_f.

Just as $J\psi$ may be expanded as a sum of ψ's so may $\partial\psi/\partial u_z$. Including the velocity normalization to convert u's to w's one finds

$$\frac{\partial\psi_\ell^r}{\partial w_z} = \frac{\ell(2r+2\ell+1)}{(2\ell+1)}\ \psi_{\ell-1}^{r} - \frac{2(\ell+1)}{2\ell+1}\ \psi_{\ell+1}^{r-1} \tag{29}$$

Combining these results yields the Boltzmann moment equation

$$\sum_s [a_{rs}(\ell)/a_{00}(1)](2\ell+1)\langle\psi_\ell^s\rangle = \gamma[\ell(2r+2\ell+1)\langle\psi_{\ell-1}^r\rangle - 2(\ell+1)\langle\psi_{\ell+1}^{r-1}\rangle] \tag{30}$$

with

$$\gamma = (m/2kT_i)^{1/2}\ [eE/mNa_{00}(1)] \tag{31}$$

The summation over s may be truncated by assuming that

$$[a_{rs}(\ell)/a_{00}(1)] \sim \Delta^{s-r} \tag{32}$$

for $s > r$, where Δ is small, and neglecting all terms with $s - r > n$ where n is some selected order of approximation. Also remember that $\langle\psi_0^0\rangle = 1$ and that T_i should be chosen so that $\langle\psi_0^1\rangle = 0$. In practice T_i is chosen and the condition $\langle\psi_0^1\rangle = 0$ determines γ, i.e., E/N is adjusted to the particular T_i under consideration. The truncated equations may be solved by various methods of sequential approximation(7)(8). In particular the first order results are those discussed earlier.

Whilst the two temperature theory (T and T_i) serves to calculate the drift velocity it is not expected to satisfactorily determine the diffusion coefficients. Lin, Viehland, and Mason(17)(18) have extended the theory to one characterized by three temperatures (T, T_L, and T_T) which assumes that the ion velocity distribution is dominated by a term of the form

$$g(\vec{u}) = (m/2\pi kT_T)(m/2\pi kT_L)^{1/2} \times \exp[-m(u_x^2 + u_y^2)/2kT_T - m(u_z-u_o)^2/2kT_L] \tag{33}$$

The complete set of orthogonal basis functions are

$$\psi_{pqr} = H_p([m/2kT_T]^{1/2}u_x) \times H_q([m/2kT_T]^{1/2}u_y) \times H_r([m/2kT_L]^{1/2}[u_z-u_o]) \tag{34}$$

where H is the Hermite polynomial(16). The Boltzmann moment equation becomes

$$N\langle J\psi\rangle = (e/m)\vec{E}\cdot\langle\vec{\nabla}_u\psi\rangle - [\langle\vec{u}\psi\rangle - \langle\vec{u}\rangle\langle\psi\rangle]\cdot\vec{\nabla}\,\ell n\, n \tag{35}$$

where the last term accomodates the diffusion effects. The development of the three temperature theory then parallels the two temperature theory in the determination of the moments, albeit with the additional features introduced by the ion density gradient. Finally the drift velocity and diffusion coefficients are given by

$$u = u_0 + \frac{1}{2}(2kT_L/m)^{1/2}\langle\psi_{001}\rangle \qquad (36)$$

$$D_L = \frac{1}{2}(2kT_L/m)^{1/2}\partial\langle\psi_{001}\rangle/\partial(\nabla_z \ln n) \qquad (37)$$

$$D_T = \frac{1}{2}(2kT_T/m)^{1/2}\partial\langle\psi_{100}\rangle/\partial(\nabla_t \ln n) \qquad (38)$$

where ∇_z and ∇_t are the longitudinal and transverse gradient operators, respectively. Unfortunately low order results do not produce simple expressions for D_L and D_T similar to those for T_f and v in the two temperature theory.

Results

The two temperature theory and, just recently, the three temperature theory have been applied to many alkali ion-rare gas and some negative halogen ion-rare gas combinations. As these are all closed shell ions and atoms the potentials are spherically symmetric. Hence the inversion procedure is available to obtain potentials determined directly from experimental data. In some cases theoretical potentials are available and may be used to obtain mobilities for comparison with experiment.

Initially the (Li^+, Na^+, K^+, Rb^+) - (He, Ne, Ar) potentials, developed by Waldman and Gordon[19] using an electron cloud-Drude model calculation, were tested[20]. The overall agreement was very good but enough discrepancies remained to indicate that the mobility data could be used to refine the potentials. Several Li^+ - He potentials, including some obtained by ab-initio calculations, were also tested and agreement to better than 1% was obtained using the CEPA potential of Hariharan and Staemmler[21] except at very high values of E/N. The directly determined Li^+ - He potential differed from the CEPA potential by about 15% in the hard core region[22]. For instance at an internuclear separation of 2 a.u. the CEPA potential value was .081 a.u., the directly determined potential was .068 a.u., and a beam scattering experiment yielded .070 a.u. at this separation distance[23]. The experimental data used in all these studies were derived from several sourses but mainly the Georgia Tech drift tube group headed by E. W. McDaniel. A compendium of ion-neutral transport properties has been published by Ellis et al.[24] The Li^+ - Ar case has also been used to test several theoretical potentials, including the CI potential of Olson and Liu[25]. A directly determined potential was also obtained but the differences between it and the CI potential are very small and of questionable significance given the experimental errors of about 2%. The short region of E/N where the mobility calculation does not fully converge, as discussed earlier, does not effect the overall result[15].

The first directly determined potentials for which no theoretical potentials were available were the cases of Cs^+ - Ar, Cs^+ - Kr, and Cs^+ - Xe. Trial potentials of the 4-6-n form

$$V(r) = -C_4/r^4 - C_6/r^6 + C_n/r^n \qquad (39)$$

were used with C_4 fixed by the polarisation limit. A variety of n, C_n, and C_6 parameters were used but the inversion procedure only converged for $n \approx 12$ and for narrow ranges of C_6 and C_n. As noted earlier the convergence is inhibited unless the well depth given by n, C_n, and C_6 is correct. The final potentials are, of course, not of the 4-6-n type and are given in tabular form by Gatland et al.[26]. Although they do not fully describe the potential two particular features are of interest: the internuclear separation distance, r_m, and the depth, ε, of the potential well. Values of these parameters for Cs^+ - Ar are r_m = 6.65 a.u. and ε = .00310 a.u.; for Cs^+ - Kr r_m = 6.58 a.u. and ε = .00445 a.u.; and for Cs^+ - Xe r_m = 7.67 a.u. and ε = .00400 a.u. (the atomic length and energy units are .529 Å and 27.2 eV, respectively). These values are also given in Table 1 along with other ion-atom combinations.

It is interesting to note that the inversion procedure only produces one potential for each of the Cs^+ cases. However the experimental error of about 2% does not carry over directly to the potentials. The mobility is relatively insensitive to a deepening of the well if the separation distance is appropriately decreased at the same time. Errors associated with r_m and ε are typically in the range from 5% to 10%.

Directly determined potentials were also obtained for several other alkali ion cases: Rb^+ - Ar, Rb^+ - Kr, Rb^+ - Xe[27], K^+ - Ar, K^+ - Kr, and K^+ - Xe[28]. In each of these cases the Gordon-Waldman potential was used as the trial potential and the values of r_m and ε for the directly determined potentials are given in Table 1. The full form of the potentials for K^+ - Ar, K^+ - Kr, and K^+ - Xe are shown in Fig. 1 and the resulting mobility curves are shown in Fig. 2, Fig. 3, and Fig. 4 along with those for the Gordon-Waldman potentials and the experimental data.

As regards the negative halogen ions four cases have been analyzed to obtain directly determined potentials: Cl^- - Xe by Thackston et al.[29], and Br^- - Ar, Br^- - Kr, and Br^- - Xe by Lamm et al.[30]. All the trial potentials were of the 4-6-n type, the values of r_m and ε for the directly determined potentials are given in Table 1, and the full tabulated potentials appear in the original articles. All the directly determined potentials discussed so far were obtained using the two temperature theory.

Several halogen ion-rare gas potentials, some calculated theoretically and others developed using data from beam experiments, have been tested against mobility data. Viehland and Mason(31) used the three temperature theory to test the F^- - Xe potential of de Vreugd et al.(32) which is a fit to differential cross sections. The depth of the well was found to be in good agreement with both mobility and diffusion data. At the same time Viehland and Mason also checked the Cl^- - Xe well depth against the limit set by de Vreugd. Viehland, Mason and Lin(33) further tested the CI and SCF H^- - He potentials, the SCF Br^- - He, and the SCF Cl^- - Ar potentials of Olson and Liu(34)(35)(36) as well as potentials for the same systems derived from beam scattering experiments(37)(38). For H^- - He the CI potential gave mobilities in good agreement with experiment, those from the SCF potential were about 3% too low and those from the beam potential (extrapolated) were nearly 10% too high. For Br^- - He the SCF potential again gave mobilities about 3% low whilst the beam results agreed well. For Cl^- - Ar both the SCF and beam potentials gave reasonable mobilities for E/N above 200 Td but were considerably too high at low E/N, i.e.,

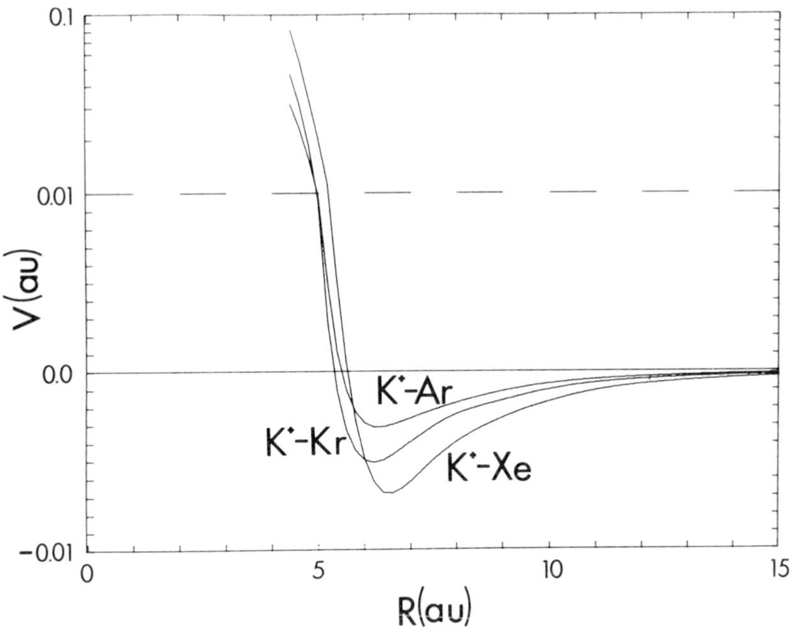

Fig. 1. The directly determined potentials for K^+ - Ar, K^+ - Kr, and K^+ - Xe. (Note logarithmic scale for energies above 0.01 a.u.)

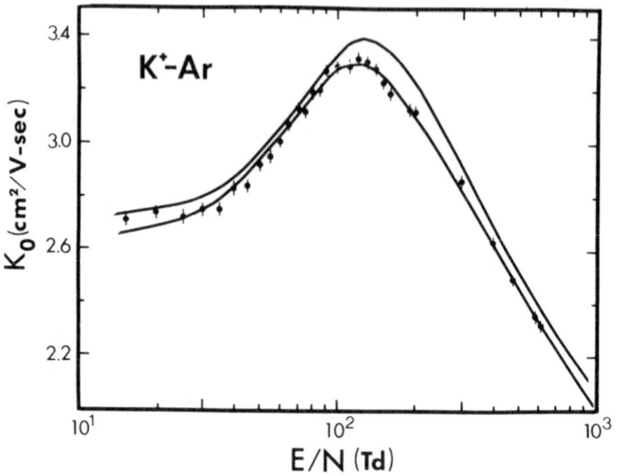

Fig. 2. The K^+ - Ar reduced mobilities from experiment (data points), Gordon-Waldman potential (upper curve), and the directly determined potential (lower curve).

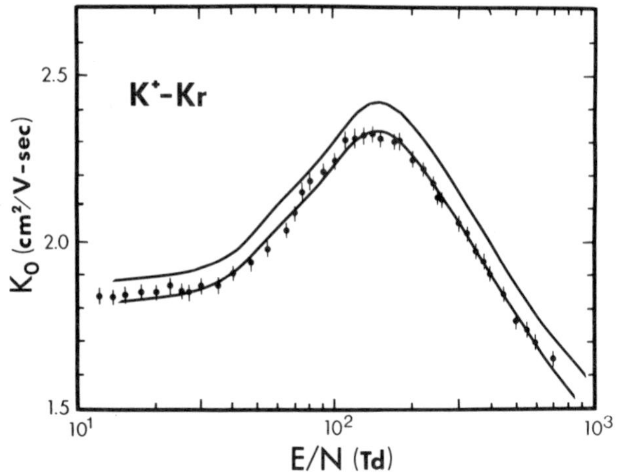

Fig. 3. The K^+ - Kr reduced mobilities from experiment (data points), Gordon-Waldman potential (upper curve), and the directly determined potential (lower curve).

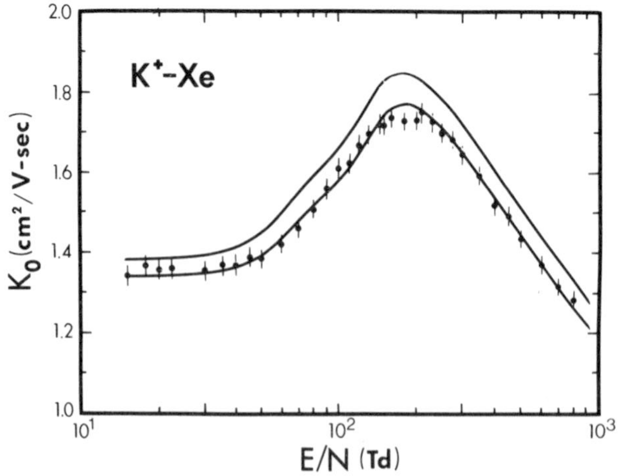

Fig. 4. The K^+ - Xe reduced mobilities from experiment (data points), Gordon-Waldman potential (upper curve), and the directly determined potential (lower curve).

below the mobility peak, a feature which could not be tested by the H^- - He or Br^- - He data.

Recently Viehland has analyzed the Li^+ - (He, Ne, Ar, Kr, Xe) cases using the three temperature theory(39). He utilized potentials based on ab-initio calculations, beam scattering experiments, particularly those of Polak-Dingels, Rajan and Gislason(40), as well as potentials determined directly from mobility data using the two temperature theory(15)(22). Both mobilities and diffusion coefficients were calculated for the various potentials and compared with several sets of experimental data, with varying results. The latest beam results(40) agreed well with the mobility data for Li^+ - Ne, Li^+ - Ar, Li^+ - Kr, and Li^+ - Xe, but disagreed somewhat in the case of Li^+ - He where the well depth is relatively shallow. Directly determined potentials were obtained for Li^+ - He, a slight modification of the two temperature result(22), Li^+ - Ne, Li^+ - Kr, and Li^+ - Xe. The Li^+ - Ar potential based on the two temperature(15) theory also gave the best results for the three temperature theory, but again convergence was poor in the E/N range from 30 Td to 80 Td. The potential well positions and depths (two temperature, three temperature, and beam) are listed in Table 1.

Discussion

In many cases the potentials determined directly from mobilities using the two temperature theory were the first experimental measurements of these potentials for internuclear separation distances at and outside the well region (apart from the polarization limit). Some beam scattering experiments, which probe the hard core, had been extended out to the well region and could be compared with mobility results but the overlap region was rather narrow.

During the last few years this situation has improved considerably. Gislason and coworkers have published potentials based on beam scattering total cross section experiments for potassium ions in the rare gases(41), lithium ions in rare gases(40), and cesium ions in rare gases(42). All of these yield positions and depths for the potential wells as listed in Table 1. Also Viehland is completing a study of all the remaining alkali ions (Na^+, K^+, Rb^+, and Cs^+) in all the rare gases (He, Ne, Ar, Kr, and Xe). In three cases, Rb^+ - He, Cs^+ - He, and Cs^+ - Ne, he finds that the mobility data obtained at a gas temperature of 300K does not provide low enough ion-atom collision energies, even at $E/N \approx 0$, to probe the potential well. For all the other cases he has obtained directly determined potentials using the three temperature theory. Preliminary results on the positions and depths of the potential wells is included in Table 1 but the reader is referred to the forthcoming paper(43) for the full details of the potentials and the comparisons with other results. Viehland finds two areas of concern. Firstly the two temperature and three temperature directly determined potentials do not agree in the cases of K^+ - Ar and Rb^+ - Xe. Secondly his potentials do not agree with those obtained by Gislason(41)(42) for the cases of Li^+ - He (as noted earlier), Na^+ - He, K^+ - He, K^+ - Ne, and Cs^+ - Ar. These are all the potentials with well depths of less than .00275 a.u. Both these areas of disagreement are being reviewed but the potentials obtained using the three temperature theory appear to be on the sounder footing in both cases.

The determination of potentials from mobility data for Rb^+ - He, Cs^+ - He, and Cs^+ - Ne will require drift tube experiments at lower temperatures. Though data on these ion-atom combinations has not yet been obtained some low temperature experiments have been performed and indicate that such data should be adequate. Koizumi, Kobayashi, and Kaneko have measured the mobilities of Ne^+ and Ar^+ in He at 82 K(44) and have found that the mobilities are close to the polarization limit for low E/N. Cassidy and Elford(45) have measured mobilities and have derived potentials for Li^+ - He and K^+ - He using data at 80K. They report that their Li^+ - He potential agrees well with that of Hariharan and Staemmler(21) but the details have not yet been published.

The overall picture now appears quite promising. The combination of low temperature mobility data, room temperature mobility data, and data from beam experiments spans a very wide range of collision energies. This in turn should lead to further refinements in the potentials and the determination of some potentials not yet amenable to the inversion procedure. Also several negative halogen ion-rare gas combinations remain to be studied. The general agreement between the potentials determined from mobilities and those from CI ab-initio calculations is encouraging and the pattern of differences between the directly determined potentials and the SCF potentials may also be of interest. The general trends in the alkali ion-rare gas well depths and positions are as expected. Only Cs^+ - Kr is slightly anomolous. Hopefully further theoretical calculations and improved experimental techniques will clarify this situation.

It is a pleasure to thank Dr. L. A. Viehland, Dr. E. Gislason, and Dr. E. W. McDaniel for suggestions, comments, and information about their latest results.

Table 1. Potential well positions and depths in a.u.*

Ion	Gas	Two Temperature		Three Temperature		Beam	
Li^+	He	3.69	.00272	3.65	.00270	3.71	.00261
	Ne	-	-	3.94	.00452	3.99	.00419
	Ar	4.55	.01030	-	-	4.57	.01151
	Kr	-	-	4.58	.01460	4.63	.01691
	Xe	-	-	4.74	.02010	4.80	.01963
Na^+	He	-	-	4.61	.00126	4.20	.00189
	Ne	-	-	4.59	.00243	4.48	.00280
	Ar	-	-	5.09	.00700	5.20	.00599
	Kr	-	-	5.50	.00810	5.42	.00772
	Xe	-	-	5.64	.00947	5.73	.00952
K^+	He	-	-	5.41	.00084	4.84	.00092
	Ne	-	-	5.55	.00146	5.07	.00173
	Ar	6.27	.00314	5.55	.00468	5.60	.00460
	Kr	6.18	.00511	6.12	.00530	6.00	.00482
	Xe	6.52	.00687	6.36	.00772	6.39	.00603
Rb^+	He	-	-	-	-	-	-
	Ne	-	-	5.97	.00123	-	-
	Ar	6.53	.00324	6.44	.00315	-	-
	Kr	6.70	.00421	6.69	.00425	-	-
	Xe	7.40	.00453	6.70	.00679	-	-
Cs^+	He	-	-	-	-	5.65	.00058
	Ne	-	-	-	-	6.27	.00089
	Ar	6.65	.00310	6.57	.00313	6.99	.00233
	Kr	6.58	.00445	6.58	.00433	7.05	.00371
	Xe	7.67	.00400	7.76	.00417	7.59	.00439
Cl^-	Xe	7.13	.00496	-	-	-	-
Br^-	Ar	7.06	.00217	-	-	-	-
	Kr	7.01	.00320	-	-	-	-
	Xe	6.85	.00534	-	-	-	-

*References. Two temperature: 15, 22, 26, 27, 28, 29 and 30. Three temperature: 39 and 43. Beam 40, 41, 42, and 46.

References

(1) G. H. Wannier, Bell System Technical Journal 32, 170 (1953).

(2) E. W. McDaniel and E. A. Mason, "The Mobility and Diffusion of Ions in Gases", Wiley, 1973.

(3) H. R. Skullerud, J. Phys. B. 6, 728 (1973).

(4) H. R. Skullerud, J. Phys. B. 6, 918 (1973).

(5) S. L. Lin and J. N. Bardsley, J. Chem. Phys. 66, 436 (1977).

(6) H. R. Skullerud, Brit. J. Appl. Phys. 1, 1567 (1968).

(7) L. A. Viehland and E. A. Mason, Ann. Phys. (NY) 91, 499 (1975).

(8) L. A. Viehland and E. A. Mason, Ann. Phys. (NY) 110, 287 (1978).

(9) L. A. Viehland, M. M. Harrington, and E. A. Mason, Chem. Phys. 17, 433 (1976).

(10) E. B. Smith, Physica (Utr) 73, 211 (1974).

(11) S. L. Lin, I. R. Gatland, and E. A. Mason, J. Phys. B: Atom. Molec. Phys. 12, 4179 (1979).

(12) F. Howorka, F. C. Fehsenfeld, and D. L. Albritton, J. Phys. B: Atom. Molec. Phys. 12, 4189 (1979).

(13) L. A. Viehland and E. A. Mason, private communication.

(14) H. R. Skullerud, Proc. 3rd Int. Swarm Seminar, Innsbruck, Austria, 1983.

(15) I. R. Gatland, J. Chem. Phys. 75, 4162 (1981).

(16) "Handbook of Mathematical Functions", M. Abromowitz and I. A. Stegun, Dover, 1965.

(17) S. L. Lin, L. A. Viehland, and E. A. Mason, Chem. Phys. 37, 411 (1976).

(18) L. A. Viehland and S. L. Lin, Chem. Phys. 43, 135 (1979).

(19) M. Waldman and R. G. Gordon, J. Chem. Phys. 71, 1325 (1979) and private communication.

(20) I. R. Gatland, L. A. Viehland, and E. A. Mason, J. Chem. Phys. 66, 537 (1977).

(21) P. C. Hariharan and V. Staemmler, Chem. Phys. 15, 409 (1976).

(22) I. R. Gatland, W. F. Morrison, H. W. Ellis, M. G. Thackston, E. W. McDaniel, M. H. Alexander, L. A. Viehland, and E. A. Mason, J. Chem. Phys. 66, 5121 (1977).

(23) H. Inouye and S. Kita, J. Chem. Phys. 57, 1301 (1972).

(24) H. W. Ellis, R. Y. Pai, E. W. McDaniel, E. A. Mason, and L. A. Viehland, Atomic Data and Nuclear Data Tables 17, 177 (1976) and 22, 179 (1978).

(25) R. E. Olson and B. Liu, Chem. Phys. Lett. 62, 242 (1979).

(26) I. R. Gatland, M. G. Thackston, W. M. Pope, F. L. Eisele, H. W. Ellis, and E. W. McDaniel, J. Chem. Phys. 68, 2775 (1978).

(27) I. R. Gatland, D. R. Lamm, M. G. Thackston, W. M. Pope, F. L. Eisele, H. W. Ellis, and E. W. McDaniel, J. Chem. Phys. 69, 4951 (1978).

(28) D. R. Lamm, M. G. Thackston, F. L. Eisele, H. W. Ellis, J. R. Twist, W. M. Pope, I. R. Gatland and E. W. McDaniel, J. Chem. Phys. 74, 3042 (1981).

(29) M. G. Thackston, F. L. Eisele, W. M. Pope, H. W. Ellis, E. W. McDaniel, and I. R. Gatland, J. Chem. Phys. 73, 3183 (1980).

(30) D. R. Lamm, R. D. Chelf, J. R. Twist, F. B. Holleman, M. G. Thackston, F. L. Eisele, W. M. Pope, I. R. Gatland, and E. W. McDaniel, J. Chem. Phys. 79, 1965 (1983).

(31) L. A. Viehland and E. A. Mason, Chem. Phys. Lett. 83, 298 (1981).

(32) C. de Vreugd, R. W. Wijnaendts van Resandt, and J. Los, Chem. Phys. Lett. 65, 93 (1979).

(33) L. A. Viehland, E. A. Mason, and S. L. Lin, Phys. Rev. A 24, 3004 (1981).

(34) R. E. Olson and B. Liu, Phys. Rev. A 17, 1568 (1978).

(35) R. E. Olson and B. Liu, Phys. Rev. A 20, 1344 (1979).

(36) R. E. Olson and B. Liu, Phys. Rev. A 22, 1389 (1980).

(37) T. L. Bailey, C. J. May, and E. E. Muschlitz, J. Chem. Phys. 26, 1446 (1957).

(38) S. Kita, K. Noda, and H. Inouye, J. Chem. Phys. 64, 3446 (1976).

(39) L. A. Viehland, Chem. Phys. 78, 279 (1983).

(40) P. Polak-Dingels, M. S. Rajan, and E. A. Gislason, J. Chem. Phys. 77, 3982 (1982).

(41) F. E. Budenholzer, E. A. Gislason, and A. D. Jorgensen, J. Chem. Phys. 78, 5279 (1983).

(42) M. S. Rajan and E. A. Gislason, J. Chem. Phys. 78, 2426 (1983).

(43) L. A. Viehland, to be submitted to Chem. Phys.

(44) T. Koizumi, N. Kobayashi, and Y. Kaneko, J. Phys. Soc. Japan 48, 1678 (1980).

(45) R. A. Cassidy and M. T. Elford, Proc. 3rd Int. Swarm Seminar, Innsbruck, Austria, 1983.

(46) E. A. Gislason, private communication.

Transverse Ion Diffusion in Gases

E. Märk[1] and T. D. Märk[2]

[1] Höhere Technische Bundeslehr- und Versuchsanstalt, A-6020 Innsbruck, Austria

[2] Institut für Experimentalphysik, Universität Innsbruck, A-6020 Innsbruck, Austria

1. Introduction

Transport properties of ions in gases (i.e. ion mobilities and diffusion coefficients) are of intrinsic, fundamental and applied interest /1/. On the one hand they can give information about the ion-neutral interaction potential; on the other hand they can be used to describe quantitatively the behavior of ions moving in a neutral buffer gas and related charge transport phenomena and are required together with ionization cross section data /2/ for a quantitative understanding of electrical discharges.

Research on ionic transport properties began almost 100 years ago /3-6/. Early research has been reviewed by Loeb /7/ and Massey /8/. With recent improvements in experimental and theoretical techniques ion mobilities and longitudinal diffusion coefficients in gases of good accuracy have become available for a number of atomic, molecular and cluster ions in various buffer gases, and surveys of experimental and theoretical results have been given by Mc Daniel and Mason /1/ and others /9-14/.

Meaningful measurements of transverse diffusion coefficients were first made only in the late 1960's and comparatively few data are available /15-25/. Moreover, recent studies have indicated, that some of the old data are either incorrect or refer to ions with unknown identity. Mc Daniel and Mason /1/ have prepared a comprehensive discussion and review of the history of experimentation in this field through the end of 1972 with some updates of data in later compilations /9-11/. New experimental techniques for obtaining reliable transverse diffusion coefficients have only been developed very recently and are discussed in the present review /26-37/.

Additionally, the reader is referred to a number of recent improved theoretical studies on ion diffusion, some including the effect of inelastic collision processes, e.g. see /38-58, 12,13/. It is interesting to point out in this conjunction that Waldman et al. /57/ recently concluded that the generalized Nernst-Townsend relations (see also experimental tests /38, 59-64/), providing approximate connections between ion mobility and longitudinal and transverse diffusion coefficients, remain valid even when resonant charge transfer is the dominant scattering mechanism. Moreover, they showed that diffusion coefficients of systems dominated by resonant charge transfer contain additional information on the ion-neutral interaction potential, not contained in the mobility data. Thus it might be possible to combine both, mobility and diffusion data, into an inversion scheme to determine the multiple ion-neutral interaction potentials, similar to the procedure now being used to determine ion-neutral interaction potentials from mobility data for non resonant charge transfer systems /65-75/.

2. Experimental

Measurements of the transverse diffusion coefficient D_T as a function of the reduced field strength E/N are made in drift tubes. To date three experimental techniques have been utilized to measure D_T (E/N), i.e. the Townsend method, the attenuation method and the radial ion distribution (RID) method. Descriptions of some of the instruments used earlier appear in previous reviews /1,8/ and will not be repeated here. Desirable features for all of these techniques, if accurate data are to be obtained, are low and stable ion swarm current densities, negligible space charge, no distortions of the electric drift field, ion identification and separation, ultra high vacuum conditions and accurate pressure determination.

2.1. Townsend Method

This method dates back to Townsend /5/ and Llewellyn-Jones /76/ and variants of this method have been used recently for measurements of the ratio of the transverse diffusion coefficient to the mobility, D_T/K, /15-21,26,27,37/. In this method ions from an ion source (electron impact, glow discharge, thermionic filament) enter the diffusion chamber (drift tube) through a small hole or a slit and travel under the action of a uniform electric field (in direction of the axis of the drift tube) in a neutral buffer gas to an end plate. This end plate consists of two or more ion collecting electrodes. The spreading of the ion swarm as it travels through the diffusion chamber is directly related to D_T/K and E/N and can be determined indirectly by measuring the ratio of ion currents to these ion collecting electrodes /16,17/. In order to increase the accuracy this ratio is also measured as a function of off-axis displacement /18-21,26.27,37/ and of drift distance /37/ (this, however, requires stable ion source outputs, which is not always fulfilled). The success of this method depends on the ability to ascertain that only one ion species is present, that this ion species can be identified and that ion molecule reactions do not occur to an appreciable degree within the diffusion region. In an attempt to overcome some of these problems drift-selected ions have been used /16,17,26, 27/ and/or mass spectrometers have been added behind the end plate /26,27,37/. Despite these improvements, these apparatus will only provide reliable data if only one ion species is present in the diffusion region. Moreover, there seems to be a longstanding problem with the interpretation of the results of the Townsend-Huxley diffusion experiments /77/.

2.2. Attenuation Method

This method was introduced by the Georgia Tech group in 1968 /22-25/ using a pulsed drift tube mass spectrometer apparatus. Descriptions of their apparatus have been given in several reviews /1,7,78,79/ and will therefore not be repeated here. The method used consists of measuring the total number of ions composing the arrival spectra at various drift distances and comparing this intensity with the integrated intensity on the axis calculated for the corresponding conditions. Because the total ion intensity depends on both, transverse diffusion and loss ion molecule reactions, an iterative fitting procedure has to be used assuming that the ion molecule reaction constants vary fairly slowly with E/N. Moreover, this analysis depends on the assumption that the ion species under study is only produced in the ion source. According to /1/ even in the simple situation of no ion molecule reactions the accuracy achieved by this methods is typically 20% due to the fact that it is difficult to obtain stable ion source outputs as the ion source is moved along the axis of the drift tube.

Recently, Varney et al. /28/ reported a new variant to measure transverse diffusion coefficients with a drift tube mass spectrometer. This method consists of measuring the count rate of ions traversing a drift tube from a point source to a point exit aperture as a function of E/N, thus avoiding the necessity to vary the drift distance. On the other hand, it is necessary to know the reaction rate constants for loss ion molecule reactions as a function of E/N and the obtained D_T/K values have to be normalized by some additional means.

2.3. Radial Ion Distribution (RID) Method

This method was introduced recently by the Innsbruck group /29-36/ (see also an earlier study by Rees and Alger /80/) and belongs to the Townsend /5/ tradition in that we measure the lateral (radial) distribution of the swarm directly. It differs, however, in two important aspects from the Townsend method: the ion current density is sensed at individual points, rather than as integrals over two (or more) ion collecting electrodes, and the ions are mass-analyzed to select the particular species we wish to study.

2.3.1. Mathematical Analysis

The starting point in the analysis of the space-time behavior of a drifting ion swarm in a static gas target under the influence of a uniform field is the modified diffusion equation /1,81/. For cylindrical symmetry, we may write

$$\frac{\partial n}{\partial t} = D_L \frac{\partial^2 n}{\partial z^2} + D_T \frac{1}{r} \frac{\partial}{\partial r} r \frac{\partial}{\partial r} n - w \frac{\partial n}{\partial z} - \alpha n + \beta \qquad (1)$$

where $n = n(r,z,t)$ is the number density of a single species of ions at z,r (axial and radial coordinates, respectively) at time t. α is the rate of loss of these ions per unit time due to reactions with reactants which are assumed to be uniformly dispersed throughout the volume. D_L and D_T are the longitudinal and transverse diffusion coefficients, respectively, of the ions. w is the drift velocity of the ions through the gas under the influence of constant, uniform electric field, E. β is a source term to represent an input of ions at the entrance of the drift space.

The solution to this diffusion equation for a point source of ions located at z=r=o which emits a vanishingly narrow pulse of n_o ions at time t=0, where all boundaries are ignored, is /1/

$$n(r,z,t) = \frac{n_o}{(4\pi t)^{3/2} D_T D_L{}^{1/2}} \exp\left(-\alpha t - \frac{r^2}{4D_T t} - \frac{(z-wt)^2}{4D_L t}\right). \qquad (2)$$

A steady-state solution for the case when the point source emits a steady stream of ions, rather than a singular pulse, can be constructed from the above time-dependent solution /34/.

$$n(r,z) = \frac{n'_o}{4\pi D_T z} \exp(-r^2 w/4zD_T - \alpha z/w). \qquad (3)$$

where n_o' is a constant rate of ions (number per unit time), and $4D_L\alpha/w^2 < 1$ and $r^2D_L/z^2D_T < 1$.

We measure particle current density rather than ion volume density. If the sampling aperture does not unduly distort the drift field in its vicinity, we expect the particle current density, j, to be related to the ion volume density, n, by the formula /1/

$$j = wn - D_L \frac{\partial}{\partial z} n. \tag{4}$$

With the above approximations, and remembering that the transverse diffusion coefficient, D_T, in the present method is determined from only the dependence of j upon r, we can take for our further analysis, the particle current density

$$j(r) = j(o) \exp \quad (-r^2w/4D_Tz) \tag{5}$$

where $j(o) = (n_o' \; w/4\pi D_T z) \exp \; (-\alpha z/w)$.
This formula has to be modified if a slit source of finite width is used. If the x dependence of the ion current is determined keeping y constant (that is, moving perpendicular to the narrow slit), we can write with very good approximation,

$$j(x) = j'(o) \exp \quad (- x^2w/4D_Tz) \tag{6}$$

This equation, then, is the basis for our further analysis of the experimental data; the ratio D_T/K is determined by measuring a number of points of the radial ion distribution and then fitting the experimental data with a gaussian shape function.

2.3.2. Apparatus

Figure 1 presents a schematic view of the apparatus used in our measurements of D_T/K. The key elements are the ion source IQ which can be moved laterally, normal to the entrance slit I with a precision stage S, the ion diffusion region D in which a very uniform, variable electric field can be maintained, the exit aperture E fixed on the axis of symmetry and the quadrupole mass filter QMF which can be tuned to pass only the desired ion species into the channeltron detector C.

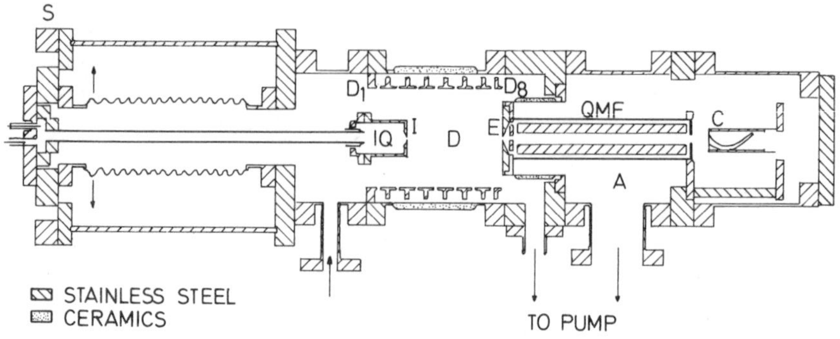

Fig. 1 Schematic view of the transverse diffusion apparatus after Sejkora and Märk /33/. The ratio D_T/K is measured directly by the radial ion distribution method.

Special care was taken so that the gases used were of very high purity. For this reason the gas inlet and mixing system of glass and nonmagnetic stainless steel can be baked at 300°C and can be pumped to high vacuum. Separate spectrographic quality gases or gas mixtures at different pressures can be fed into the ion source and diffusion drift region; the working pressure (0.1 to 1 Torr) in these regions is measured with precise capacitance gauges.

Different ion sources can be interchangably attached to the ion source mounting stage with its associated gas and electrical feed-throughs. One source is based on α particle emission. It follows an earlier conception by Crompton and Elford /82/ and has been described in detail by Hilchenbach /83/ and briefly by Sejkora et al. /29,30/. A cross section through the longitudinal axis of the source is displayed in Fig. 2.

The source operates as follows: Alpha particles (5.5 MeV) from the 9.7 x 10^7 Bq ^{241}Am foil enter region A, where the gas pressure is typically 0.3 Torr. The potential between foil and

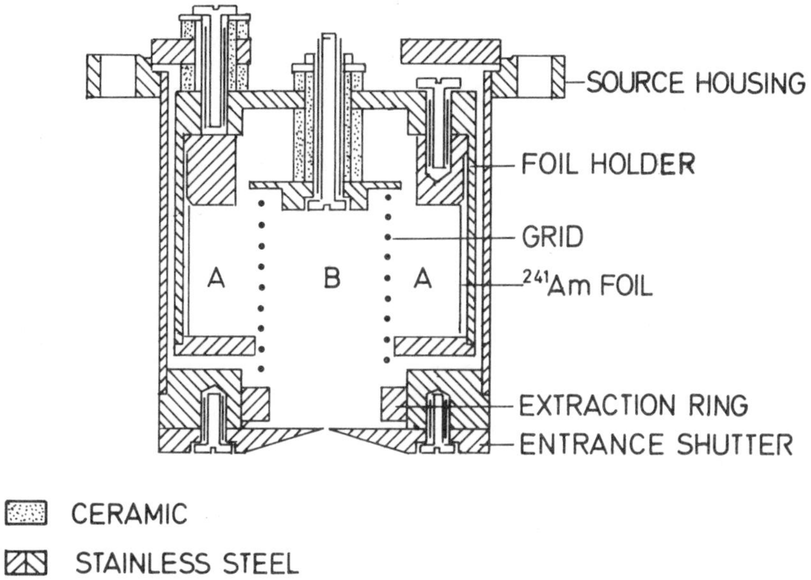

Fig. 2. Schematic view of the low current-high stability ion source after Sejkora et al. /29,30/ and Hilchenbach /83/. The ionization within the source is produced by Townsend avalanches initiated by α particles.

grid, U_{GC}, is adjusted (150-250 V) so that Townsend avalanches occur in the radial field in A thereby greatly enhancing the ionization occurring in the gas (e.g. see Fig. 3). The electrons produced in region A then enter the nearly field-free region B where ions are produced by electron impact, as well as by the original α particles. Approximately 2% of the resulting ions are extracted through the entrance slit with the help of the potential between foil and entrance shutter, U_{GE}, (typically 70 Volts). Since the half life of ^{241}Am is ~ 460 years, the current output from this source is quite steady, if other conditions are maintained.

The properties of this source which are especially important for our measurements are: 1. High stability for ion currents even below 10 pA, 2. working pressures above 0.1 Torr, and 3. no temperature gradients.

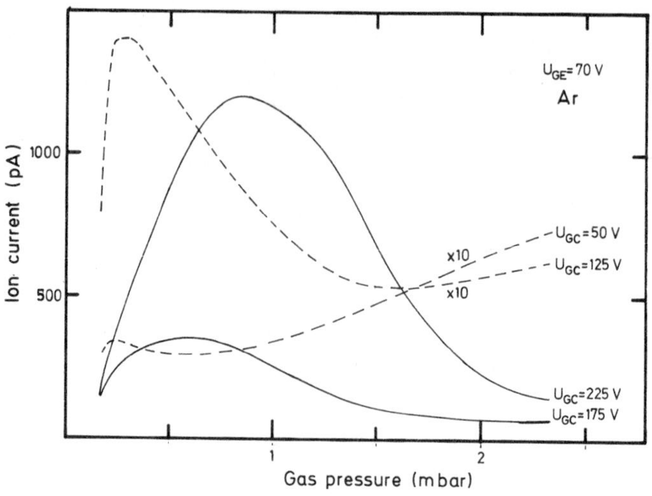

Fig. 3. Ion current-pressure characteristic of the ion source shown in Fig. 2. The current shown is measured at the entrance shutter.

The ion source mount is attached to a dove-tail stage which can be driven by a precision screw to provide carefully controlled translation perpendicular to the axis of symmetry of the drift tube. The dove-tail mechanism prevents any tilting of the source mount. Bearing surfaces are ground flat to 0.001 mm. The stepping motor, which is computer controlled, operates at a frequency of 50 Hz. The displacement of the entrance slit, mounted on the source holder, can be controlled with an accuracy of at least 0.01 mm.

The ions enter the diffusion region from the ion source under the influence of the extraction potential, U_{GE}, discussed above. The entrance slit formed by two knife edges of stainless steel, and whose length 2a = 20 mm, is adjusted to have a precise width of 2b = 0.2 mm. The shape of this entrance slit shutter has been recently improved in order to minimize field distortion in the diffusion region /35,36/.

The diffusion region consists of a cylindrical drift tube 161 mm long and 114 mm in diameter (see Fig. 1). The uniform electric field is applied through a series of 8 "drift rings" whose potential differences are maintained by a chain of resistors with 500 Ω trimmers, across a doubly-stabilized, voltage-regulated power supply. With a center-to-center spacing of the drift rings of 23 mm, an interring gap of 3 mm and an inside diameter of the rings of 114 mm, it can be shown /84/ that a constant field region is thereby produced such that in a cylinder 40 mm in diameter, centered on the axis of symmetry, the largest field deviation at any point is no larger than 0.1 % of the average field. However, when the cylindrical source is inserted into this uniform field, slight distortion occurs. This distortion can be compensated for as follows: the entrance shutter is maintained at the same potential as the 3rd drift ring, as the two have the same axial position. The first two drift rings, D_1 and D_2, are maintained with a larger potential differences than the others. By a semiempirical method, using He^+ in He, as well as by direct computation, one can demonstrate that, if the gap potential for D_1, D_2 and D_3 is maintained at 2.0 times the gap potential for the remaining rings, a reasonably uniform field results /32/.

As with the entrance slit, special care must be taken with the exit aperture to insure that the current density measurement is not distorted by inhomogeneous fields and contaminants. The exit aperture, a circular hole 0.15 mm in diameter, is machined from stainless steel. The surface of this exit aperture was machined as flat and scratch-free as possible, with the edges of the circular hole almost knifelike (the edge thickness is 0.05 mm) to minimize transition losses /85-87/. Great care was taken to insure that the exit shutter, which could be heated by means of a "thermocoax", was free of surface contamination. The exit shutter, although electrically isolated to allow the measurement of the total ion current, is maintained at ground potential as is the 8th drift ring, D_8. Ions passing through the exit aperture from the diffusion region into an analyzer region, are accelerated by an extraction electrode into a quadrupole mass filter. The ions transmitted by the mass filter enter a Channeltron detector, C, whose entrance aperture is biased at - 3kV, where current pulses are formed and passed on for electronic analysis.

A thorough discussion of the microprocessor-controlled data acquisition is presented by Sejkora /84/. At the heart of the system is a microcomputer whose CPU is the Intel 8080 chip. Operator access to the system is trough a visual display terminal. Working storage is a 48 kbyte RAM; data storage is upon "mini floppy disks" with a storage capacity of 100 kbyte/disk through a floppy disk drive. Since the ion source with the stepping motor and its control logic reside at a substantial positive potential, while the control electronics are kept at ground, it is necessary to optically couple the stepping motor control logic to the computer interface carrying instructions to the stepping motor.

At a given E/N and for a specified set of operating conditions, we have the raw data on a run over the radial ion density profile for a number of distances from the axis of symmetry of the diffusion chamber. Figure 4 shows a typical ion density profile for Ar^+ in Ar. The solid line through the points is the result of fitting to eq. 6. The value of the quantity D_T/K is extracted from the best fit /34/.

A thorough discussion of the statistic (fitting procedure) and possible systematic errors are presented by Sejkora et al. /34/. Figs. 5 and 6 show checks for possible space charge effects and pressure effects. Moreover, the apparatus is presently under reconstruction to provide ion source displacement in the z-direction, i.e. variation of the drift distance. This allows additional checks for possible end effects and field distortion (see above).

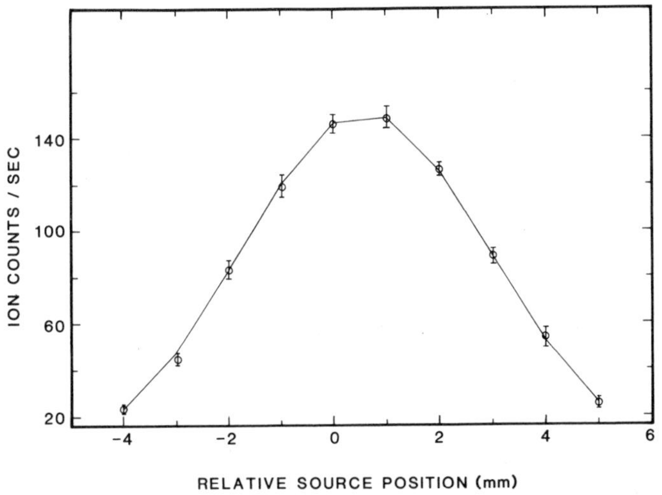

Fig. 4. Typical Ar^+ ion density profile in argon. Solid line joins fitted points after Sejkora et al. /34/.

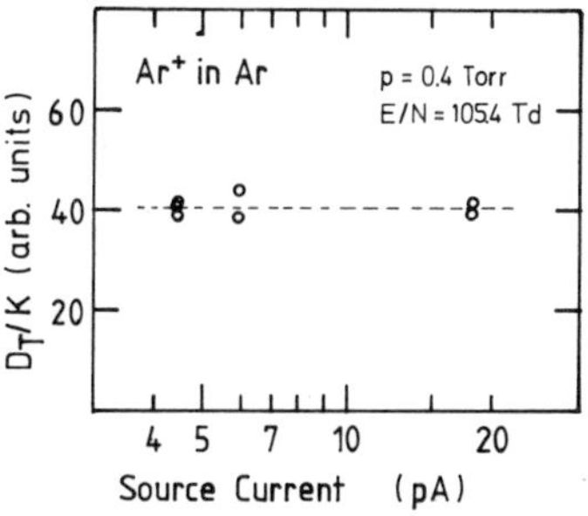

Fig. 5. Check for possible space charge effects after Sejkora et al. /34/.

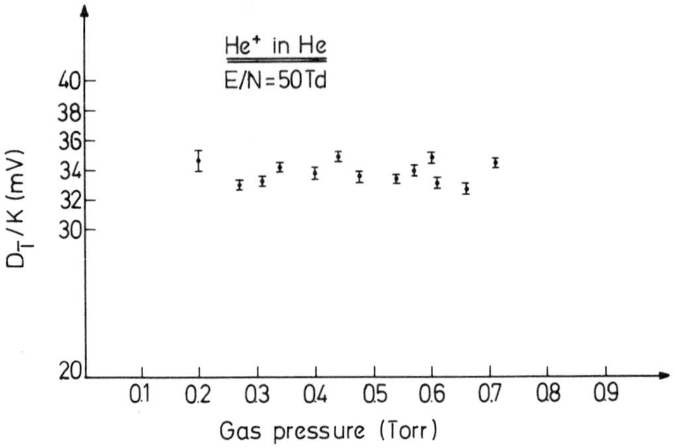

Fig. 6. Check for possible pressure effects after Märk et al. /35,36/.

3. Results

Earlier experimental results have been reviewed by Mc Daniel and Mason /1/ in 1973 and summarized by Ellis et al. /11/ presenting data up to 1978. The present review brings results on transverse diffusion coefficients up to date as of December 1983. The scientific motivation for undertaking precise measurements of D_T, beyond its intrinsic interest, is that D_T, in conjunction with D_L, is particularly sensitive to the long range part of the ion-neutral interaction potential, which is diffucult to study in other ways /1,46/. Specifically, D_T is very sensitive to the large angle scattering cross section /34/. The usual interpretation of measured D_T/K characteristics by experimentalists is as an approximate expression due to Wannier /88/ and discussed by Skullerud /46/

$$D_T/K = a + b \quad w^2 \qquad (7)$$

From very basic considerations one can demonstrate that

$$a = kT/q \qquad (8)$$

a relationship variously known as the "Nernst-Townsend" /3,4/ or "Einstein" relation, and /43,46/

$$b = (1 + m/M) R\ (4m/M + 3R)^{-1}\ M/q \qquad (9)$$

where M and m are the masses of the buffer gas and ion, respectively, and R is identified as the ratio of the viscosity cross section to the momentum cross section, σ_2/σ_1. One of the reason for measuring D_T/K as a function of E/N is to obtain useful information about the ion-neutral interaction potential. Several authors /27,46/, however, stress the point that the semi-quantitative expression given in equ. (7) must be examined further before any firm conclusions can be drawn. On the other hand more elaborate theories are available for some cases /51,57,58/, and comparison between experiment and theory can reveal information about the respective ion-neutral interaction potential /32,37/.

3.1. Atomic Ions

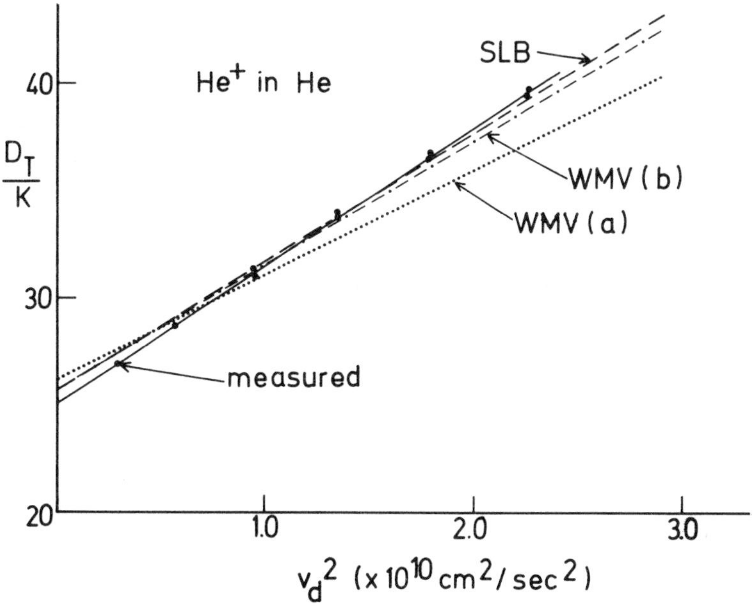

Fig. 7. D_T/K for He^+ in He versus drift velocity squared. Solid line is linear fit to the data (solid points) of Sejkora et al. /32/ yielding a = (25.14 ± 0.26) mV. Also shown are linear fits to two sets of theoretical

calculations, designated SLB (see /51/), designated WMV (a) (see /57/), and designated WMV (b) (see /57/ kT_T versus w^2), yielding a = 25.75 mV, a = 26.10 mV and a = 25.8 mV respectively.

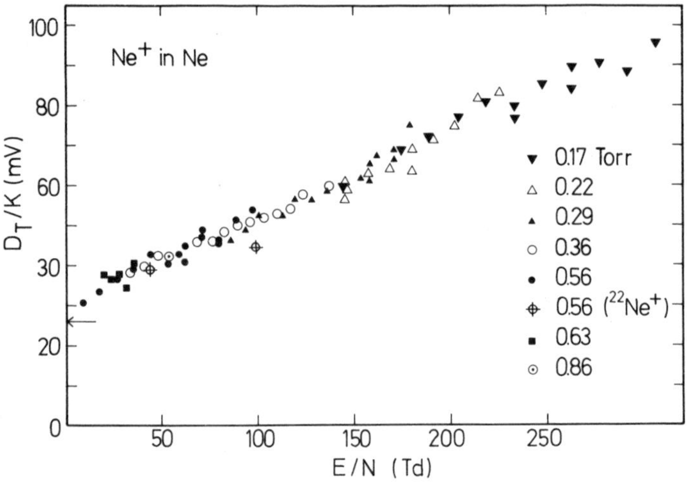

Fig. 8. D_T/K for Ne^+ in Ne versus E/N after Märk et al. /35, 36/ (preliminary data obtained without field correction, see discussion in text).

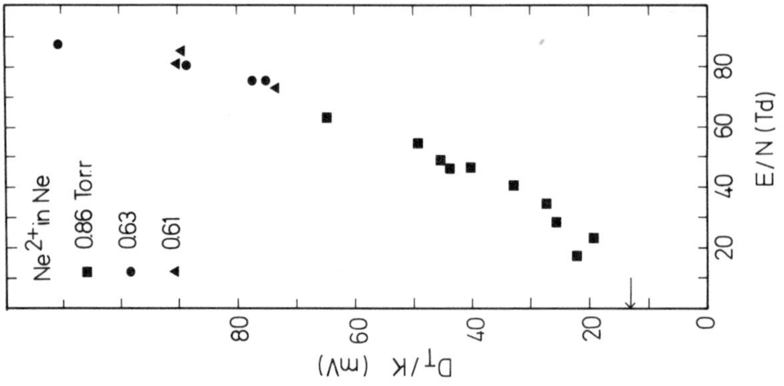

Fig. 9. D_T/K for Ne^{2+} in Ne versus E/N after Märk et al./35,36/ (preliminary data obtained without field correction, see discussion in text).

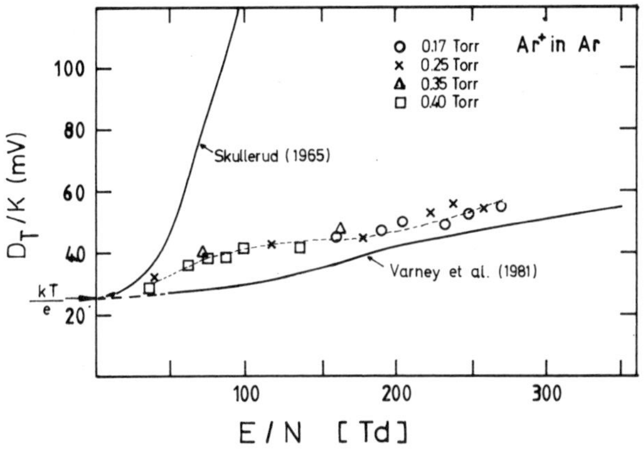

Fig. 10. D_T/K for Ar^+ in Ar versus E/N. Solid line data by Varney et al. /28/, and by Skullerud /18/, respectively. Data points preliminary results obtained in Innsbruck /83/ without field correction (see discussion in text).

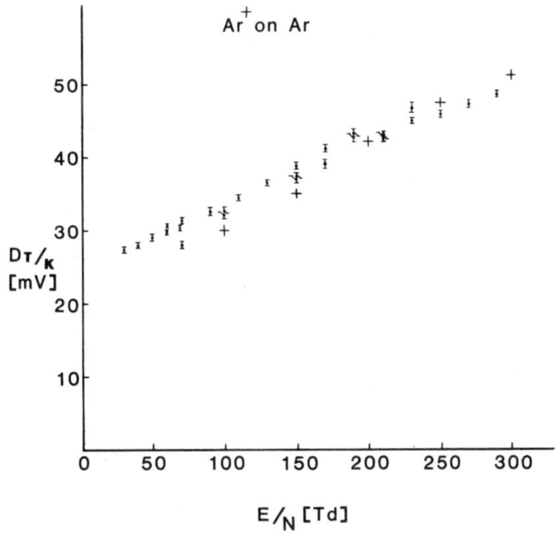

Fig. 11. D_T/K for Ar^+ in Ar versus E/N. Full dots by Sejkora et al. /34/ and crosses by Varney et al. /28/.

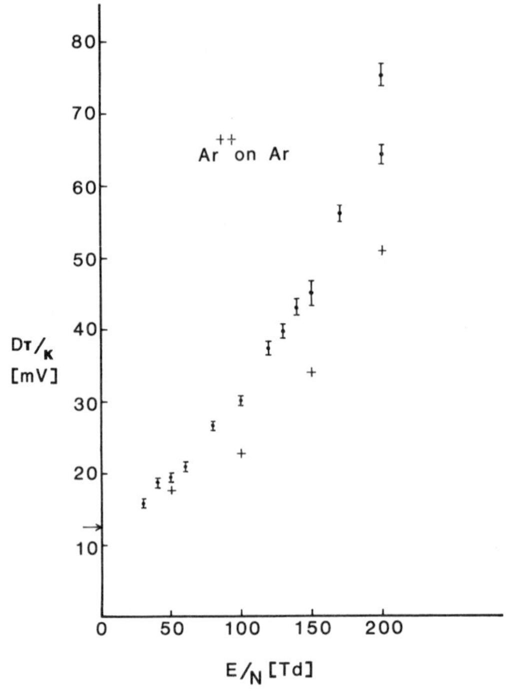

Fig. 12. D_T/K for Ar^{2+} in Ar versus E/N. Full dots by Sejkora et al. /34/ and crosses by Varney et al. /28/.

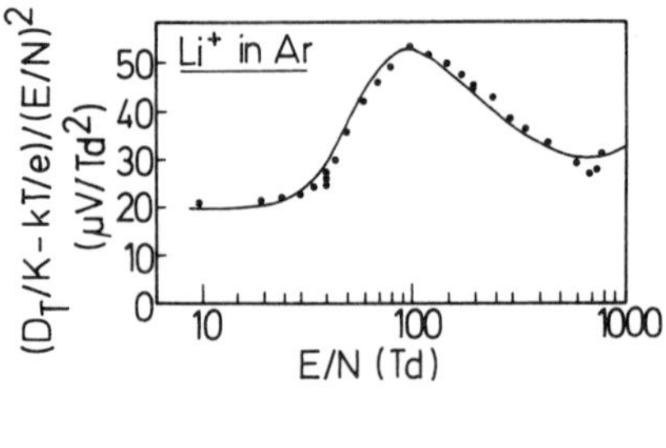

Fig. 13. The reduced transverse diffusion coefficient $(D_T/K - kT/e) / (E/N)^2$ versus E/N for Li^+ in Ar. Full dots are data by Stefannson /37/ and full line calculation by Skullerud /58/.

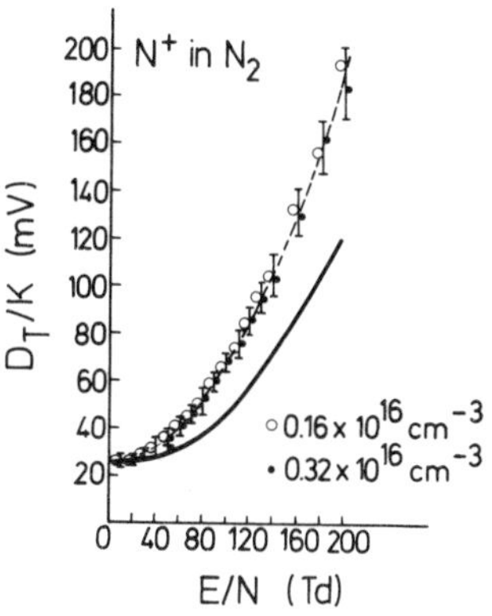

Fig. 14. D_T/K for N^+ in N_2 versus E/N. Full line experimental results by Moseley et al. /23,25/, full dots, experimental results by Rees and Alger /80/ and dashed line calculated function with R = 0.57 (see equ. 7 to 9).

3.2. Molecular Ions

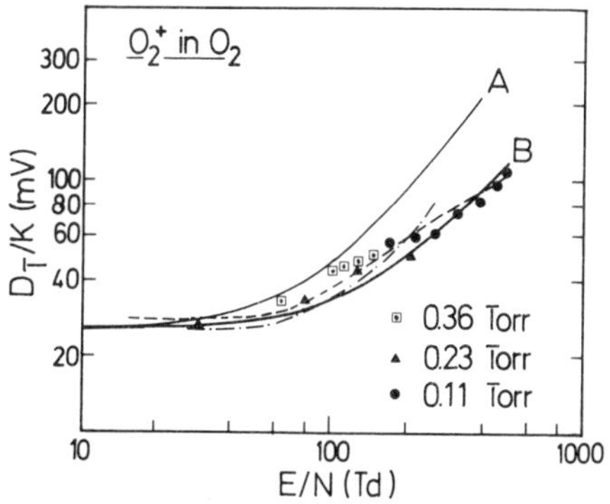

Fig. 15. D_T/K for O_2^+ in O_2 versus E/N. Line designated -.-.- experimental results by Dutton and Howells /17/, line designated - - - corrected experimental results by Gray and Rees /26/, points experimental results by Alger et al. /27/, curve A calculated function with R = 0.697, and curve B calculated function with R = 0.18 (see equ. 7 to 9).

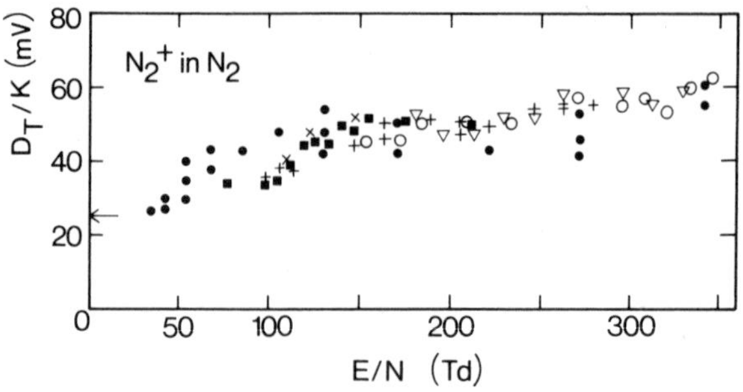

Fig. 16. D_T/K for N_2^+ in N_2 versus E/N. Full dots: results of Moseley et al. /25/. Other points results of Sejkora and Märk /33/ measured at various pressures in the drift tube (preliminary results obtained without field correction, see discussion in text).

3.3. Cluster Ions

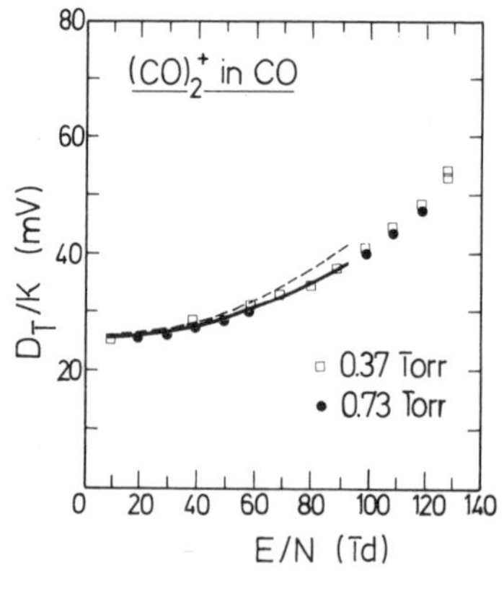

Fig. 17. D_T/K for $(CO)_2^+$ in CO versus E/N. Points experimental results of Alger et al. /27/, full line calculated function with R = 0.54, dashed line calculated function with R = 0.697 (see equ. 7 to 9). See also recent results by Rees and Alger /80/, which are in good agreement with the results of Alger et al. /27/.

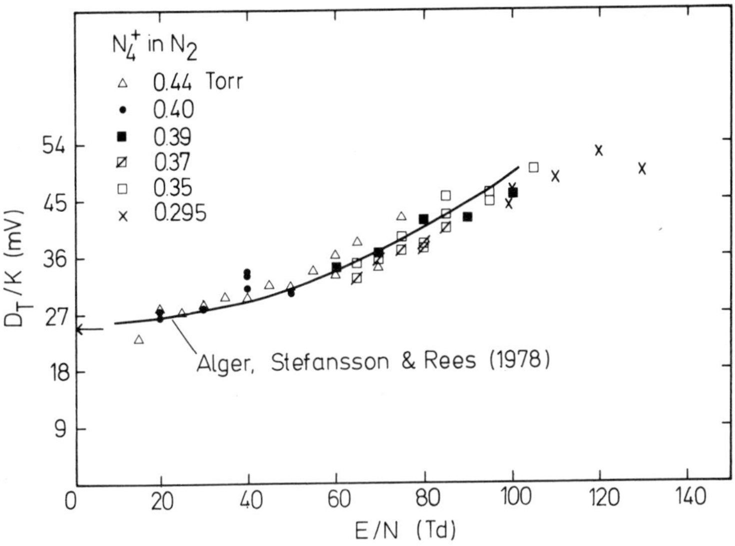

Fig. 18. D_T/K for $(N_2)_2^+$ in N_2 versus E/N. Full line experimental results of Alger et al. /27/, points results of Märk et al. /35,36/ measured at various pressures between 0.295 and 0.44 Torr. It is interesting to note that Alger et al. /27/ have also determined D_T/K for N_3^+, however, data points for N_3^+ and $(N_2)_2^+$ for a given E/N were found to lie within $\pm$ 2% of each other. Furthermore, Sejkora et al. /32/ have studied D_T/K for a given E/N over a wide pressure range. Preliminary results indicate a pressure dependence due to the fact that the apparent diffusion coefficient, measured by the detector tuned to $(N_2)_2^+$, is in fact the diffusion of a positive particle which spends pressure-dependent fractions of its transit time as N_2^+, N_4^+ and N_4^{+*} /89/.

3.4. Negative Ions

The available data for transverse diffusion of negative ions are not extensive. They include D_T/K for O^- and CO_3^- in CO_2, SF_6^- and $(SF_6)_2^-$ in SF_6 (e.g. see Figs. 19-21), and have been recently summarized by Rees /90/ and Rees and Alger /80/.

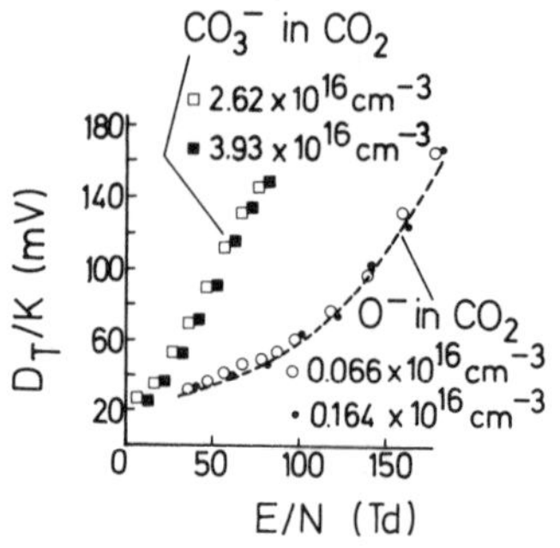

Fig. 19. D_T/K for O^- and CO_3^- in CO_2 versus E/N after Rees and Alger /80/. Dashed line calculated values for O^- with R = 0.79 (see exq. 7 to 9).

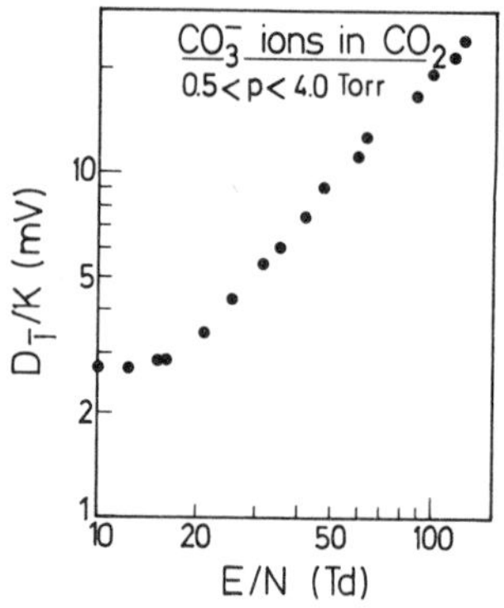

Fig. 20. D_T/K for CO_3^- in CO_2 versus E/N after Rees /90/.

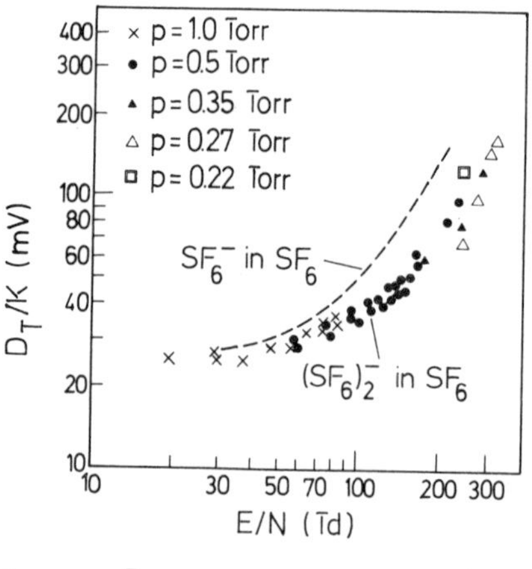

Fig. 21. D_T/K for SF_6^- in SF_6 (dashed line /91/) and $(SF_6)_2^-$ in SF_6 (points) after Rees /90/.

References

1. Mc Daniel, E.W., Mason, E.A.: The mobility and diffusion of ions in gases. New York: Wiley 1973.
2. Märk, T.D., Dunn, G.H.: Electron impact ionization. Wien: Springer. 1984.
3. Nernst, W.: Zur Kinetik der in Lösung befindlicher Körper. Z. Phys. Chem. 2, 613-637 (1888).
4. Townsend, J.S.: The diffusion of ions into gases. Phil. Trans. A 193, 129-158 (1900).
5. Townsend, J.S.: The charges on positive and negative ions in gases. Proc. Roy. Soc. A 80, 207-211 (1908), and, The charges on ions in gases, and the effect of water vapor on the motion of negative ions. Proc. Roy. Soc. A 81, 464-471 (1908).
6. Thomson, J.J., Thomson, G.P.: Conduction of electricity through gases, Vol. I (unrevised reprint of 1928 3rd edition). New York: Dover. 1969.
7. Loeb, L.B.: Basic processes of gaseous electronics. Berkeley: Univ. Calif. Press. 1960.
8. Massey, H.S.W.: Electronic and ionic impact phenomena, Vol. III. Oxford: Clarendon. 1971
9. Viehland, L.A., Mason, E.A., Morrison, W.F., Flannery, M.R.: Tables of transport collision integrals for (n, 6,4) ion-neutral potentials. Atom. Data Nucl. Data Tabl. 16, 495-514 (1975)
10. Ellis, H.W., Pai, R.Y., Mc Daniel, E.W., Mason, E.A., Viehland, L.A.: Transport properties of gaseous ions over a wide energy range. Atom. Data Nucl. Data Tabl. 17, 177-210 (1976).
11. Ellis, H.W., Mc Daniel, E.W., Albritton, D.L., Viehland, L.A., Lin, S.L., Mason, E.A.: Transport properties of gaseous ions over a wide energy range. Part II. Atom. Data Nucl. Data Tabl. 22, 179-217 (1978).
12. Viehland, L.A., Mason, E.A.: Gaseous ion mobility and diffusion in electric fields of arbitrary strength. Ann. Phys. 110, 287-328 (1978).
13. Kumar, K., Skullerud, H.R., Robson, R.E.: Kinetic theory of charged particle swarms in neutral gases. Aust. J. Phys. 33, 343-448 (1980).
14. Märk, T.D., Castleman, Jr. A.W.: Experimental studies on cluster ions. Adv. Atom. Molec. Phys. 20 (1984).
15. Crompton, R.W., Elford, M.T., Gascoigne, J.: Precision measurements of the Townsend energy ratio for electron swarms in highly uniform electric fields. Austral. J. Phys. 18, 409-436 (1965)
16. Dutton, J., Llewellyn Jones, F., Rees, W.D., Williams, E.M.: Drift and diffusion of ions in hydrogen. Phil. Trans. Roy., A 259, 339-354 (1966).
17. Dutton, J., Howells, P.: The motion of oxygen ion in oxygen. J. Phys. B1, 1160-1170 (1968).
18. Skullerud, H.R.: Measurement of positive ion diffusion in argon and hydrogen. In: Proc. 7th Int. Conf. Phenom. ionized Gases, p. 50-53 Belgrade: Gradevinska Knjiga. 1965.
19. Tunnicliffe, R.J., Rees, J.A.: The determination of Townsend energy ratio (k_1) for positive potassium ions in hdydrogen and nitrogen. In: Proc. 8th Int. Conf. Phenom. Ionized Gases, p. 14. Wien: Springer. 1967.
20. Fleming, I.A., Tunnicliffe, R.J., Rees, J.A.: The drift and diffusion of potassium ions in nitrogen. J. Phys.

D2, 551-556 (1969).

21. Fleming, I.A., Tunnicliffe, R.J., Rees, J.A.: Concerning determinations of the drift velocity and lateral diffusion of positive ions in hydrogen. J.Phys. B2, 780-789 (1969).

22. Miller, T.M., Moseley, J.T., Martin, D.W., Mc Daniel, E.W.: Reactions of H^+ in H_2 and D^+ in D_2; Mobilities of hydrogen and alkali ions in H_2 and D_2 gases. Phys. Rev. 173, 115-122 (1968).

23. Moseley, J.T., Snuggs, R.M., Martin, D.W., Mc Daniel, E.W.: Longitudinal and transverse diffusion coefficients of mass identified N^+ and N_2^+ ions in nitrogen. Phys. Rev. Lett. 21, 873-878 (1968).

24. Moseley, J.T., Gatland, I.R., Martin, D.W., Mc Daniel, E.W.: Measurement of transport properties of ions in gases: Results for K^+ ions in N_2. Phys. Rev. 178, 234-239 (1969).

25. Moseley, J.T., Snuggs, R.M., Martin, D.W., Mc Daniel, E.W.: Mobilities, diffusion coefficients, and reaction rates of mass-identified nitrogen ions in nitrogen. Phys. Rev. 178, 240-248 (1969).

26. Gray, D.R., Rees, J.A.: The lateral diffusion of mass identified positive ions in oxygen. J. Phys. B5, 1048-1055 (1972).

27. Alger, S.R., Stefansson, T., Rees, J.A.: Measurements of the lateral diffusion of O_2^+ ions in oxygen, N_3^+ and N_4^+ ions in nitrogen, and $CO^+.CO$ ions in carbon monoxide. J. Phys. B11, 3289-3297 (1978).

28. Varney, R.N., Helm, H., Alge, E., Störi, H., Lindinger, W.: Transverse diffusion of Ar^+ and Ar^{2+} in Ar at 298 K. J. Phys. B14, 1695-1705 (1981)

29. Sejkora, G., Hilchenbach, M., Grössl, M., Helm, H., Elford, M.T., Lindinger, W., Märk, T.D.: Lateral diffusion of mass identified ions: experimental. In: Proc. 3rd Symp. Atom. Surf. Phys. (Lindinger, W., et al., Eds.) p. 322-326. Maria Alm. 1982.

30. Sejkora, G., Hilchenbach, M., Elford, M.T., Märk, T.D.: Lateral diffusion of mass identified ions; low current stable ion source. Europhys. Conf. Abstr. 6D, 143 ,1982)

31. Girstmair, P., Sejkora, G., Bryant, H.C., Märk, T.D.: Transverse diffusion of the $(N_2)_2^+$ cluster ion. In: Proc. XIIIth Int. Conf. Phys. Electr. Atom. Coll., post deadline paper, p. 11, Berlin 1983.

32. Sejkora, G., Bryant, H.C., Girstmair, P., Hesche, M., Djuric, N., Märk, T.D.: Transverse diffusion of mass identified ions in their parent gas. In: Proc. 3rd Intern. Swarm Seminar (Lindinger, W., et al., Eds.) p. 201-205. Innsbruck. 1983.

33. Sejkora, G., Märk, T.D.: Transverse diffusion of mass identified N_2^+ ions in nitrogen. Chem. Phys. Lett. 97, 123-126 (1983).

34. Sejkora, G., Girstmair, P., Bryant, H.C., Märk, T.D.: The transverse diffusion of Ar^+ and Ar^{2+} in Ar. Phys. Rev. A, in print (1984).

35. Märk, T.D., Sejkora, G., Girstmair, P., Hesche, M., Märk, E., Bryant, H.C., Elford, M.T.: Transverse ion diffusion studied with the radial ion distribution method. In: Proc. 4th Symp. Atom. Surf. Phys. (Howorka, F., et al., Eds.), 144-154, Maria Alm (1984).

36. Märk, T.D., Girstmair, P., Sejkora, G., Hesche, M., Märk, E., Bryant, H.C.: Transverse ion diffusion in rare gases. In: Proc. 7th ESCAMPIG, Bari. 1984.
37. Stefannson, T.: Diffusion of lithium ions in argon. In: Proc. 3rd Intern. Swarm Seminar (Lindinger, W., et al., Eds) p. 227-233. Innsbruck. 1983.
38. Mc Daniel, E.W., Moseley, J.T.: Tests of the Wannier expressions for diffusion coefficients of gaseous ions in electric fields. Phys. Rev. A3, 1040-1044 (1971).
39. Robson, R.E.: A thermodynamic treatment of anisotropic diffusion in an electric field. Aust. J. Phys. 25, 685-693 (1972).
40. Wannier, G.H.: On a conjecture about diffusion of gaseous ions. Austr. J. Phys. 26, 897-900 (1973).
41. Skullerud, H.R.: Monte-Carlo investigations of the motion of gaseous ions in electrostatic fields. J. Phys. B6, 728-742 (1973).
42. Whealton, J.H., Mason, E.A., Robson, R.E.: Composition dependence of ion-transport coefficients in gas mixtures. Phys. Rev. A9, 1017-1020 (1974).
43. Whealton, J.H., Mason, E.A.: Transport coefficients of gaseous ions in an electric field. Ann. Phys. 84, 8-38 (1974).
44. Viehland, L.A., Mason, E.A.: On the relation between gaseous ion mobility and diffusion coefficients at arbitrary electric field strengths. J. Chem. Phys. 63, 2913-2915 (1975).
45. Viehland, L.A., Mason, E.A.: Gaseous ion mobility in electric fields of arbitrary strength. Ann. Phys. 91, 499-533 (1975).
46. Skullerud, H.R.: On the relation between the diffusion and mobility of gaseous ions moving in strong electric fields. J. Phys. B9, 535-546 (1976).
47. Robson, R.E.: On the generalized Einstein relation for gaseous ions in an electrostatic field. J. Phys. B9, L 337- L 339 (1976).
48. Skullerud, H.R.: Progress in the transport theory of weakly ionized gases. In: Proc. 13th Int. Conf. Phys. Ionized Gases, p. 303-319. Berlin, 1977.
49. Viehland, L.A., Lin, S.L.: Application of the three-temperature theory of gaseous ion transport. Chem. Phys. 43, 135-144 (1979).
50. Lin, S.L., Viehland, L.A., Mason, E.A.: Three-temperature theory of gaseous ion transport. Chem. Phys. 37, 411-424 (1979).
51. Sinha, S., Lin,S.L., Bardsley, J.N.: The mobility of He^+ in He. J. Phys. B12, 1613-1622 (1979).
52. Skullerud, H.R., Forsth, L.R.: Perturbation treatment of thermal motions in gaseous ion-transport theory. J. Phys. B12, 1881-1888 (1979).
53. Waldman, M., Mason, E.A.: Generalized Einstein relations from a three-temperature theory of gasous ion transport. Chem. Phys. 58, 121-144 (1981)
54. Viehland, L.A., Mason, E.A., Lin, S.L.: Test of the interaction potentials of H^- and Br^- ions with He atoms and of Cl^- ions with Ar atoms. Phys. Rev. A 24, 3004-3009 (1981).

55. Viehland, L.A., Mason, E.A.: Well depths of XeF^- and $XeCl^-$ from ion transport data. Chem. Phys. Lett. 83, 298-300 (1981).
56. Viehland, L.A.: Gaseous ion transport coefficients. Chem. Phys. 70, 149-156 (1982).
57. Waldman, M., Mason, E.A., Viehland, L.A.: Influence of resonant charge transfer on ion diffusion and generalized Einstein relations. Chem. Phys. 66, 339-349 (1982).
58. Skullerud, H.R.: Calculation of ion drift and diffusion basis set expansions with non-gaussian weight functions. In: Proc. 3rd Intern. Swarm Seminar (Lindinger, W., et al., Eds) p. 212-217. Innsbruck. 1983.
59. Pai, R.Y., Ellis, H.W., Akridge, G.R., Mc Daniel, E.W.: Longitudinal diffusion coefficients of Li^+ and Na^+ ions in He, Ne and Ar: Experimental test of the generalized Einstein relation. J. Chem. Phys. 63, 2916-2918 (1975).
60. Pai, R.Y., Ellis, H.W., Akridge, G.R., Mc Daniel, E.W.: Generalized Einstein relation: Application to ions in molecular gases. Phys. Rev. A12, 1781-1784 (1975).
61. Ellis, H.W., Thackston, M.G., Pai, R.Y., Mc Daniel, E.W.: Longitudinal diffusion coefficients of Rb^+ ions in He, Ne, Ar, H_2, N_2, O_2 and CO_2. J. Chem. Phys. 65, 3390-3391 (1976).
62. Pai, R.Y., Ellis, H.W., Mc Daniel, E.W.: The generalized Einstein relation-application to Li^+ and Na^+ ions in hydrogen gas. J. Chem. Phys. 64, 4238-4239 (1976).
63. Eisele, F.L., Thackston, M.G., Pope, W.M., Gatland, I.R., Ellis, H.W., Mc Daniel, E.W.: Experimental test of the generalized Einstein relation for Cs^+ ions in molecular gases: H_2, N_2, O_2, CO, and CO_2. J. Chem. Phys. 67, 1278-1279 (1977).
64. Thackston, M.G., Eisele, F.L., Pope, W.M., Ellis, H.W., Mc Daniel, E.W.: Further tests of the generalized Einstein relation: Cs^+ ions in Ar, Kr, and Xe. J. Chem. Phys. 68, 3950-3951 (1978).
65. Morrison, W.F., Akridge, G.R., Ellis, H.W., Pai, R.Y., Mc Daniel, E.W., Viehland, L.A., Mason, E.A.: Test of the Li^+-He interaction potential. J. Chem. Phys. 63, 2238-2241 (1975).
66. Viehland, L.A., Mason, E.A., Stevens, T.H., Monchik, L.: Test of the H_2^+ + He interaction potential. Comparison of the interactions of He with H^+, H_2^+ and H_3^+. Chem. Phys. Lett. 44, 360-362 (1976).
67. Viehland, L.A., Harrington, M.M., Mason, E.A.: Direct determination of ion-neutral molecule interaction potentials from gaseous ion mobility measurements. Chem. Phys. 17, 433-441 (1976).
68. Gatland, I.R., Morrison, W.F., Ellis, H.W., Thackston, M.G., Mc Daniel, E.W., Alexander, M.H., Viehland, L.A., Mason, E.A.: The Li^+-He interaction potential. J. Chem. Phys. 66, 5121-5125 (1977).
69. Gatland, I.R., Viehland, L.A., Mason, E.A.: Tests of alkali ion-inert gas interaction potentials by gaseous ion mobility experiments. J. Chem. Phys. 66, 537-541 (1977).
70. Gatland, I.R., Thackston, M.G., Pope, W.M., Eisele, F.L., Ellis, H.W., Mc Daniel, E.W.: Mobilities and interaction potentials for Cs^+-Ar, Cs^+-Kr, and Cs^+-Xe. J. Chem. Phys. 68, 2775-2778 (1978).

71. Gatland, I.R., Lamm, D.R., Thackston, M.G., Pope, W.M., Eisele, F.L., Ellis, H.W., Mc Daniel, E.W.: Mobilities and interaction potentials for Rb^+-Ar, Rb^+-Kr, and Rb^+-Xe. J. Chem. Phys. 69, 4951-4954 (1978).
72. Thackston, M.G., Eisele, F.L., Pope, W.M., Ellis, H.W., Mc Daniel, E.W., Gatland, I.R.: Mobility of Cl^- ions in Xe gas and the Cl^--Xe interaction potential. J. Chem. Phys. 73, 3183-3185 (1980).
73. Lamm, D.R., Thackston, M.G., Eisele, F.L., Ellis, H.W., Twist, J.R., Pope, W.M., Gatland, I.R., Mc Daniel, E.W.: Mobilities and interaction potentials for K^+-Ar, K^+-Kr, and K^+-Xe. J. chem. Phys. 74, 3042-3045 (1981).
74. Takebe, M.: The generalized mobility curve for alkali-ions in rare gases: Clustering reactions and mobility curves. J. Chem. Phys. 78, 7223-7226 (1983).
75. Cassidy, R.A., Elford, M.T.: The mobility of Li^+ and K^+ ions in helium and argon at 294 and 80 K and derived interaction potentials. In: Proc. 3rd Intern. Swarm Seminar (Lindinger, W., et al., Eds) p. 222-226. Innsbruck. 1983.
76. Llewellyn Jones, F.: The energy of agitation of positive ions in argon. Proc. Phys. Soc. 47, 74-85 (1935).
77. Crompton,R.W.: The present status of the Townsend-Huxley experiment in theory and practice. Aust. J. Phys. 25, 409-417 (1972).
78. Mc Daniel, E.W., Cermak, V., Dalgarno, A., Ferguson, E.E., Friedman, L.: Ion-molecule reactions. New York: Wiley. 1969.
79. Mc Daniel, E.W., Mc Dowell, M.R.C.: Case Studies in Atomic Collision Physics I (Eds.). Amsterdam: North Holland. 1969.
80. Rees, J.A., Alger, B.A.: Transport properties of mass-identified ions in oxygen, nitrogen, carbon dioxide and carbon monoxide. Proc. IEE 126 (1979) 356-360.
81. Huxley, L.G.H., Crompton, R.W.: The diffusion and drift of electrons in gases. New York: Wiley. 1974.
82. Elford, M.T.: Private communication. 1981.
83. Hilchenbach, M.: Radioaktive Ionenquelle mit Townsend-lawinenverstärkung. Diplomarbeit. Universität Innsbruck. 1982.
84. Sejkora, G.: Bestimmung transversaler Diffusionskoeffizienten als Funktion der reduzierten Feldstärke. Aufbau einer mikroprozessorgesteuerten Apparatur und Messungen in Argon und Stickstoff. Dissertation. Universität Innsbruck. 1982.
85. Märk, T.D., Helm, H.: Mass spectrometry as a technique for studying atomic properties of low pressure plasmas (particle extraction and detection system). Acta Physica Austriaca 40, 158-180 (1974).
86. Helm, H., Märk, T.D., Lindinger, W.: Plasma sampling - a versatile tool in plasma chemistry. J. Pure Appl. Chem. 52, 1739-1757 (1980).
87. Grössl, M., Langenwalter, M., Helm, H., Märk, T.D.: Molecular ion formation in decaying plasmas produced in pure argon and krypton, J. Chem. Phys. 74, 1728-1735 (1981)
88. Wannier, G.H.: Motion of gaseous ions in strong electric fields. Bell Syst. Techn. J. 32, 170- 254 (1953)

89. Varney, R.N., Pahl, M., Märk, T.D.: Properties of the ionic systems N_4^+, O_4^+ and O_4^-. Acta Physica Austriaca 38, 287-294 (1973).
90. Rees, J.A.: Transport properties of ions in electronegative gases. Vacuum 24, 603-607 (1974).
91. Naidu, M.S., Prasad, A.N.: Mobility and diffusion of negative ions in sulphur hexafluoride. J. Phys. D3, 951-956 (1970).

4. Acknowledgements

Work partially supported by Österreichischer Fonds zur Förderung der wissenschaftlichen Forschung under Projekt S-18/05, S-18/08 and P 5148. Part of this review is a summary of experiments conducted in our laboratory. The work was done in most part in collaboration with G. Sejkora, P. Girstmair, M. Hesche and H.C. Bryant, and has been published elsewhere as noted.

Runaway Mobilities of Ions in Helium

F. Howorka

Institut für Experimentalphysik, Universität Innsbruck, A-6020 Innsbruck, Austria

1. Introduction

The term "runaway mobility" of ions in a buffer gas like helium refers to a non-equilibrium condition: it is impossible for the ions to attain a defined drift velocity in an applied field, they are rather accelerated all the time. In contrast the usual mobility concept demands the condition of equilibrium where every momentum gain in the field is compensated by an average momentum loss over a long time. Microscopically the ions - even in the case of a well-established equilibrium - never have a constant speed: only on the average over many ions the whole group has a mean drift velocity and a certain velocity distribution. Mobility is defined as the ratio of the mean velocity attained in the buffer gas under a certain field strenght and the value of the field strenght. It is usually found to be a function of the reduced field strength E/N, where E is the field strength as measured in Vcm^{-1} (or Vm^{-1}) and N is the gas number density in cm^{-3} (or m^{-3}). The unit is Townsends (1 Townsend = 10^{-17}V cm^2 or 10^{-21}V m^2; its symbol is Td).

When and where does such a non-equilibrium condition occur? The first time a runaway phenomenon for ions was predicted when Lin, Gatland and Mason /1/ tried to calculate the mobility of protons and deuterons from an accurate ab initio potential of the HeH^+ (or HeD^+) molecular ion as given by Kołos and Peek /2/. They used the two-temperature theory of Viehland and Mason /3/ to calculate the dependence of the mobility of protons and deuterons as a function of E/N. To their surprise they discovered that from a certain reduced field strength on (E/N greater than 25 Td), the concept of mobility broke down and an erratic behavior of the apparent mobility set in, depending on the degree of approximation of the set of moment equations used. They immediately concluded that a real runaway in the sense of non-equilibrium between the momentum acquired in the field and that lost by collisions must be the reason, because the ion-atom momentum transfer cross section decreases with increasing relative speed so quickly that the ions cannot lose enough energy to achieve a steady-state drift velocity.

Measurements of the mobilities of protons and deuterons in helium did exist at that time but did not extend to a high enough reduced field strength or - in one case where they did - did not show any peculiarities. In that stage of development of the theoretical calculations, Howorka, Fehsenfeld and Albritton /4/ tried to experimentally verify the predicted runaway phenomenon. Only after a substantial change of the ion production part of the Boulder drift tube (National Oceanic and Atmospheric Administration, NOAA) they did succeed in measuring the region of E/N values where the runaway should occur and really found the mobility to obey the predicted behaviour.

A further prediction of the theory, that the onset of runaway should depend on the field strenght and the gas number density separately rather than on the ratio of these quantities, was verified by Moruzzi and Kondo /5/. They found that at given E/N, the dependence of the apparent mobility on the gas number density was even stronger than expected (proportional to N rather than to $N^{1/2}$). This effect could be explained by a model by Waldman and Mason /6/ in which a cutoff in the collision frequency function leads to a set of dependences of the apparent mobility on the gas number density which show linear regions as intermediate cases between the "accelerator" (or runaway) regime and the "drift" region where a mobility concept exists. Very similar results have been obtained by Gatland (private communication), using a truncated Coulomb potential to calculate the ions still drifting after some time. Here again, a strong pressure dependence of the runaway onset is predicted. Runaway, like thermalisation, occurs earlier at higher pressures.

A survey over these calculations and experimental verifications has been given by Mason and Waldman /7/. In the meantime, a further case of the runaway mobility is to be expected: for H^- in helium, runaway is predicted by Viehland, Mason and Lin /8/ for a reduced field strength above 50 Td. Here again, measurements exist only up to 32 Td which is too low for the onset of runaway.

Due to the connection of the mobility with the transversal and longitudinal diffusion coefficients via generalized Einstein relations, the latter ones also are expected to show a runaway behaviour. Direct measurements of the runaway diffusion coefficients have never been undertaken, but from the mobility measurements (or rather the arrival time distributions) they can at least be qualitatively inferred. This will be shown in the following sections.

The aim of the present paper is to explore the runaway phenomenon in an experimental way, to understand phenomena in the experiments and to look at consequences this phenomenon has on the measurements of other magnitudes.

2. Experimental Setup

Mobilities are usually studied in static or flow drift tubes. In the case of H^+ and D^+ in helium, the NOAA flow drift tube /9/ was utilized for measuring the mobilities. Former measurements using the same tube extended only to an E/N value of 20 Td /9/. Using an electron impact ion source in the flow tube (upstream of the drift tube section), the result shown in Fig.1 was obtained.

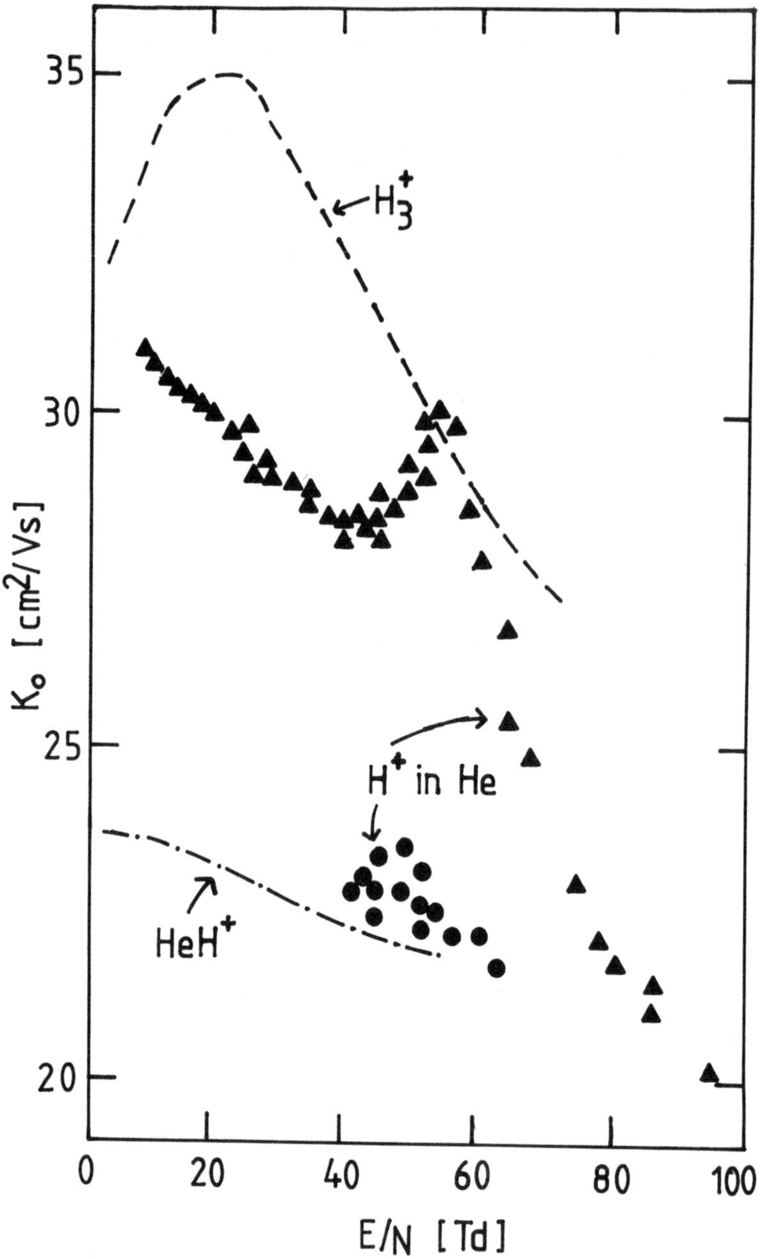

Figure 1. The mobility of protons as measured in a flow drift tube without selected ion injection. Around 50 Td, two distinctly different mobility peaks in the arrival time histograms point to the collisional breakup of H_3^+ and HeH^+, respectively. These two peaks merge above 70 Td. Mobility values of H_3^+ and HeH^+ are taken from refs. 10 and 13.

Fig. 1 shows the mobility of H^+ ions in helium in the E/N range 0 - 100 Td. As can be seen, the mobility first declines to about 40 Td, then it rises to a maximum at 60 Td. This could correspond to the predicted sharp rise in the apparent mobility as predicted by Lin, et al. /1/. Then, however, a sharp decline of the mobility follows up to the highest E/N values attainable. This behaviour can only be understood by the presence of other ions in the tube: HeH^+ and H_3^+. The production of H^+ ions from H_2, regardless of the means, generally also produces H_2^+ ions, which then rapidly form H_3^+ ions via the fast reaction

$$H_2^+ + H_2 \rightarrow H_3^+ + H. \qquad (1)$$

Thus, without selected-mass injection, the ion entrance shutter would pass an ion pulse consisting of H^+, H_2^+ and H_3^+ ions, which then would drift separately with their characteristic velocities. Normally for a drift tube followed by a mass spectrometer, the co-existence of multiple ion species need not pose problems in mobility measurements. Indeed, the mobilities of H_2^+ and H_3^+ in He have been measured straightforwardly over a wide range of E/N in this way and similar for H^+ in He at low E/N (Ellis et al., /10/). However, measurement of the latter mobility at high E/N is complicated by the collision-induced dissociation of H_3^+ into H^+ by energetic collisions with the helium buffer gas atoms.

Usually in such cases, the H_3^+ concentration is much larger than that of the primary H^+ ions; hence, even a small collision-induced source of H^+ ions can overwhelm the primary H^+ signal. Since these collision-induced H^+ ions will have spent part of their time drifting as H_3^+ ions, the H^+ arrival-time spectrum at high E/N will have a mixture of the characteristics of the H^+-in-He and H_3^+-in-He mobilities, which are sufficiently close in magnitude in the range of E/N where the H^+ runaway is believed to occur. Therefore, a decision is not possible whether the rise in the apparent mobility at 60 Td is really due to a runaway effect or due to the fact that more and more H^+ ions in the bulk are produced in the tube by kinetic breakup of H_3^+ ions which are produced in the source region.

Therefore, the flow drift tube has to be altered by adding a mass-selected venturi inlet ion injection according to Adams and Smith /11/. Now only H^+ ions can enter the tube.

The H^+ ions are created in a DC-discharge hollow-cathode ion source at 0.1 Torr H_2 pressure. The H^+, H_2^+ and H_3^+ ions leave the source through a cluster of five 0.05 mm diameter holes and are directed toward the entrance of a 10 cm quadrupole mass spectrometer, which passes only the H^+ component to the venturi-inlet aperture of the flow tube. The pressures in the injection mass spectrometer chamber and in the flow tube are typically 2×10^{-4} and 0.5 Torr, respectively. The low-energy (approximately 25 eV) passage of the ions from the low to the high pressures is made possible by the venturi pumping effect of the helium buffer gas expanding into the flow tube in the immediate vicinity of the aperture, as described by Smith and Adams /12/. The inlet used here has a 2 mm diameter aperture that is immediately surrounded by a concentric annular opening of 0.4 mm width through which typically 180 STP cm^3 of helium enters per second. The entering H^+ are carried down the flow tube by this gas flow at a flow velocity of 10^4 cm s^{-1} to the ion entrance shutter, which is electrically pulsed open briefly ($\leq$ 10 μs) to let a thin slice of H^+ ions enter the drift tube section of the apparatus.

The DC drift field is established and maintained sufficiently uniform by a series of electrically separated, vacuum-tight guard rings. The magnitude of the drift field can be varied up to about 10 V cm^{-1}, which corresponds to a 600 V potential on the upstream end of the apparatus. The ion swarm flows and drifts down the remainder of the tube and a fraction is sampled by a 1 mm aperture in the earthed cone protruding into the flowing buffer gas. The sampled ions are mass analysed by a 10 cm quadrupole mass spectrometer and are detected individually using a strip secondary-electron multiplier. The corresponding signal pulse denoting the arrival of each ion is stored by its arrival time, where zero time corresponds to the trigger signal that pulses the ion shutter open briefly.

When the entrance ion shutter is held open continuously, the H^+ ion signals are about 10^3 counts s^{-1}. Such a DC level yields a workable pulsed signal. For example, about 10^5 pulsed openings of the entrance shutter (30s data acquisition time) produces a smooth histogram of arrival times, from which the mean transit time can be established. This total time includes contributions from several effects. The fraction that corresponds only to the field-induced drift over the 55.9 cm length of the drift tube is determined using the separately measured flow time at zero field and mass-analysis time (~2 μs for H^+), the latter obtained by using the exit shutter. The mean drift velocity v_d corresponding to the established E/N follows directly and, from the mobility K,

$$v_d = KE \qquad (2)$$

defined here in the usual way. As is customary, the data reported here are reduced mobilities K_o, which are normalized to STP conditions:

$$K_o = (p/760)(273.16/T)K \qquad (3)$$

where p and T are the helium buffer gas pressure (Torr) and temperature (K), respectively.

3. Results and discussion

The measurements that have been made of the H^+ mobilities in He at 300 K are shown in Fig. 2.

Of main interest are the results at higher E/N, where Lin et al /1/ predict a breakdown of the mobility concept for H^+ and D^+ in helium. Figure 2 shows that, as E/N is increased, the H^+ and D^+ mobilities intitially decrease. However, at E/N = 37 ± 3Td for H^+ and at E/N = 48 ± 3Td for D^+, the observed mobilities begin what evolves into a rather precipitous increase with further increases in E/N. This is interpreted as the detection of the runaway mobility predicted by Lin et al. /1/. This interpretation is supported by several features of the observations and the theory.

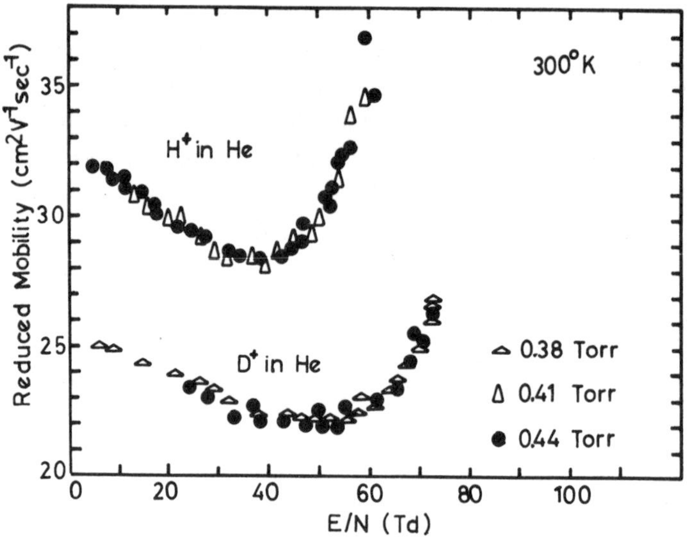

Figure 2. The mobilities of H^+ and D^+ in helium as measured in the SIFDT (selected ion flow drift tube) of the National Oceanic and Atmospheric Administration in Boulder, Colo., USA, at three different pressures and at 300°K. The steep onset of runaway at 37 and 48 Td, respectively, is clearly observable. The figure is taken from ref. 4.

Firstly, this increase is most unusual and is thought to be unique. No mobility of ions in helium exhibits such a rapid increase, nor has such been reported for any other ion-neutral atom combinations (see the extensive tabulations of Ellis et al/10,13/ or the figures given by Lindinger and Albritton /14/). The interpretation here is that, as E/N is increased, the average motion of the ions in the swarm becomes more and more a mean acceleration, rather than a mean drift velocity. The "apparent" mobilities, which have been calculated in the usual way from the observed mean transit time and the drift-tube length, will increase rapidly with E/N, since the increasing mean acceleration of the growing number of runaway ions will yield smaller and smaller mean transit times.

Secondly, the E/N dependence of the observed mobilities in

figure 2 are consistent with that predicted by Lin et al /1/. However, theory and experiment detect the runaway with different sensitivities; hence, the comparison can only be semi-quantitative. In the calculations, the higher orders of the moment expansion that yields the mobility are progressively more sensitive to the runawy effect. Thus, the early growth of the high-energy tail of the ion velocity distribution, where the runaway ions originate, can be detected quite sensitively. At about 27 Td for H^+ and 35 Td for D^+, Lin et al /1/ found that the fifth-order expansion, when compared with the fourth-order expansion, yields mobilities that are increasing sharply with increaseing E/N, which is attributed to the runaway effect. In the experiment, on the other hand, there must be a sufficient number of runaway ions to alter the mean arrival time of all ions, implying that the increase in the observed "apparent" mobilities should occur at higher E/N than those at which runaway is detectable with the theory. A comparison of the data in figure 2 with the predictions from the theory (see figure 2 of Lin et al /1/) shows that this is indeed the case.

Thus, while a direct comparison of theoretical and experimental "apparent" mobilities is not possible here, the difference in the observed 'onsets' of runaway for H^+ and D^+ should provide a much more straightforward comparison of theory and experiment. Lin et al /1/ predict that, in addition to the H^+ and D^+ mobilities scaling by $m_r^{-1/2}$, the E/N axis should scale by $m_r^{1/2}$, i.e. $m_r^{1/2}K_o$ versus $m_r^{-1/2}$ E/N should be a single curve, even when runaway is occurring. Their figure 2, which contains the calculated and measured scaled mobilities versus scaled E/N, demonstrates that this scaling is what is observed. For example, the ratio of the E/N values at which the minimum in the H^+ and D^+ mobilities occur in figure 2 is 0.77, which is practically identical with the predicted ratio of 0.775.

Thirdly, Lin et al /1/ point out that the runaway effect should also occur in the H^+ and D^+ diffusion coefficients in helium at elevated E/N. Their calculations detect the runaway of both the transverse and longitudinal diffusion coefficient components, D_T and D_L, at roughly the same E/N values at which runaway was detected for the mobilities. In an experiment, the longitudinal ion diffusion coefficient, D_L, is the major factor in determining the width of the arrival-time histogram of the ions; thus, this width can serve as an indicator of runaway diffusion coefficients. Indeed, the high-E/N histograms obtained in the present measurements can be interpreted in terms of such effects.

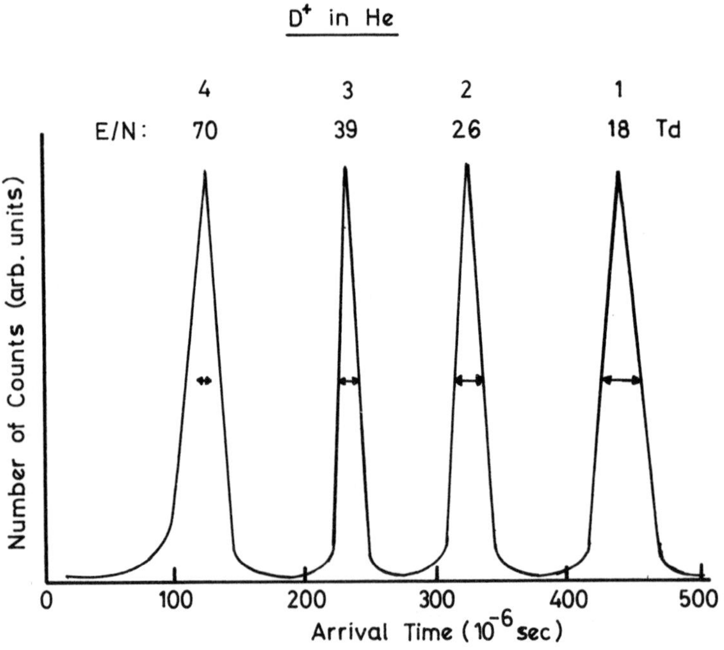

Figure 3. Observed arrival-time histograms recorded for D^+ ions in 0.43 Torr as a function of E/N. The arrows denote the full widths at half maximum calculated from a flow-drift model of normal ion transport. The abnormal width of the high-E/N histogram is interpreted as the 'runaway diffusion coefficients' predicted by Lin et al /1/. The 'early toe' on this histogram is attributed to the accelerating 'runaway' ions. From ref. 4.

This is demonstrated in figure 3, which gives four histograms recorded especially for the purpose of illustrating this point. The histograms in figure 3 are normalised to the same height to facilitate comparison. An ion entrance shutter pulse width of 4 μs was used in each case. The four histograms were recorded at different E/N values, with E/N increasing from right to left in figure 3. The first three of these E/N values are sufficiently small that runaway effects would not be expected to be observed. For these three, the diffusively broadened widths are progressively narrower with increasing E/N, which is what is normally observed in drift tubes, since the decreasing residence time affords less and less time for diffusion, overcoming even the normal increase of D_T and D_L with increasing E/N. The arrows in figure 3 show quantitatively the full widths at half maximum expected by the transport model of Gatland /15/, modified to include drift, diffusion and flow, and with D_T and D_L given by McDaniel and Mason /16/. The agreement between observed and calculated widths for these three histograms is very good, which is consistent with the picture that the transport coefficients K_o, D_T and D_L are essentially normal at these E/N values.

Histogram 4, however, was recorded at an E/N at which the runaway ions should constitute a substantial fraction of the total and at which the 'apparent' mobilities are observed to be steeply increasing with increasing E/N. Figure 3 shows that this histogram is considerably broadened, having a width that is almost three times as wide as that expected from normal transport. This abnormal width no doubt arises from the fact that the longitudinal diffusion coefficient is experiencing a runaway effect also, as Lin et al /1/ predict, and/or from the fact that a large fraction of the ions in the swarm are experiencing a mean acceleration and hence have earlier times than the drifting fraction.

Finally, this high-E/N histogram exhibits an anomalous 'early toe'. Normally, arrival-time histograms are slightly skewed to later times, simply because the trailing members of the ion swarm remain in the drift space longer that the leading members and hence have more time to diffuse. When observed, such an early toe is normally attributed to the production of the observed ion in the drift space by a faster ion, either from ion-neutral reactions or collision-induced dissociation. In the present study, it was easy to demonstrate that D^+ was the fastest ion species in the drift tube. Thus, the early toe on histogram 4, which is unique in the observations of arrival-time histograms under such circumstances, is attributed here to the runaway ions, since their continued mean acceleration implies such early arrival times.

The influence of the runaway in the transverse diffusion coefficient, D_T, cannot be directly inferred from Fig. 3 but manifests itself in the recording times of the histograms. Whereas at low E/N values with no runaway present the time to record a histogram is about 30 sec, at high E/N (runaway regime) the recording time is much larger - up to an hour or more. It took about an hour to record histogram no. 4 as compared to 1 minute (or less) for histograms 1-3. This effect has to be borne in mind when studying the partial runaway of ions other that H^+ or D^+ which might occur at much higher field stengths. Weak ion signals might simply disappear due to the D_T runaway, without piling up to give a suitable arrival-time histogram.

4. Consequences

Up to now, the runaway of protons and deuterons has not yet found practical applications. It is, however, detectable in ion-molecule reactions. Some reactions of ions with hydrogen-containing molecules branch off into several channels, one of which might be proton production. Examples for such reactions are $Ne^+ + H_2O$ or $He^+ + H_2O$. In such cases the branching ratios are often independent of the kinetic energy of the ions. If, however, protons are produced, this channel seemingly disappears in a helium buffer at an energy corresponding to the onset of runaway. Fig. 4 shows as an example the branching ratios of the reaction $Ne^+ + H_2O$ which produces (at low energy) H^+, OH^+, H_2O^+ and NeH^+ as studied in a helium buffered drift tube /17/. Due to the disappearance of H^+ above 0.6 eV, the remaining ionic product channels seem to increase; this, however, is an artefact attributable to the D_T runaway of the protons produced. The case is even more dramatic in $He^+ + H_2O$. Here again, a proton channel exists but these protons are produced at an energy of 4 eV in the reaction which is far above the onset of runaway in D_T. It is, therefore, extremely difficult to observe these ultrafast H^+ particles directly; they only manifest themselves as secondary products. Only a very careful investigation with minute additions of H_2O (an experimental difficulty) or the study of the reaction in a different buffer gas (where no runawy exists) enables one to obtain the branching ratios correctly as a function of energy.

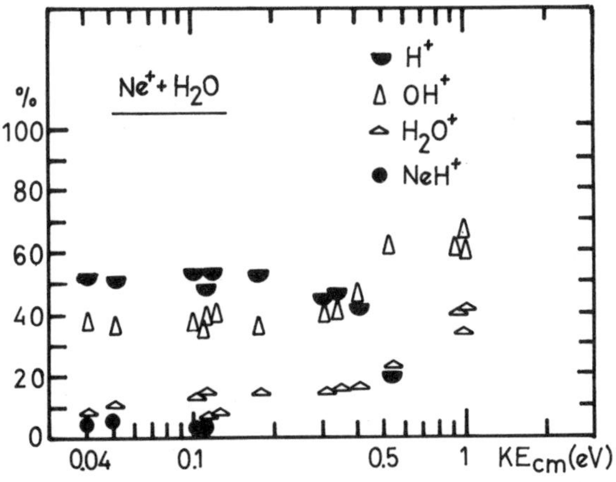

Figure 4. Branching ratio of the reaction $Ne^+ + H_2O$ as a function of energy, measured in a helium buffer in the Innsbruck selected ion flow drift tube (ref. 17). The distribution of the reaction products H^+, OH^+, H_2O^+ and NeH^+ is constant up to 0.4 eV; above this energy, the H^+ disappears due to runaway in the transversal diffusion coefficient of H^+, apparently shifting the product distribution.

5. Conclusions

The runaway mobility has so far found interest mainly in theoretical work as a proving ground for kinetic theories connecting interaction potentials and transport coefficients. Despite the fact that moment equation methods discovered the onset of runaway, they seem to be unable to describe the speed distribution of runaway completely. Direct solutions and path-integral methods on the other hand could provide this information (see Skullerud and Kuhn, this book). In addition to H^+ and D^+ in helium, there are predicted runaway mobilities for H^- in helium (onset 50 Td, see ref. 8), Li^+ in helium (above 200 Td) and Li^+ in argon (above 600 Td, see Viehland in ref. /18/). For Li^+ in argon, the diffusion coefficients as calculated by Skullerud /19/ and as measured by Stefánsson /20/ show a rise in the ratio $(D_T/K - kT/e)/(E/N)^2$ vs. E/N around 700 Td. Whether this is a real runaway or determined by end effects is still an open question.

Experimental difficulties are to be expected for the runaway of H^- because of the competition of runaway mobility, collisional detachment of the electron and D_T runaway.

The disappearance of reaction-produced H^+ in a helium buffer due to D_T runaway is more a nuisance than a useful phenomenon; it may have consequences for the sampling of ions from discharges and plasmas which contain helium as the majority gas. Detailed information of the influence of fast runaway particles on the energy transport in plasmas, stars, interstellar media etc. is still missing. The direct measurement of the diffusion coefficient (D_L and D_T) runaway is desirable but has not yet been undertaken for the most important cases, H^+ and D^+ in helium.

6. Acknowledgements. The author thanks the Fonds zur Förderung der wissenschaftlichen Forschung in Österreich for support and the colleagues in Boulder for their kind hospitality.

7. References

1. Lin, S.L., Gatland, I.R. and Mason, E.A.: Mobility and Diffusion of Protons and Deuterons in Helium - a Runaway Effect. J. Phys. B: Atom. Molec. Phys. 12, 4179-88 (1979)
2. Kołos, W. and Peek, J.M: New ab initio Potential Curve and Quasibound States of HeH^+. Chem. Phys. 12, 381-6 (1976)
3. Viehland, L.A. and Mason, E.A.: Gaseous Ion Mobility in Electric Fields of Abitrary Strength. Ann. Phys., NY. 91, 499-533 (1975)
4. Howorka, F., Fehsenfeld, F.C., Albritton, D.L.: H^+ and D^+ Ions in He: Observations of a Runaway Mobility. J. Phys. B12, 4189-4197 (1979)
5. Moruzzi, J.L., Kondo, Y.: The Mobility of H^+ Ions in Helium. Jap. J. Appl. Phys. 19, 1411-1412 (1980)
6. Waldman, M., Mason, E.A.: to be publ. (1983)
7. Mason, E.A., Waldman, M.: Theory of Ion Transport in Gases - Runaway Ions. Electron and ion swarms, ed. by L.G. Christoporou, Pergamon press, pp. 147-156 (1981)
8. Viehland, L.A., Mason, E.A., and Lin, S.L.: Test of the Interaction Potentials of H^- and Br^- Ions with He Atoms and of Cl^- Ions with Ar Atoms. Phys. Rev. A24, 3004-9 (1981)
9. Mc Farland, M., Albritton, D.L., Fehsenfeld, F.C., Ferguson, E.E. and Schmeltekopf, A.L.: Flow-Drift Technique for Ion Mobility and Ion-Molecule Reaction Rate Constant Measurements. J. Chem. Phys. 59, 6610-9, 6620-8 (1973)
10. Ellis, H.W., Pai, R.Y., Mc Daniel, E.W., Mason, E.A. and Viehland, L.A.: Transport Properties of Gaseous Ions Over a Wide Energy Range, I. Atom. Data Nucl. Data Tables 17, 177-210 (1976)
11. Adams, N.G. and Smith, D.: The Selected Ion Flow Tube (SIFT): a Technique for Studying Ion-Neutral Reactions. Int. J. Mass Spectrom. Ion Phys. 21, 349-59 (1976)
12. Smith, D. and Adams, N.G.: Gas Phase Ion Chemistry, ed. M.T. Bowers (New York: Academic), pp. 1-44 (1979)
13. Ellis, H.W., Mc Daniel, E.W., Albritton, D.L., Viehland L.A., Lin, S.L. and Mason, E.A.: Transport Properties of Gaseous Ions over a Wide Energy Range, II. Atom. Data Nucl. Data Tables 22, 179-217 (1978)

14. Lindinger, W. and Albritton, D.L.: Mobilities of Various Mass-Identified Positive Ions in Helium and Argon. J. Chem. Phys. 62, 3517-22 (1975)

15. Gatland, I.R.: Case Studies in Atomic Physics vol. 4, ed. E.W. Mc Daniel and M.R.C. Mc Dowell (Amsterdam: North Holland), pp. 369-437 (1974)

16. Mc Daniel, E.W. and Mason, E.A.: The Mobility and Diffusion of Ions in Gases (New York: Wiley) (1973)

17. Dobler, W., Lindinger, W., Howorka, F.: Kinetic Energy Dependences of the Branching Ratios of the Reactions He^+, Ne^+ and Ar^+ + H_2O. 3rd Symp. on Atomic and Surface Physics; ed. W. Lindinger, Innsbruck, pp. 299-303 (1982)

18. Viehland, L.A.; Interaction Potentials for Li^+ - Rare Gas Systems. Chem. Phys. 78, 279-294 (1983)

19. Skullerud, H.R.: Calculation of Ion Drift and Diffusion: Basis Set Expansions with Non-Gaussian Weight Functions. Proc. 3rd Int. Swarm Seminar, W. Lindinger (ed.), Innsbruck, pp. 212-217 (1983)

20. Stefánsson, T.: Diffusion of Lithium Ions in Argon. Proc. 3rd Int. Swarm Seminar. W. Lindinger (ed.), Innsbruck, pp. 227-233 (1983)

Theory of Ion-Molecule Collisions at (1 eV–5 keV)/AMU

M. R. Flannery

School of Physics, Georgia Institute of Technology,
Atlanta, GA 30332, U.S.A.

1. Various Aspects

In ion-molecule collisions from thermal energies, to a few eV, up to a few keV, various processes such as chemical reactions (nuclear rearrangements), charge transfer (electronic rearrangement) and rotational, vibrational and electronic transitions occur, rarely in isolation except over limited energy ranges. Inelastic scattering changes the internal state by rotational, vibrational and electronic excitation of one or both collision parteners while preserving the identity of each partner, in contrast to reactive scattering (chemical reactions, three-body ion-molecule association, charge transfer, dissociation, ionization) which changes the chemical identities. There are few theories which elucidate the coupling between reactive and inelastic mechanisms, since development of feasible theoretical treatments of each of the processes in isolation presents, in itself, a considerable challenge.

The electrostatic interaction $V(R,r,\gamma)$ between an ion A^+ at relative separation $\underset{\sim}{R}$ from the center of mass of the molecule BC with internuclear vector $\underset{\sim}{r}$, depends on r which, in general, causes vibrational excitation via collision induced forces along $\underset{\sim}{r}$; on the angle γ between $\underset{\sim}{R}$ and $\underset{\sim}{r}$, which induces (via a torque) rotational excitation, and on R which mainly controls the general variation of the corresponding inelastic differential cross sections with scattering angle. We shall see, however, that vibrational excitation can be strongly enhanced when coupled to charge-transfer channels. Within various regions of the kinetic energy E of relative motion (or center-of-mass energy), chemical reactions, charge transfer, rotational and vibrational excitation may be all or partially strongly coupled, or else be all effectively decoupled over a small ($\sim$10eV) energy range. While strong couplings may exist only within all reactive mechanisms (chemical reaction and charge transfer for $E \lesssim 1$ eV) or only within all inelastic mechanisms (rotational and vibrational excitation) in range $1 \lesssim E$ (eV) $\lesssim$ 10-50, i.e., essentially in mutually exclusive energy ranges, one inelastic mechanism can be strongly coupled, for a wide range of interesting cases, with only one strong reactive scattering mechanism (as in vibrational excitation caused directly by charge transfer for $10 \lesssim E$ (eV) $\lesssim$ few keV).

A plausible upper limit to reaction at kinetic energy E of relative motion is the Langevin cross section, $\sigma_L(\text{Å}^2) = 17\alpha^{1/2}[E(\text{eV})]^{-1/2}$, for close spiralling encounters within the orbiting radius, $\sim$(4-6) Å at thermal energies, centered about the center of mass of the molecule of averaged polarizability $\alpha(\text{Å}^3)$. At thermal energies, the cross section σ_C for production [1] of ArH^+ in $(Ar^+\text{-}H_2)$ collisions is $\sigma_C(\text{Å}^2) \sim 200$. Also HD_2^+ and D_3^+ are each formed in merging beam $(D_2^+\text{-}HD)$ collisions [2] with $\sigma_C(\text{Å}^2) \sim 100$, 10 and 0.1 at kinetic energies E of relative motion of 0.002, 1 and 8 eV, respectively. The E-variation of σ_C becomes quite precipitous for $E \gtrsim 1$ eV, relative to the much slower E-variation of σ_L. Except at very low energies $E \lesssim 1$ eV, charge transfer cross sections σ_X become much larger than σ_L, since attraction can occur from overlap forces, in addition to the weaker polarization R^{-4} attraction implicit in σ_L.

For $(H^+, D_2) \to (D^+, HD)$, reaction and charge-transfer can occur for E (eV) $\gtrsim 0.04$ assisted by direct rotational and possible vibrational excitation. For $(H^+\text{-}H_2(v=0, J=0))$ collisions, in contrast, rotational excitation (with thresholds $\Delta E(J=0 \to 2,4) = 0.04, 0.14$ eV) and vibrational excitation (with $\Delta E(v=0 \to 1,2,3) = 0.516, 1.003, 1.461$ eV) are alone possible, until E reaches the threshold $E_X \sim 1.83$ eV for charge-transfer and reaction. The cross section σ_X for charge-transfer [3,4] then slowly rises to a peak ~ 10 Å^2 at $E \sim 1$ keV. Above $E_D = 4.48$ eV direct dissociation, and above $E_i \sim 11$ eV dissociation via electronic excitation, can occur, but again both peak in the keV energy region. For $Li^+\text{-}H_2$ collisions at $E \lesssim E_X = 10.2$ eV, only rotational and vibrational transitions and weak direct dissociation are possible.

Since the strong valence forces in H_3^+ tend to expand the H_2 bond length, vibrational excitation of H_2 by H^+ is much stronger than in $Li^+\text{-}H_2$ collisions where short range (forced oscillator) repulsive forces are mainly responsible for the small vibrational excitation with associated excitation of high J' levels. Rotational excitation in the lighter $H^+\text{-}H_2$ system is, however, weaker by comparison, since the torque is less and the triangular geometry of the intermediate H_3^+ complex permits less flexibility for rotation of H_2.

As the ion energy or kinetic energy of relative motion E increases from $\sim$ several eV, rotational excitation decreases since the rotational period $\tau_{rot} \sim (10^{-11}$ for O_2 -2 10^{-12} for H_2)s becomes much greater than the duration $\tau_{coll} \sim R(\AA)\ [M(A.M.U.)/2E(eV)]^{1/2}\ 10^{-14}$s of the collision between species of reduced mass M(A.M.U.) with characteristic length R(Å) for the particular process. The cross sections σ_R for rotational transitions (in Li^+-H_2 collisions) increase from threshold, exhibit a peak ~ 10 Å^2 at 1-5 eV and then decrease [5] to ~ 1 Å^2 at 100 eV. Since the vibrational period $\tau_{vib} \sim (10^{-13}$ for Br_2 - 10^{-14} for H_2)s, σ_v for vibrational excitation in Li^+-H_2 collisions, will increase from thresholds $\sim$1 eV mainly via v,J $\rightarrow$ v',J' transitions say to $\sim$0.1 Å^2 and 0.03 Å^2 for v = 0 $\rightarrow$ 1,2 respectively [6] at E$\sim$100-200 eV, and then decrease thereafter. The H^+-H_2 (v=0 $\rightarrow$ v') system involves larger vibrational cross sections with maxima, σ_v (Å^2) $\sim$ 6, 2.3, and 0.8 for v' = 1, 2 and 3, respectively, which occur [7] at E $\sim$ 30 eV.

Charge transfer collisions [8] between similar (AB^+-AB) systems are characterized by much larger cross sections σ_x which decrease only very slowly as E is raised from thermal energies to a few keV; for N_2^+-N_2, σ_x decreases [9] from $\sim$70 Å^2 at E $\sim$ 1 eV to $\sim$25 Å^2 at a few keV. Since the direct and charge transfer channels can be in near energy-resonance for symmetric systems and are closely coupled, direct vibrational excitation [8-10] is therefore characterized by large cross sections, $\sigma_v \sim$(10-20) Å^2, in contrast to weaker vibrational excitation via forced oscillator collisions, which are characterized by relatively smaller cross sections $\sim 10^{-2}$ Å^2.

The key difference between symmetric (AB^+-AB) and asymmetric (A^+-BC) charge transfer collisions is that there are initially two (elastic) paths of approach in the former case, and charge transfer then arises from the phase-change caused by the different elastic scattering potentials associated with the bonding and antibonding states of the $[AB]_2^+$-complex (states which reduce to the gerade and ungerade states of A_2^+ in A^+-A collisions). Charge-transfer in the asymmetric (A^+-BC) case involves an electronic transition between states with different asymptotic energies ($\sim$ 1.83 eV for H^+-H_2 and 11 eV for Li^+-H_2 collisions), and is therefore characterized by a threshold dependent cross section which increases to somewhat lower cross section maxima ($\sigma_x \sim 10$ Å^2 at E $\sim$ 1 keV for H^+-H_2 collisions [3]).

The variation with E of many ion-molecule processes appears as if the process were energy resonant. This is because the manifold of vibrational channels (in charge-transfer, for example) offers a band of pathways which are in near-energy resonance either with the incident channel or with a particular product (charge-transfer) channel which is strongly coupled to the incident channel.

The dominant processes for unlike systems (A^+-BC) in the (1-100) eV range of center-of-mass energy E are elastic and rotational excitation, while for $E \sim$(1 eV - several keV), the dominant process for like (AB^+-AB), apart from direct elastic scattering, is charge-transfer with its associated direct vibrational excitation. For $E \gtrsim$ few keV, direct electronic excitation to bound electronic and to dissociative electronic states occurs with relatively small cross sections $\sigma_E \sim 1$ Å^2. For lower $E \lesssim 1$ eV, chemical reactions will compete quite strongly with any of the above processes when operative.

In the following sections, feasible and tractable semiclassical (classical path, quantal target) theories for inelastic (rotational excitation) processes alone, and for reactive processes (symmetric and asymmetric charge transfer) coupled to direct inelastic processes (vibrational excitation) will be outlined and discussed. The theories are particularly appropriate for the energy range (1 eV - several keV), since chemical reactions are specifically excluded. Full quantal treatments are not presented since their application is feasible to at most 10 eV, and since they are well documented elsewhere [11-15]. Space allows only the mention of quasi-classical trajectory methods [11,15,17], valuable for highly averaged quantities in rotational excitation in non-reactive systems, and of semi-classical S-matrix theory [13,16] and classical perturbation theory [17,18], which are all based on classical relative and internal motions and which are valuable towards realization of the conceptual power of classical mechanics. Apart from vibrational excitation in collinear collisions, the S-matrix theory, however, becomes computationally unfeasible [11,18] for three dimensional systems, even in comparison with full quantal methods.

2. Rotational and Vibrational Transitions in Non-Reactive Collisions

2.1 Quantal Methods

The traditional full quantal close coupling (CC) treatment of rovibrational transitions is the generalization [19] of the formulation [20] of Arthurs and Dalgarno for scattering of a rigid rotor (RR) by a structureless projectile to vibrating rotor (VR) collisions, and is, by now, well established. It is, however, practical only for $E \lesssim$ few eV, since the inherent partial wave expansion for the relative motion converges only very slowly as E is increased, and excessive amounts of computer time are required. It has, as yet, only been applied [19] to rotational and vibrational excitation in Li^+-H_2 collisions at 0.6 eV and 1.2 eV where coupling with the two excited vibrational channels (v = 1,2) is extremely weak, and vibrational excitation is then at most incidental.

Many simplified quantal schemes (J_z-coupled states (CS) [21], infinite-order sudden (IOS) [22], effective potentials [23], L-dominant [24]) based on CC have therefore been introduced for $E \lesssim 10$ eV. These methods seek to reduce the (rotational) dimensionality of CC by assuming, for example, that the Z-axis of a rotating frame always lies along $\underset{\sim}{R}$ (as in CS [21] and in earlier rotating-atom approximation of Bates [25], where transitions are therefore restricted to zero change ΔM_J in the Z-components of the rotational quantum number J) or by assuming that the orientation $\underset{\sim}{r}$ of the molecule can be considered fixed during the collision (as in the sudden or adiabatic-nuclei approximation [26]). In IOS, a combination of both sudden and CS is adopted. At low energies ($\lesssim$ few eV), vibration may either be ignored, as in rigid rotor (RR) approximations, or be treated exactly (VR), within the above methods by full solution of the vibrationally coupled equations.

Since these dimensionality-reducing (DR) schemes remain based on a partial-wave analysis they are also very time-consuming. For the transition $J = 0 \to 2$ in Li^+-N_2 collisions at 4.23 eV, which involves close-coupling of many J-levels, 10^4 partial waves were required in a recent IOS treatment [27]. Under the added requirement that reliable interaction potentials, in general, ab-initio, are needed as input, it is therefore not too surprising to note that only a few special systems have been studied in detail by the above quantal methods, at only a few energies chosen to coincide with measurement. For example, pure rotational and rovibrational transitions have been studied in the collisions: H^+-H_2 by CS-RR [28] at 3.7 eV, by IOS-VR [29,30] at 4.67, 6 and 10 eV; Li^+-H_2 by CC-RR [31], CC-VR [19], CS [32] for $E \lesssim 1.2$ eV and by IOS-VR [33] at 5.54 and 8.8 eV; Li^+-N_2 by IOS-VR [27] at 4.23 eV.

Comparison of the IOS-VR cross sections [29] with rotationally resolved measurements [34] for H^+-H_2 ($v=0$, $J=0 \to v'(0,1,2,3)$, J') indicates good to excellent agreement at 10 eV for the variation with scattering angle of the vibrational transition probabilities (ratio of cross sections) when summed over final levels J', and excellent agreement for the relative differential cross sections for vibrational excitation. Only scattered agreement [30] is noted, however, for rotational excitation ($0,0 \to 0,J'$) at 4.67 and 6 eV.

The agreement between IOS-VR [33] and measured [35] vibrational transition probabilities (summed over J') for Li^+-H_2 collisions is reasonable only at 8.8 eV and very poor at 5.54 eV. When compared with the only measured absolute integral cross sections (~ 4.2 Å^2 [5]) for rotational excitation in Li^+-H_2 ($0,0 \to 0,2$) collisions at 12 eV, (and beyond), the IOS-VR cross section 19.8 Å^2 [33] at 8.8 eV is a factor of 5 higher; and the IOS-VR cross section ~ 0.3 Å^2 [33] for vibrational excitation ($v=0 \to 1$) at 8.8 eV is much in excess of the measured [6] peak values ~ 0.08 Å^2 which are spread over a wide range around 150 eV.

Many of the above reduced-dimensionality quantal schemes have varying degrees of success e.g., the most accurate of these, the J_Z-coupled states (CS) method [32] when compared with CC [19] predicts σ_R for Li^+-H_2 ($J=0 \rightarrow 2$) collisions to within only 33% at 0.6 - 1.2 eV, (in contrast to its success with atom-molecule collisions), and differential cross sections in substantial disagreement at intermediate angles. Also the great sensitivity [29,28] exhibited by any method to the choice of various (reasonably accurate) interactions $V(R,r,\gamma)$ places a great premium for highly accurate (time consuming) calculations of these anisotropic surfaces [19,27,29] over an extensive range of (R,r,γ) parameters.

2.2 Semiclassical Theory (Eikonal path-Quantal target)

There is therefore some need for more accurate yet feasible treatments of the energy region, $1 \lesssim E$ (eV) $\lesssim 10$, and great need for treatments feasible in the region 10 ev $\lesssim E \lesssim 1$ keV. For low energies $\lesssim 10$ eV only relative differential cross sections have been measured [34-36]. Some measurements [5,6] of integral cross sections exist for rotational and vibrational excitation in Li^+-H_2 collisions at center-of-mass energies E in the range (12-90 eV) for which accurate theory has not yet been applied.

For the above reasons, and in order to explore the strengths of semiclassical methods which naturally focus attention on collisional time-scales and which naturally provide greater physical insight to various levels of approximation, McCann and Flannery [37-39] have developed a multistate orbital (or energy-conserving) treatment of rotational and vibrational transitions in ion-molecule and neutral-molecule collisions at intermediate and high impact energies. Rather than seeking schemes which reduce the dimensionality of the quantal CC within a phase shift analysis of the relative motion, the basis of their semiclassical treatment is initially a semiclassical description (via an eikonal or JWKB wavefunction) of the relative motion, which is then coupled to the full quantal response of the closely coupled internal modes of the collision partners, via the introduction at each $\underset{\sim}{R}$ of an interaction averaged over the quantal response of all of the internal modes. Even at low energies ($\sim$0.6 eV) the theory realized considerable success [39] when compared with full quantal CC differential and integral cross sections [19],(Figure 1), and with measured relative differential cross sections. The procedure apparently introduces less error than does the best (CS) [32] dimensionality-reducing (DR) quantal scheme, as evidenced by comparison of CS [32] and EC [39] with CC [19] for Li^+-H_2 at 0.6 eV. Within the EC method, DR approximations may in addition be made. Figure 1 indicates the excellent agreement between the semiclassical

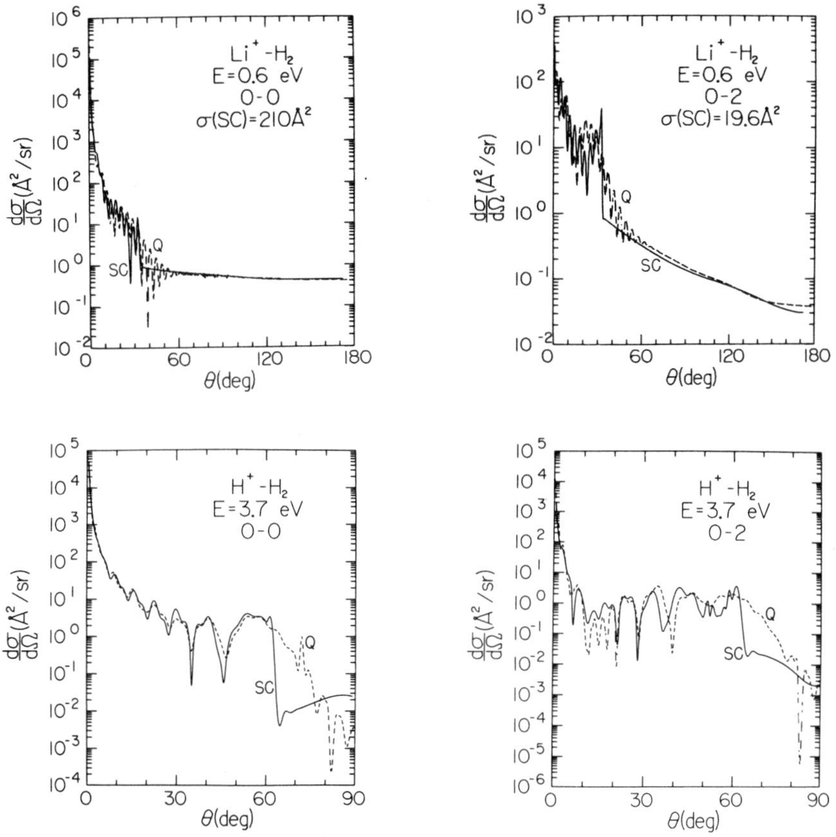

Fig. 1. Quantal [19,28] and semiclassical [39] differential cross sections for elastic ($0 \rightarrow 0$) and inelastic ($J=0 \rightarrow 2$) scattering in Li^+-H_2 and H^+-H_2 collisions.

---

[39] and quantal methods [28] under the common CS approximations for H^+-H_2 collisions.

The method is extremely efficient and accurate in that transition amplitudes converge much faster in the impact-parameter representation than they do in the angular momentum (partial-wave) representation. It is therefore quite a natural method for (1 eV - 1 keV) inelastic collisions. It has recently been shown [40,41] to be also highly reliable for reactive collisions for $E \lesssim 10$ eV.

2.3 Multistate Orbital (Energy Conserving) Method

In contrast to customary semiclassical treatments based on a straight line, the classical path $\underset{\sim}{R}(t)$ for relative motion between species A (ion or atom) and BC is obtained [37] by solving Hamilton's equations

$$\partial Q_i/\partial t = \partial\langle H\rangle/\partial P_i \quad ; \quad \partial P_i/\partial t = - \partial\langle H\rangle/\partial Q_i \tag{1}$$

directly for the variation with time t of the generalized coordinates Q_j (≡ X,Y,Z) for $\underset{\sim}{R}(t)$, the vector position of A relative to the center-of-mass of BC, and of the associated generalized momentum $P_j(t)$ for motion under the (Ehrenfest) averaged Hamiltonian [37],

$$\langle H(\underset{\sim}{R}(t))\rangle = \sum_{j=1}^{3} P_j^2(\underset{\sim}{R}(t))/2\mu + \langle\Psi(\underset{\sim}{r},\underset{\sim}{R}(t)|H_0(\underset{\sim}{r}) + V(\underset{\sim}{r},\underset{\sim}{R}(t))|\Psi(\underset{\sim}{r},\underset{\sim}{R}(t))\rangle, \quad (2)$$

where μ is the reduced mass of the (A-BC) system. The system wavefunction is therefore expanded in terms of the complete set of eigenfunctions $\phi_n(\underset{\sim}{r})$ of the unperturbed Hamiltonian H_0 for the internal states of the isolated species A and BC of energy E_n, $(H_0\phi_n = E_n\phi_n)$, as

$$\Psi(\underset{\sim}{r},\underset{\sim}{R}) = \sum_n A_n(\underset{\sim}{R})\ \phi_n(\underset{\sim}{r})\ \exp[iS_n(\underset{\sim}{R}) - X_n(\underset{\sim}{R})] \quad (3)$$

where $\underset{\sim}{r}$ denotes the collective internal coordinates of A and BC in the absence of interaction, and where $\phi_n(\underset{\sim}{r})$ separates, via the Born-Oppenheimer approximation, into associated electronic, vibrational and rotational parts. The relative motion is described by the above combination of functions A_n, S_n and X_n which possess the following physical characteristics. The eikonal, or classical action, is the solution of the Hamilton-Jacobi equation [37-39],

$$(\underset{\sim}{\nabla}S_n)^2 = K_n^2(\underset{\sim}{R}) = k_n^2 - (2\mu/\hbar^2)\ V_{nn}(\underset{\sim}{R}) \quad (4)$$

which, of course, is the heavy-particle limit to Schrödinger's equation,

$$[\ -\nabla_{\underset{\sim}{R}}^2 + (2\mu/\hbar^2)\ V_{nn}(\underset{\sim}{R})]\psi(\underset{\sim}{R}) = [(\nabla S_n)^2 - i(\nabla^2 S_n)]\psi(\underset{\sim}{R}) \equiv K_n^2\psi(\underset{\sim}{R}) \quad (5)$$

The wavefunction, $\psi(\underset{\sim}{R}) = \exp(iS_n)$, describes relative motion in the static channel n under interaction V_{nn}, which is a diagonal element of

$$V_{ij}(\underset{\sim}{R}) = \langle\phi_i(\underset{\sim}{r})|V(\underset{\sim}{r},\underset{\sim}{R})|\phi_j(\underset{\sim}{r})\rangle \quad (6)$$

where $V(\underset{\sim}{r},\underset{\sim}{R})$ is the instantaneous electrostatic interaction between A and BC at separation $\underset{\sim}{R}$. The wavenumber and kinetic energy of relative motion at asymptotic R is k_n and $\hbar^2 k_n^2/2\mu$ and the local momentum at $\underset{\sim}{R}$ is $\hbar\underset{\sim}{K}_n(\underset{\sim}{R})$. The phase X_n is the solution of

$$\nabla_{\underset{\sim}{R}}^2 S_n - 2(\nabla_{\underset{\sim}{R}} S_n)\cdot(\nabla_{\underset{\sim}{R}} X_n) = 0 \quad (7)$$

which represents flux conservation, $(\underset{\sim}{\nabla}\cdot\underset{\sim}{J}_n = 0)$, in static channel n, since the current $\underset{\sim}{J}_n$ associated with the exponential factor in (3), for real S_n and X_n, is

$$\underset{\sim}{J}_n = \exp(-2X_n)\underset{\sim}{v}_n(\underset{\sim}{R}) \equiv (\hbar/\mu)\ \exp(-2X_n)\ \underset{\sim}{\nabla}S_n \quad (8)$$

when (4) is used for the local velocity $\underset{\sim}{v}_n$. Thus, flux associated with channel n of (3) is changed only via transitions to other states f, and A_n is then the probability amplitude of remaining in state n. On assuming that the eikonal $S_n(R)$ contains the major R-variation of (3), the stationary-state Schrödinger equation with (3) inserted then yields the three-dimensional equation [37-39],

$$i\underset{\sim}{K}_f \cdot \underset{\sim}{\nabla}_R A_f(\underset{\sim}{R}) = (\mu/\hbar^2) \underset{n\neq f}{S} A_n(\underset{\sim}{R})\, V_{fn}(\underset{\sim}{R}) \exp i(S_n - S_f) \exp(X_f - X_n) \qquad (9)$$

The first term in (2) is the kinetic energy of relative motion and, with (3) inserted, the second term reduces under the assumption that the orbits in the different channels do not differ appreciably (i.e., $X_n \approx X_f$), to the $\underset{\sim}{r}$-averaged interaction [37-39],

$$\langle V(\underset{\sim}{R})\rangle = \underset{n}{S} [|A_n(\underset{\sim}{R})|^2 \varepsilon_n + \underset{f}{S} A_f^*(\underset{\sim}{R})\, A_n(\underset{\sim}{R})\, V_{fn}(\underset{\sim}{R}) \exp i(S_n - S_f)] \qquad (10)$$

The time t ($\equiv s/v(\underset{\sim}{R})$) is introduced as a (dummy) variable to measure distance s along the common trajectory i.e., $\underset{\sim}{K}_f \cdot \underset{\sim}{\nabla}_R A_f = K_f \partial A_f/\partial s \equiv (\mu/\hbar)\, \partial A_f/\partial t$, and the mean momentum is

$$\frac{1}{2}\hbar(K_n + K_f) = P(\underset{\sim}{R}) = \mu|d\underset{\sim}{R}/dt| \qquad (11)$$

so that the evolution of the eikonal difference in (9) and (10) can be written as

$$S_n(\underset{\sim}{R}) - S_f(\underset{\sim}{R}) = [\varepsilon_{fn} t + \int_{t_0}^{t} \{V_{ff}(\underset{\sim}{R}(t)) - V_{ii}(\underset{\sim}{R}(t))\}\, dt]/\hbar \qquad (12)$$

where $\varepsilon_{fn} = \varepsilon_f - \varepsilon_n$. The amplitudes

$$C_f(t) = A_f(t) \exp[-(i/\hbar) \int_{t_0}^{t} V_{ff}(\underset{\sim}{R}(t))dt] \qquad (13)$$

when inserted in (9), therefore satisfy the usual set of coupled equations,

$$i\hbar\, \partial C_f/\partial t = \underset{n}{S} C_n(t)\, V_{fn}(\underset{\sim}{R}(t)) \exp(i\varepsilon_{fn} t/\hbar) \qquad (14)$$

derived for the internal response of a system to a time dependent perturbation $V(\underset{\sim}{r},\underset{\sim}{R}(t))$. The implicit time dependence for motion under the averaged interaction (10), which reduces with the aid of (12) and (13) to,[37-39],

$$\langle V(\underset{\sim}{R}(t))\rangle = \underset{n}{S} [|C_n(t)|^2 \varepsilon_n + \underset{f}{S} C_f^*(t) C_n(t) V_{fn}(\underset{\sim}{R}(t)) \exp(i\varepsilon_{fn} t/\hbar)] \qquad (15)$$

is determined from Hamilton's equations (1), which now reduce to [37-39]

$$\partial Q_i/\partial t = P_i(t)/\mu \quad ; \quad \partial P_i/\partial t = - \underset{n}{S} \underset{f}{S} C_f^*(t)\, C_n(t) [\partial V_{fn}/\partial Q_i] \exp(i\varepsilon_{fn} t/\hbar) \qquad (16)$$

The external relative motion (16) is therefore solved self consistently with (14), which characterizes the quantal internal response of the molecule, at all times. Two important consequences [37,39] of this procedure are that total energy is conserved at all times during the collision, ($d\langle H\rangle/dt = \partial\langle H\rangle/\partial t = 0$) - energy is therefore being continually redistributed between the internal modes and the relative motion (a valuable asset) - and that total probability $S_f |C_f(t)|^2$ is naturally conserved at each point on the common trajectory provided by (15) and (16).

The differential cross section per steradian is

$$\sigma_{if}(\theta,\phi) = (k_f/k_i)|f_{if}(\theta,\phi)|^2$$
$$= (k_f/k_i)|(2\mu/4\pi\hbar^2)\langle\phi_f(\underset{\sim}{r})\exp(i\underset{\sim}{k}_f\cdot\underset{\sim}{R})|V(\underset{\sim}{r},\underset{\sim}{R})|\Psi(\underset{\sim}{r},\underset{\sim}{R})\rangle_{\underset{\sim}{r},\underset{\sim}{R}}|^2 \quad (17)$$

for scattering of an incident beam directed along $\underset{\sim}{k}_i(0,0)$ into unit solid angle about $\underset{\sim}{k}_f(\theta,\phi)$. When the main contribution to (17) arises from the stationary phase which produces the classical orbit $\underset{\sim}{R}(\theta,\phi)$, then the scattering amplitude $f_{if}(\theta,\phi)$ and hence differential cross section $\sigma_{if}(\theta,\phi)$ is easily deduced from flux considerations, as [37,39]

$$f_{if}(\theta,\phi) = \lim_{R\to\infty} [A_n(\underset{\sim}{R}) \exp iS_n'(\underset{\sim}{R})][\sigma_{el}^n(\theta,\phi)]^{1/2} \quad (18)$$

where $\sigma_{el}^n(\theta,\phi)$, the differential cross section for elastic scattering by $V_{ff}(\underset{\sim}{R})$ in channel f, or by $\langle V(R)\rangle$ of (15) for <u>common</u> trajectories $\underset{\sim}{R}_{i,f}(\theta,\phi)$, can be determined either from classical or quantal procedures, and where

$$S_n'(\underset{\sim}{R}) = S_n(\underset{\sim}{R}) - \underset{\sim}{K}_n\cdot\underset{\sim}{R} \quad (19)$$

is the classical action along the path measured relative to the action for undeflected particles with the same wavenumber. When more than one trajectory contributes to a given scattering angle θ, then the primitive result [37] can be suitably generalized [39] to yield (uniform) differential cross sections which exhibit the customary oscillatory structure (Fig. 1) due to interference between the classical actions $S_n'(\underset{\sim}{R})$ associated with each classical path which contributes to a specific scattering angle θ. Representative comparison with full CC quantum treatment at <1 eV is shown in Figure 1. Implementation to higher energies (1 eV → 1 keV), for which the above treatment was mainly designed (since full CC quantal treatments then become prohibitive), is underway.

In the limit of high impact energy, the scattering becomes increasingly concentrated in the forward direction, and the classical cross section $\sigma_{el}^n(\theta,\phi)$ if used, diverges (albeit extremely slowly). The cross section for transitions $(J_iM_i \to J_fM_f)$ between sublevels M_i and M_f associated with rotational levels J_i and J_f, respectively, can then be obtained directly from (17) to yield [42,43],

$$\sigma_{if}(\theta,\phi) = (k_f/k_i)|k_i \int_0^\infty J_\Delta(K'\rho)[C_f^{(SL)}(\rho,\infty) - \delta_{if}]\rho d\rho|^2 \quad (20)$$

in the heavy-particle/high-energy limit, where J_Δ is the spherical Bessel function of order $\Delta = (M_f - M_i)$, where $K'^2 = (k_i - k_f)^2 - \varepsilon_{fi}/\hbar^2 v_i^2$, and where $C_f^{SL}(\rho,t)$ are the solutions of (14) evaluated using a straight line orbit, and trajectory,

$$\underset{\sim}{\rho}(\rho,\phi) = \text{constant} \quad ; \quad \underset{\sim}{R}(t) = \underset{\sim}{\rho} + \underset{\sim}{v}_i\, t \quad (21a)$$

in terms of the impact parameter $\underset{\sim}{\rho}$. The integral cross section then reduces to [42,43]

$$\sigma_{if}(v_i) = 2\pi(k_f/k_i) \int_0^\infty |C_f^{(SL)}(\rho,\infty) - \delta_{if}|^2 \rho d\rho \tag{21b}$$

which permits, in contrast to the usual expression without δ_{if}, the evaluation also of elastic scattering cross sections.

When the coupling between the classical path and quantal target can be ignored i.e., when energy conservation which results from (15) is therefore violated, the classical trajectory R(t), $\theta(t)$ and orbit $\underset{\sim}{R}(\theta,\phi)$ at energy $E = \frac{1}{2}\mu v_i^2$ can be assumed planar, and is determined by integration of,

$$\frac{1}{2}\mu(dR/dt)^2 = E - [\overline{V}_0(R) + E\rho^2/R^2] \quad ; \quad d\theta/dt = -\rho v_i^2/R^2 \tag{22}$$

where $\overline{V}_0$ is either some averaged spherical interaction independent of the transition amplitudes $C_n(t)$, or is the spherical part of the surface $V(\underset{\sim}{r},\underset{\sim}{R})$, so that ϕ remains constant.

The above energy conserving (EC) classical path (CP) method reproduces the rainbow oscillatory structure and is capable of very effective description of inelastic rotational rainbows [15] characterized as structures in the final rotational distribution of differential cross sections at fixed angles for those molecular systems (N_2, CO, Na_2) associated with small rotational constants so that many rotational levels must be included. These rotational rainbows arise from coherent superposition of scattering from different orientation angles γ. As the final level J_f is increased the rotational rainbows in $J_i \rightarrow J_f$ transitions shift to larger angles due to hard collisions and have wider widths than the usual diffraction oscillations which occur from near forward scattering by the repulsive part of the interaction. The analysis of these rainbows by the pure CC quantal method is not at all feasible, and reduced dimensionality simplifications as CS and IOS may be too inaccurate.

The above EC-CP method can also be extended [44] to cover vibrational transitions and vibrational rainbows. For vibrational excitation simpler time-dependent semiclassical schemes based on an uncoupled classical path (21a), (22) or its three-dimensional equivalent, have been adopted [45-48]. Billing [49] has treated vibrational excitation by assuming that rotation and relative motion can be determined classically from various Hamiltonians which include an asserted average over the one-dimensional quantal vibrational response, but which are not, however, consistent with energy conservation.

3. Charge-Transfer and Associated Vibrational Excitation

In contrast to charge transfer in (B^+-C) collisions [50] involving atomic species, theoretical development of viable descriptions of AB^+-AB, A^+-BC, BC^+-A, AB^+-CD charge-transfer collisions are relatively scarce [8-10,51-61]. This is mainly due to great difficulty of accurate determination of the various electronic interaction matrix elements which are, in general, anisotropic and which involve orbitals based on three and four centers. The added complexity introduced by acknowledgement of vibrational and even rotational transitions (which proceed via the anisotropic part of the interaction) is however tractable, and given the electronic interactions, semiclassical multistate calculations, although time-consuming, are feasible. Bates and Reid [8] have realized the first tractable and simplified multi-vibrational level treatment of charge-transfer between like systems (AB^+-AB) which do not involve an electronic transition. Hedrick et al.[60] have provided the first detailed account of A^+-BC and BC^+-A charge-transfer collisions which involve an electronic transition and have shown (as with Bates and Reid [8]) that direct vibrational excitation is greatly enhanced by coupling with the charge-transfer channel.

In the collision,

$$AB^+(v_i^+,J_i^+) + AB(v_i,J_i) \rightarrow AB^+(v_f^+,J_f^+) + AB(v_f,J_f) \quad ; \quad (D) \qquad (23a)$$

$$\rightarrow AB(v_f,J_f^+) + AB^+(v_f^+,J_f) \quad ; \quad (X) \qquad (23b)$$

at low and intermediate $E \lesssim$ few keV, it can be assumed that only the ground electronic states of each species participate, so that the $[AB]_2^+$ complex has only two possible electronic states. These represent the active electron remaining either attached to the slow, neutral target AB - the direct (D) channel, (23a) - or attached to the (fast) projectile ion - the exchange channel (X), (23b). The kinetic energy E of relative motion of two species each of mass M (A.M.U.) and moving with relative speed u (a.u.) is $\sim 25\ M\ u^2$ (keV) such that for E sufficiently high $\gtrsim$ 10 eV when the collision duration $\tau_{coll} >> \tau_{rot}$, the rotational period, electron-transfer may be considered to occur at a fixed orientation $\underset{\sim}{r}$ of each internuclear axis, so that for spherical interactions, rotation of the target and projectile remains unaffected i.e., $J_i = J_f$ and $J_i^+ = J_f^+$. Vibrational transitions $v_i^{(+)} \rightarrow v_f^{(+)}$ are however important since $\tau_{coll} \gtrsim \tau_{vib}$, the vibrational period.

3.1 Two-State Treatment for (A^+-A) and AB^+-AB Charge-Transfer

Of all the charge transfer processes involving molecules, (23) is closest in spirit to the symmetric resonance process [50],

$$A^+ \text{ (fast)} + A \text{ (slow)} \rightarrow A^+ \text{ (fast)} + A \text{ (slow)} \quad ; \quad (D) \tag{24a}$$

$$\rightarrow A \text{ (fast)} + A^+ \text{ (slow)} \quad ; \quad (X) \tag{24b}$$

between atomic species, insofar as (23b) proceeds via the time-varying field generated by the energy splitting $\varepsilon^+(R) - \varepsilon^-(R)$ associated with the bonding and antibonding eigenfunctions, the electronic states of the $[AB^+\text{-}AB]$ complex, which become symmetric or antisymmetric with respect to the mid-point of the separation $\underset{\sim}{R}$ between the center of masses of AB^+ and AB, only when internuclear vectors $\underset{\sim}{R}_P$ and $\underset{\sim}{R}_T$ of AB^+ and AB, respectively, are equal and parallel.

Since the bonding-antibonding splitting becomes naturally decoupled, via the Born-Oppenheimer approximation, from the Franck-Condon factors which couple the vibrational states of ion and neutral, (23a) also permits feasible detailed study of characteristics of convergence in the number of vibrational states used in the wavefunction expansion. Also, direct vibrational excitation (23a) is achieved via its close coupling with the charge-transfer channel (23b) normally characterized by large cross sections ~ 10 Å^2 at ~ 100 eV.

For straight line relative motion, the cross section for (24b), which lays the conceptual groundwork for (23b) is [50,8,9]

$$\sigma^X_{el} = 2\pi \int_0^\infty \sin^2 \eta^X_{el}(\rho)\, \rho d\rho \tag{25}$$

where the phase difference between elastic scattering in each of the gerade and ungerade channels (which are acknowledged by expanding the system wavefunction as a linear combination of gerade (+) and ungerade (-) eigenfunctions of A_2^+) is [50,9]

$$\eta^X_{el} = (1/2\hbar) \int_{-\infty}^{\infty} [\varepsilon^+(R(t)) - \varepsilon^-(R(t))]dt \tag{26}$$

When either an LCAO (linear combination of atomic orbitals) is used for the gerade and ungerade states of A_2^+, or when the system wavefunctions is expanded in terms of atomic states ϕ_i^D, ϕ_f^X, based on each center, then (25) follows, with [50,9]

$$\eta^X_{el} = (1/\hbar) \int_{-\infty}^{\infty} V(R(t))dt \tag{27}$$

where the exchange interaction is given, in general, by [50,9]

$$[1-|\langle\phi_i^D|\phi_i^X\rangle|^2]V(\underset{\sim}{R}) = [\langle\phi_f^X|V^D|\phi_i^D\rangle - \langle\phi_f^X|\phi_i^D\rangle\langle\phi_i^D|V^D|\phi_i^D\rangle], \tag{28a}$$

$$= [\langle\phi_i^D|V^X|\phi_f^X\rangle - \langle\phi_i^D|\phi_f^X\rangle\langle\phi_f^X|V^X|\phi_f^X\rangle], \tag{28b}$$

The initial and final wavefunctions for the active electron attached to the target and projectile ionic cores are $\phi_i^D(\underset{\sim}{r}_1)$ and $\phi_f^X(\underset{\sim}{r}_2)$ respectively, with

$\underset{\sim}{r}_2 = \underset{\sim}{R} + \underset{\sim}{r}_1$, and i = f for energy resonance. The electrostatic interaction for the direct (D) channels associated with the fast ion and slow target is $V^D(\underset{\sim}{r}_1,\underset{\sim}{R})$ and, for the exchange (X) channels associated with a fast neutral and slow ion, is $V^X(\underset{\sim}{r}_2,\underset{\sim}{R})$. Since the bra-ket $<|>$ denotes integration over the coordinate $\underset{\sim}{r}_1$ (or $\underset{\sim}{r}_2$) of the active electron, the integrations in (28) involve, in general, one-electron two center integrals. For symmetric resonance processes, the diagonal terms in (28a) and (28b) are equal.

By comparison of (26) and (27), the gerade-ungerade splitting for electronic states in the LCAO approximation may be approximated as

$$\varepsilon^+(R) - \varepsilon^-(R) = 2V(R) \qquad (29)$$

On assuming in (23) that only one pair ($v_i^+ = v_f^+$, $v_i = v_f$) of vibrational levels participate, then on separating the wavefunction for the isolated neutral and ionic molecular systems into their electronic, vibrational and rotational parts (via the Born-Oppenheimer approximation), the charge transfer cross section for (23b) is [52]

$$\sigma^X = 2\pi \int_0^\infty \sin^2 [P_{ii}\eta_{el}^X]\, \rho d\rho \qquad (30)$$

where P_{ii} is the diagonal element of the matrix $\underset{\sim}{P}$ with element

$$P_{fn} = <v_f^+|v_n><v_n^+|v_f> \equiv F_{fn}\, F_{nf} \qquad (31)$$

where each (Franck-Condon) positive or negative amplitude F_{in} is the vibrational overlap for the $AB^+(v_i^+) \rightarrow AB(v_n)$ transition. The electronic interaction $V(\underset{\sim}{R})$ in (27) now involves, in general, orbitals on four centers.

The above expression (30), derived originally by Gurnee and Magee [52], is valid in the low energy limit only when the ground vibrational levels of both ion and neutral alone participate in an energy-resonant process. It is the natural generalization of the electronic result (25) for this case. For vibrationally excited ions and ground neutrals, initially, accidental degeneracy can occur when a band of various pairs of (v_f^+, v_f) vibrational states is in close energy-resonance with the original pair (v_i^+, $v_i = 0$) so that (30) is inadequate, even at low energies E. Moreover, as E is raised, an increasing number of vibrational levels participate in the processes (23).

3.2 Multilevel Treatment for (AB^+-AB) Charge-Transfer

The wavefunction for the time dependent response of the internal modes (represented collectively by $\underset{\sim}{r}$) of the (AB^+-AB) system under the influence of their mutual electrostatic interaction $V(\underset{\sim}{r},\underset{\sim}{R}(t))$ is expanded as [8]

$$\Psi(\underset{\sim}{r},t) = \sum_{\alpha=D,X} \sum_n A_n^\alpha(t)\, \Phi_n^\alpha(\underset{\sim}{r}) \exp(-iE_n^\alpha t/\hbar) \qquad (32)$$

in terms of $\Phi_n(\underset{\sim}{r})$ which form a complete set of molecular eigenfunctions (with electronic, vibrational and rotational parts) of the unperturbed

Hamiltonian H_0 for the isolated molecular systems with total internal energy E_n at infinite center-of-mass separation R. The index α denotes whether the labelled quantities are associated with the direct (D) channels (23a) or with the charge-transfer (X) channels (23b). Insert (32) into the time-dependent Schrodinger equation, and assume that electronically averaged quantities such as the overlap

$$S(\underset{\sim}{R}) = \langle\psi_{el}^{D}|\psi_{el}^{X}\rangle_{\underset{\sim}{r}} \tag{33}$$

between the electronic parts $\psi_{el}^{D,X}$ of the wavefunctions $\Phi_n^{D,X}$ for the D and X channels, and the diagonal distortion (secular) terms

$$V^{DIS}(\underset{\sim}{R}) = \langle\psi_{el}^{D,X}|V(\underset{\sim}{r},\underset{\sim}{R})|\psi_{el}^{D,X}\rangle_{\underset{\sim}{r}} \tag{34}$$

are all independent of each of the internal nuclear separations $\underset{\sim}{R}_P$ and $\underset{\sim}{R}_T$ of the projectile P and target molecules T i.e., S and V^{DIS} are spherical in R so that rotational state of P and T remains unchanged. By working to lowest order in S, Bates and Reid [8] then found that the amplitudes,

$$C_n^{\alpha}(t) = A_n^{\alpha}(t)\exp(i/\hbar)\int_{-\infty}^{t} V^{DIS}(\underset{\sim}{R}(t))dt \tag{35}$$

satisfied the following set of coupled equations [8]

$$i\hbar\,\partial C_f^{\alpha}/\partial t = V(R)\sum_n P_{fn}\,C_n^{\bar{\alpha}}(t)\exp(i\varepsilon_{fn}t/\hbar) \tag{36}$$

where

$$V(\underset{\sim}{R}) = \langle\psi_{el}^{D}|V(\underset{\sim}{r},\underset{\sim}{R})|\psi_{el}^{X}\rangle - S(\underset{\sim}{R})\,V^{DIS}(\underset{\sim}{R}) \tag{37}$$

is assumed spherical, where P_{fn} is the product (31) of Franck-Condon amplitudes F_{fn}, and where ε_{fn} is the difference $(\varepsilon_f-\varepsilon_n)$ between the combined energies ε_f of the vibrational levels of the ion and neutral (the small difference between the rotational energies in the D and X channels being neglected). The label $\bar{\alpha}$ in (36) is D when α is X and vice versa, i.e., the D channel cannot proceed without participation of the X channels and vice versa. For straight line relative motion, (21a) along the Z-axis, the cross section for charge-transfer (α = X)

$$\sigma_{if}^{\alpha}(E) = 2\pi\int_0^{\infty}|C_f^{\alpha}(\rho,t\to\infty)|^2\,\rho d\rho \tag{38}$$

reduces to (30) for the energy resonant case, $(v_i^+ = v_f^+,\ v_i = v_f)$.

Once the solutions $C_n^{\bar{\alpha}}$ are obtained from (36) subject to $C_f^{D}(\rho,t\to-\infty) = \delta_{ni}$, $C_f^{X}(\rho,t\to-\infty) = 0$, differential cross sections are given by (20) with $\Delta = 0$. The use of (21b), rather than (38), permits the additional evaluation of direct elastic integral cross sections. Also the assumption of straight-line relative motion may be removed by the orbital (or energy-conserving method) of McCann and Flannery [37] so that extension to the lower (eV) energies can be obtained [56,40]. It is worth noting that the above treatment ignores electron translational factors $\exp(i\underset{\sim}{u}\cdot\underset{\sim}{r})$, which arise from the momentum change introduced as the active electron jumps from the target to the projectile ion and which become important mainly in the keV region.

At low impact speeds u, when only these levels n which form a band in near energy resonance with the initial channel need be included, (since, otherwise the exponent in (36) oscillates rapidly to yield negligible contribution), and at high u, $\exp(i\varepsilon_{fn}\, Z/\hbar u)$ in (36) is effectively unity, the set (36) may then be solved analytically by diagonalization techniques [8].

At high energies this procedure yields [8],

$$\sigma^X_{if} = P^2_{fi}\,\sigma^x_{el}(E) \quad ; \quad \sigma^D_{if} = 0\ (i \neq f) \quad ; \tag{39}$$

such that the charge transfer cross section σ^X_{if} when summed over all final vibrational levels f is simply σ^X_{el}, as in the atomic case (25) - (27). Thus, near energy-resonant channels ($\varepsilon_{fi} \approx 0$) with reasonable vibrational overlaps P_{fi} control the process, at low E, while σ^X_{if} at large E is mainly determined by P_{fi}; dependence at intermediate E results from coupling between energy defects ε_{fi} and P_{fi}.

The main limitation to application of this theory [(36) and (39)] is the inherent complexity of calculation of the exchange interaction (37) which involves orbitals based on four centers and anisotropic V. Bates and Reid [8] extended the method of Firsov for the splitting (29) and calculated the D and X cross sections for $H_2^+(v)$ - $H_2(v = 0)$ collisions in the (C.M.) energy range 1 eV $\lesssim$ E $\lesssim$ 25 keV. Flannery and associates [9,10,51,53-56] investigated, in detail, N_2^+-N_2, CO^+-CO, NO^+-NO and O_2^+-O_2 collisions by adopting a prescription due to Sato for (29); and H_2^+-H_2, D_2^+-D_2, T_2^+-T_2 collisions by using the Bates-Reid interaction. Up to 100 vibrational channels may be coupled. The similar vibrational frequencies of the neutral and ionic species (N_2, CO) results in a grouping of the energy defects for different vibrational channels into different "bands" separated by approximately one vibrational quanta. The largest vibrational overlaps F_{ij} are in the near resonant band. Dissimilar vibrational frequencies for the ionic and neutral species (NO, O_2) prevent this grouping, and large vibrational overlaps are spread over a moderate number of product channels. For (H_2, D_2, T_2), the ionic and neutral vibrational frequencies are dissimilar but the energy defects do, however, exhibit band structure and the vibrational overlaps as spread over a large number of product channels. These characteristics (energy defects versus vibrational overlaps) of these various systems are reflected in the variation of the calculated cross sections [9,10,53-56] with energy E.

Some calculated cross sections [54] are illustrated in Figure 2 for the first ten D and X channels with the smallest energy defects for H_2^+ ($v_i^+ = 0$) + H_2 ($v_i = 0$) collisions. At low E, the channels in close energy resonance have the largest cross sections as expected from (30). Above 100 eV-ion energy, the vibrational overlaps start to play an important role and the cross sections are influenced considerably by energy defect and overlap, until at high energies ($\sim$ 5 keV), the vibrational overlaps, in accord with (39), control the relative cross sections which increase with increasing overlap (Figure 2).

Quantative comparison of theory and measurements is difficult in that the vibrational population of the reactant ions, produced in many experiments by high-energy electron impact, can be questionable due to competition between direct ionization and autoionization. Figure 3 illustrates comparison of theoretical charge transfer cross sections for reactant ions H_2^+ ($v_i^+ = 0, 1$) with various measurements for ions produced mainly in the lower levels ($v_i^+ = 0, 1$) by 16 eV electron impact. At low energies ($\sim$ 1 ev) rearrangement channels which produce H_3^+ and which are not included in the above theory, seriously compete with charge-transfer. Comparison between theory and measurements for other systems may be found in the review by Moran [51].

Within the Bates-Reid method, Smirnov and associates [57] determined the matrix element $V(\underset{\sim}{R})$ of (37) by asymptotic methods (which are based on the assumption that electron transfer occurs at distances R large compared with molecular dimensions). They were then able to extract both the spherical and anisotropic parts of $V(\underset{\sim}{R})$ from Hartree-Fock one electron wavefunctions for a wide range of systems.

Since the main limitation to application of the Bates-Reid method is a lack of reliable data on the exchange interaction, Borkman and Cobb [69] have recently computed ab-initio potential-energy curves for the H_2^+-H_2 complex using SCF (self consistent field) and CI (configuration interaction) techniques. This is the only comprehensive study yet available for AB^+-AB system, is extremely time consuming, and indicates that the interaction $V(\underset{\sim}{R})$ is extremely anisotropic. Rotational excitation is therefore expected to be important for E $\lesssim$ 100 eV or so. Lee and De Pristo [70] have recently found agreement between their simple model of $V(\underset{\sim}{R})$ and the ab-initio results.

De Pristo [40] adopted the orbital (or energy-conserving) method of McCann and Flannery [37,56], rather than the straight-line relative motion, within the Bates-Reid treatment, and has shown good-excellent agreement with the

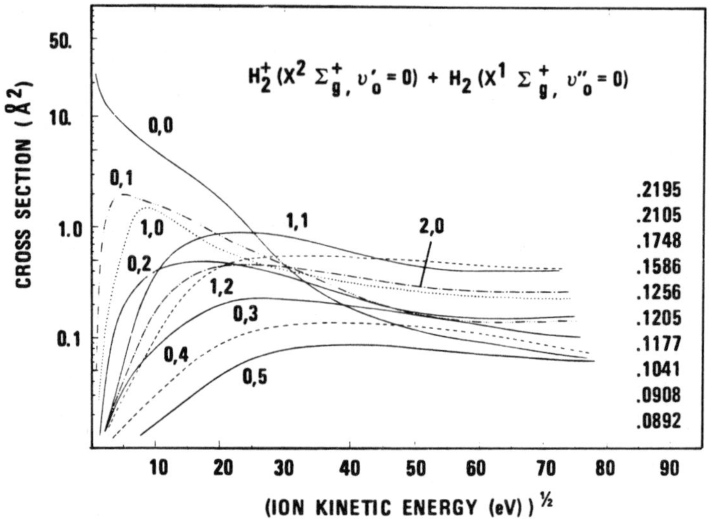

Fig. 2. State-to-state $(v_i^+ = 0, v_i = 0) \rightarrow (v_f, v_f^+)$ charge-transfer cross sections; (v_f, v_f^+) label each curve which at high energy become arranged in order of the indicated vibrational overlaps P_{fi} (from [54]).

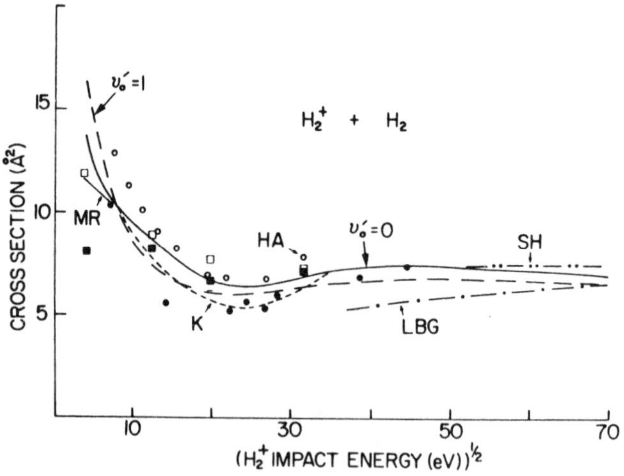

Fig. 3. Present theoretical (labelled $v_0' = 0,1$) and various experimental (SH [62], K [63], MR [64], LBG [65], HA [66], ● [67], (■, $v_0' = 0$; □, $v_0' = 1$) [68]) cross sections for $H_2^+ (X^2\Sigma_g^+, v_0') - H_2(X^1\Sigma_g^+, v_i = 0)$ charge-transfer collisions (from [51] and [54]).

only full quantal treatment performed by Becker [59] for O_2^+-O_2 charge transfer collisions at 8 and 36 eV (C.M.). Finally Sears and De Pristo [58] have recently developed a simple number scaling relationship in which they assert that the entire set of unknown state-to-state cross sections can be generated from one cross section and two parameters based on overlaps, energy defects and collision durations. It essentially represents an effort to extend the exact parametric dependence (39) at high energies down to lower energies by acknowledging interplay between ε_{fi} and P_{fi}.

3.3 Multistate Treatment of (A^+-BC) and (BC^+-A) Charge Transfer

For unlike (asymmetric) systems, Hedrick et al. [60] expand the system wavefunction as (32) and find that the amplitudes $A_n^{D,X}(t)$ satisfy the set of coupled equations,

$$i\hbar[\dot{A}_f^X + S_{21}\sum_n \dot{A}_n^D F_{fn}\exp(iE_{fn}^{XD}t/\hbar)] = V_{21}^D \sum_n A_n^D F_{fn}\exp(iE_{fn}^{XD}t/\hbar) + V_{22}^X A_f^X \tag{40a}$$

$$i\hbar[\dot{A}_f^D + S_{12}\sum_n \dot{A}_n^X F_{nf}\exp(iE_{fn}^{DX}t/\hbar)] = V_{12}^X \sum_n A_n^X F_{nf}\exp(iE_{fn}^{DX}t/\hbar) + V_{11}^D A_f^D \tag{40b}$$

where F_{fn} is the Franck-Condon amplitude $\langle v_f^+|v_n\rangle$ for $BC^+(v_f^+) \rightarrow BC(v_n)$ transitions, and where $E_{fn}^{\alpha\beta} = E_f^\alpha - E_n^\beta$, the combined electronic and vibrational energy defects between the D and X channels. The matrix elements

$$\left.\begin{aligned}
\langle\Phi_f^D|V^D(\underset{\sim}{r},\underset{\sim}{R})|\Phi_n^D\rangle_{\underset{\sim}{r}} &= [\langle\Phi|V^D|\Phi\rangle \equiv V_{11}^D(R)]\ \delta_{vib}\ \delta_{rot} \\
\langle\Phi_f^D|V^X(\underset{\sim}{r},\underset{\sim}{R})|\Phi_n^X\rangle_{\underset{\sim}{r}} &= [\langle\Phi|V^X|\psi\chi\rangle \equiv V_{12}^X(R)]\ F_{nf}\ \delta_{rot} \\
\langle\Phi_f^D|\Phi_n^X\rangle_{\underset{\sim}{r}} &= [\langle\Phi|\psi\chi\rangle \equiv S_{12}(R)]\ F_{nf}\ \delta_{rot}
\end{aligned}\right\} \tag{41a}$$

$$\left.\begin{aligned}
\langle\Phi_f^X|V^X(\underset{\sim}{r},\underset{\sim}{R})|\Phi_n^X\rangle_{\underset{\sim}{r}} &= [\langle\psi\chi|V^X|\psi\chi\rangle \equiv V_{22}^X(R)]\ \delta_{vib}\ \delta_{rot} \\
\langle\Phi_f^X|V^D(\underset{\sim}{r},\underset{\sim}{R})|\Phi_n^D\rangle_{\underset{\sim}{r}} &= [\langle\psi\chi|V^D|\Phi\rangle \equiv V_{21}^X(R)]\ F_{fn}\ \delta_{rot} \\
\langle\Phi_f^X|\Phi_n^D\rangle_{\underset{\sim}{r}} &= [\langle\psi\chi|\Phi\rangle \equiv S_{21}(R)]\ F_{fn}\ \delta_{rot}
\end{aligned}\right\} \tag{41b}$$

represent various $\underset{\sim}{r}$-averages of the direct interaction V_D between A^+ and BC, and of the exchange interactions V^X between A and BC^+ over post and prior electronic wavefunctions (Φ for BC alone, ψ for BC^+, χ for the one-electron orbital of A), and over vibrational and rotational wavefunctions for BC and BC^+. The incident speed as before (§ 3) is sufficiently large that rotational transitions are neglected (and V_{ij} are assumed to be independent of orientation of the internal internuclear vectors of BC and BC^+ and hence of the orientation of $\underset{\sim}{R}$, the relative channel vector between A^+ and BC). In contrast to the elastic bonding and antibonding channels in the AB^+-AB case (§ 3.2), charge-transfer occurs here via an electronic transition between distinct electronic states 1 and 2 of different asymptotic energies and represented by electronic wavefunctions Φ and ψ_X. Also $V_{11}^{D} \neq V_{22}^{X}$, $V_{21}^{D} \neq V_{12}^{X}$, in contrast with the symmetric case (§ 3.2), and the overlap S_{12} cannot in general be neglected. We note, with the aid of (34) that the set (40) reduces to (36) for equal distortion terms $V_{jj}^{D,X} (\equiv V^{DIS}, j = 1,2)$ in the D and X channels, when the electronic overlap S_{if} is neglected, and when F_{fn} is naturally replaced by P_{fn}. For the case of one (incident) D channel and n-X channels, (40) reduces to a simplified set [60].

Hedrick et al. [60] calculated differential and integral cross sections (20) and (21) from the solution of (40) for Ar^+-H_2 collisions at ion energies between 0.9 and 3.5 keV. The interactions in (41) were computed directly from simple analytical electronic wavefunctions for H_2 and H_2^+ and from an analytic fit to the numerical 3p-orbital for Ar. Fair agreement with experiment was obtained.

More recently, Sidis and de Bruijn [61] have given similar theory as above but used asymptotic methods to determine the various interactions for H_2^+-Mg collisions.

4. Conclusion

We have therefore discussed, in § 1 of this Chapter various aspects of ion-molecule collisions. We have then presented in § 2 a theory suitable for differential and integral cross sections for rotational transitions with extension to vibrational transitions. In § 3, a theory which couples reactive (charge-transfer) and inelastic (vibrational excitation) scattering is described for charge-transfer and vibrational excitation in like (AB^+-AB) and unlike (A^+-BC) systems. A natural combination of the treatments in § 2 and § 3 would permit the additional coupling of rotational transitions to the present theory in § 3. This coupling would be mainly important for collision energies E < 100 eV, but would, however, place great emphasis on accurate determination of ion-molecule bonding and antibonding anisotropic interactions, which at this stage are only (partially) known for the H^+-H_2 [29], Li^+-H_2 [19], Li^+-N_2 [27] and H_2^+-H_2 [69] systems.

Acknowledgement: This research is sponsored by the U. S. Air Force Office of Scientific Research under Grant No. AFOSR-80-0055.

References

1. Albritton, D. L., *Ion-Neutral Reaction-Rate Constants Measured in Flow Reactors through 1977.* Atom. Data Nucl. Data Tables *22*, 1-101 (1978).
2. Douglas, C. H., McClure, D. J., and Gentry, W. R., J. Chem. Phys. *67*, 4931 (1977).
3. Nakai, Y., Kikuchi, A., Shirai, T., and Sataka, M., *Data on Collisions of Hydrogen Atoms and Ions with Atoms and Molecules.* Japan Atomic Energy Research Institute. Report JAERI-M83-013, February 1983.
4. Tawara, H., Kato, T., Nakai, Y., *Electron Capture and loss cross sections for collisions between heavy ions and hydrogen molecules.* Institute of Plasma Physics. Report IPPJ-AM-28, June 1983.
5. Itoh, Y., Kobayashi, N., and Kaneko, Y., J. Phys. B. *14*, 679 (1981).
6. Kobayashi, N., Itoh, Y., and Kaneko, Y., J. Phys. Soc. Jap. *46*, 208 (1979).
7. Gentry, W. R., and Giese, C. F., Phys. Rev. A *11*, 90 (1975).
8. Bates, D. R., and Reid, R. H. G., Proc. Roy. Soc. A *310*, 1 (1969).
9. Flannery, M. R., Cosby, P. C., and Moran, T. F., J. Chem. Phys. *59*, 5494 (1973).
10. Flannery, M. R., McCann, K. J., and Moran, T. F., J. Chem. Phys. *63*, 1462, 3857 (1975).
11. Bernstein, R. B., (ed.), *Atom-Molecule Collision Theory.* (Chap. 8-12) New York, Plenum, 1979.
12. Dickinson, A. S., *Non-Reactive Heavy Particle Collision Calculations.* Comput. Phys. Commun. *17*, 51-80 (1979).
13. Miller, W. H., (ed.), *Dynamics of Molecular Collisions: Part A, Part B.* New York, Plenum, 1976.
14. Lawley, K. P., (ed.), *Potential Energy Surfaces.* New York, Wiley, 1980.
15. Bowman, J. M., (ed.), *Molecular Collision Dynamics.* Berlin, Springer-Verlag, 1983.
16. Miller, W. H., *Classical-Limit Quantum Mechanics and the Theory of Molecular Collisions.* Adv. Chem. Phys. *25*, 69-177 (1974).
17. Child, M. S., (ed.), *Semiclassical Methods in Molecular Scattering and Spectroscopy.* Holland, Reidel, 1980.
18. Dickinson, A. S., and Richards, D., *Classical and Semiclassical Methods in Inelastic Heavy-Particle Collisions.* In: Adv. Atom. Molec. Phys. *18*, 165-205 (1982).
19. Lester, W. A., and Shaefer, J., J. Chem. Phys. *62*, 1913 (1975).
20. Arthurs, A. M., and Dalgarno, A., Proc. R. Soc. A *256*, 540 (1960).

21. McGuire, P., Chem. Phys. Letts. *23*, 575 (1973); McGuire, P., Kouri, D. J., J. Chem. Phys. *60*, 2488 (1974); Pack, R. T., J. Chem. Phys. *60*, 633 (1974).
22. Tsien, T. P., and Pack, R. T., Chem. Phys. Lett. *6*, 54 (1970).
23. Rabitz, H., J. Chem. Phys. *57*, 1718 (1972).
24. DePristo, A. E., and Alexander, M. H., J. Chem. Phys. *63*, 3552 (1975).
25. Bates, D. R., Proc. Roy. Soc. A *243*, 15 (1957).
26. Chase, D. M., Phys. Rev. *104*, 839 (1956).
27. Pfeffer, G. A., and Secrest, D., J. Chem. Phys. *78*, 3052 (1983).
28. McGuire, P., J. Chem. Phys. *65*, 3275 (1976).
29. Schinke, R., Dupuis, M., and Lester, W. A., J. Chem. Phys. *72*, 3909-3916 (1983).
30. Schinke, R., J. Chem. Phys. *72*, 3917-3922 (1983), and references therein.
31. McGuire, P., Chem. Phys. *4*, 249 (1974).
32. Kouri, D. J., and McGuire, P., Chem. Phys. Letts. *29*, 414 (1974).
33. Schinke, R., Chem. Phys. *34*, 65 (1978).
34. Hermann, V., Schmidt, H., and Linder, F., J. Phys. B. *11*, 493-506 (1978).
35. David, R., Faubel, M., and Toennies, J. P., Chem. Phys. Letts. *18*, 87 (1973).
36. Faubel, M., and Toennies, J. P., *Scattering Studies of Rotational and Vibrational excitation of Molecules*. In: Adv. Atom. Molec. Phys. Vol. 13, 229-314 (1977).
37. McCann, K. J., and Flannery, M. R., J. Chem. Phys. *63*, 4695 (1975).
38. McCann, K. J., and Flannery, M. R., Chem. Phys. Letts. *35*, 124 (1975).
39. McCann, K. J., and Flannery, M. R., J. Chem. Phys. *69*, 5275-5287 (1978).
40. DePristo, A. E., J. Chem. Phys. *78*, 1237 (1983).
41. Micha, D., J. Chem. Phys. (in press).
42. Flannery, M. R., and McCann, K. J., J. Phys. B *7*, 2518 (1974).
43. Flannery, M. R., and McCann, K. J., Phys. Rev. A *9*, 1947 (1974).
44. Flannery, M. R., (in preparation).
45. Kruger, H., and Schinke, R., J. Chem. Phys. *66*, 5087 (1977).
46. Schinke, R., Chem. Phys. *24*, 379 (1977); *47*, 287 (1980).
47. McKenzie, R. L., J. Chem. Phys. *66*, 1457 (1977).
48. Iwamatsu, M., Onodera, Y., Itoh, Y., Kobayashi, H., and Kaneko, Y., Chem. Phys. Lett. *77*, 585 (1981).
49. Billing, G. D., J. Chem. Phys. *61*, 3340 (1974).
50. Bates, D. R., and McCarroll, R., *Charge Transfer*. Phil. Mag. Suppl. *11*, 39-81 (1962).

51. Moran, T. F., *Electron Transfer Reactions*, Chap. 1: *Electron-Molecule Collisions and Their Applications*. Christophorou, L. G. (ed.): Vol. 2. New York, Academic, 1984.
52. Gurnee, E. F., and Magee, J. L., J. Chem. Phys. *26*, 1237 (1957).
53. Moran, T. F., Flannery, M. R., and Cosby, P. C., J. Chem. Phys. *61*, 1261 (1974).
54. Moran, T. F., Flannery, M. R., and Albritton, D. L., J. Chem. Phys. *62*, 2869 (1975).
55. Moran, T. F., McCann, K. J., Flannery, M. R., and Albritton, D. L., J. Chem. Phys. *65*, 3172 (1976).
56. Moran, T. F., McCann, K. J., Cobb, M., Borkman, R. F., and Flannery, M. R., J. Chem. Phys. *74*, 2325 (1981).
57. Yeseyev, A. V., Radtsig, A. A., and Smirnov, B. M., J. Phys. B. *15*, 4437 (1982), and references therein.
58. Sears, S. B., and DePristo, A. E., J. Chem. Phys. *77*, 290, 298 (1983).
59. Becker, C., J. Chem. Phys. *76*, 5928 (1982).
60. Hedrick, A. F., Moran, T. F., McCann, K. J., and Flannery, M. R., J. Chem. Phys. *66*, 24 (1977).
61. Sidis, V., and Bruijn, D. P., Chem. Phys. (in press).
62. Stedeford, J. B. H., and Hasted, J. B., Proc. Roy. Soc. A *227*, 466 (1955).
63. Koopman, D. W., Phys. Rev. *154*, 79 (1967).
64. Moran, T. F., and Roberts, J. R., J. Chem. Phys. *49*, 3411 (1968).
65. Latimer, C. J., Browning, R., and Gilbody, H. B., J. Phys. B. *2*, 1055 (1969).
66. Hayden, H. C., and Amme, R. C., Phys. Rev. *172*, 104 (1968).
67. Rothwell, H. R., Van Zyl, B., and Amme, R. C., J. Chem. Phys. *61*, 3851 (1974).
68. Campbell, F. M., Browning, R., and Latimer, J. C., J. Phys. B. *14*, 3493 (1981).
69. Borkman, R. F., and Cobb, M., J. Chem. Phys. *74*, 2920 (1981).
70. Lee, C-Y., and DePristo, A. R., J. Amer. Chem. Soc. (in press).

Vibrational Excitation and De-Excitation and Charge-Transfer of Molecular Ions in Drift Tubes

E. E. Ferguson

Aeronomy Laboratory, NOAA, Boulder, CO 80303, U.S.A.

Introduction

The combination of a drift tube and a flowing afterglow constructed by McFarland for his Ph.D. thesis[1] opened up a wide range of experiments which can be carried out. The addition of the selected ion feature by Adams and Smith[2] has substantially enhanced the versatility of this device, now called the selected ion flow drift tube (SIFDT). Theoretical advances in understanding the velocity (or energy) distributions of ions in drift tubes have greatly enhanced their utility. Lin and Bardsley[3] and Viehland and Mason[4] have successfully treated the problem of atomic ions in atomic buffer gases so that cross-sections or thermodynamic rate constants can be obtained for atomic ion reactions even though their velocity distributions are not Maxwellian. This allows drift tubes to be used for rate constant (or cross-section) measurements between thermal to a few eV, closing the gap between thermal energy and beam measurements. Thermodynamic as well as kinetic data can be obtained. Recently Viehland, Lin and Mason[5] have shown that molecular ions in atomic buffer gases in drift tubes will have a Boltzmann distribution of internal energy state populations at steady state, characterized by the ion-buffer gas kinetic energy. This encourages the use of drift tubes to study reaction rate constant dependences on internal energy, since the kinetic and internal energy of ions can be varied independently by varying the rare gas buffer mass. We will describe some recent measurements of vibrational enhancement of charge-transfer reactions and then discuss some direct measurements of vibrational relaxation of molecular ions in flow drift tubes and present some unifying concepts on molecular ion relaxation.

Effect of Ion Vibrational Excitation on Charge-Transfer Reactions

A number of exothermic charge-transfer reactions of triatomic ions with diatomic neutrals have recently been shown to have strong vibrational state dependences.[6] An example is shown in Fig. 1, the reaction $N_2O^+ + NO \rightarrow NO^+ + N_2O$ measured as a function of average center-of-mass kinetic energy in He, Ne, and Ar buffer gases.

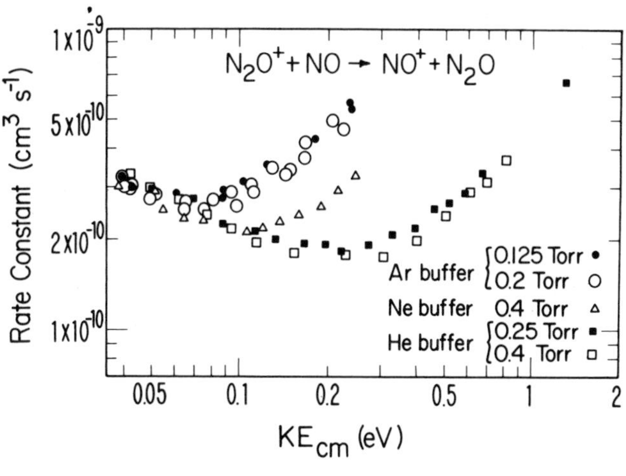

Figure 1. Rate constant for $N_2O^+ + NO \rightarrow NO^+ + N_2O$ as a function of kinetic energy in He, Ne and Ar buffer gases.

The rate constant is larger in the heavier buffer gases as a consequence of the higher N_2O^+ vibrational temperature in the heavier buffer gases.[5] The departure occurs at approximately the energy of the N_2O^+ bending mode, 0.057 eV, and much below the energy of the stretching modes, 0.215 and 0.140 eV. Fig. 2 shows the same results (solid lines) plus a number of added points representing the addition of N_2O neutral to the Ar buffer. Small amounts of N_2O reduce the rate constant by vibrational relaxation of the N_2O^+ by charge-transfer with N_2O, which occurs at near the collision rate.

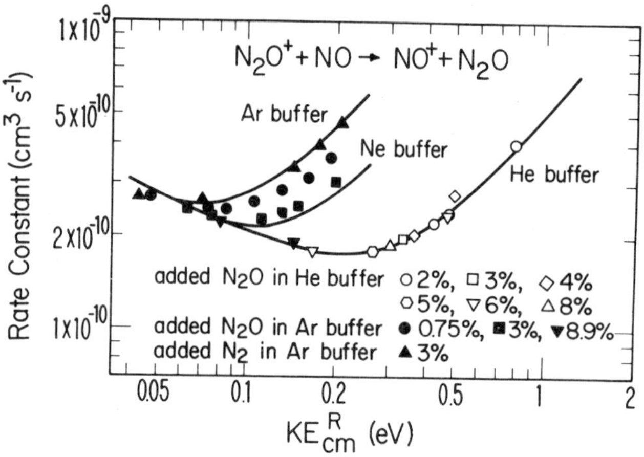

Figure 2. Same Data as Figure 1 with trace quantities of N_2O added to the He and Ar buffer gases.

Fig. 2 shows that there is no vibrational excitation in He, since there is no change in k with added N_2O. In the Viehland, Lin and Mason[5] framework this means that the vibrational excitation rate is too low for a steady state population to be reached in the experiments. There are about 10^3 collisions with the buffer so that $P_{ex} < 10^{-3}$ or $k_{ex} < 10^{-12}$ cm^3s^{-1}. On the other hand for Ne and Ar near threshold and above $P \gtrsim 10^{-3}$ and $k_{ex} \gtrsim 10^{-12}$ cm^3s^{-1}. From detailed balance it follows that $N_2O^+(v)$ vibrational deexcitation rate constants would have similar constraints, i.e. $k_d < 10^{-12}$ cm^3s^{-1} for He and $k_d \gtrsim 10^{-12}$ cm^3s^{-1} for Ar and Ne.

Similar limits have also been found for the ions CO_2^+, NO_2^+, H_2O^+ and SO_2^+ in which vibrational enhancements of charge transfer with NO have been observed,[6] i.e. $k_{ex} \sim k_d < 10^{-12}\ cm^3s^{-1}$ for He in every case and $k_{ex} \sim k_d \geq 10^{-12}\ cm^3s^{-1}$ for either Ne or Ar and sometimes both. Fig. 2 also shows that rotational excitation has no effect on the charge-transfer rate constant since the Ar data is quenched to the He data by addition of 8.9% N_2O. Rotational excitation cannot be quenched by N_2O (which is heavier than Ar) and the average rotational energy at the common point in Fig. 2 is three times greater in the Ar-N_2O buffer, 0.19 eV compared to 0.06 eV in the He buffer. We have also found rotational (and vibrational energy) to be ineffective in a number of O^- transfer negative ions reactions,[7] one of which is shown in Fig. 3, the reaction $CO_3^- + SO_2 \rightarrow SO_3^- + CO_2$, which has the same rate constant versus kinetic energy in both He and Ar buffer gases from thermal energy to 0.4 eV which corresponds to a center-of-mass energy four times greater in Ar than He, 0.32 eV compared to 0.08 eV, which should correspond closely to the rotational energies in the buffers.

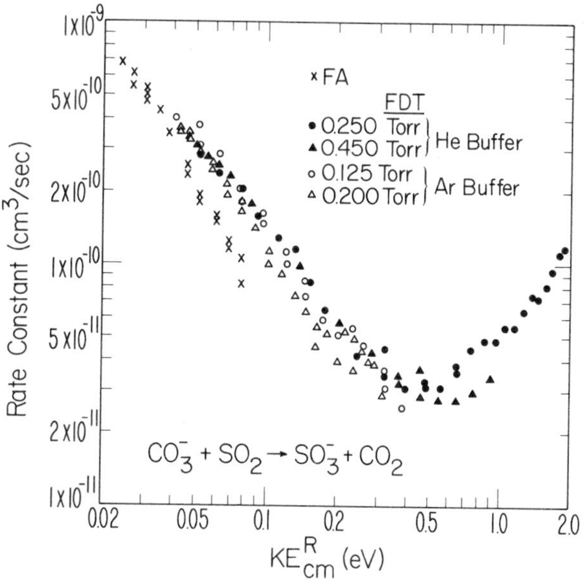

Figure 3. Rate constant for $CO_3^- + SO_2 \rightarrow SO_3^- + CO_2$ as a function of KE in He and Ar buffer gases and as a function of T (crosses).

The vibrational energy ratio is at least this large, and even larger if He has a very low vibrational excitation probability ($<10^{-3}$) as it does for positive ions.

The decrease of k with KE at low KE in Fig. 1 and 2 and 3 is a general feature of slow ($k<<k_c$) reactions and has been taken as evidence that the reactions go through intermediate complexes, where the probability of reaction is <<1 and this probability decreases with decreasing complex lifetime, which in turn decreases with increased energy content.[8] At higher energies a direct mechanism inevitably takes over and is presumed to be responsible for the rate constant increase with KE. Vibrational energy enhances the rate constant at all energies for the reaction shown in Fig. 1 (and for the other triatomic ion charge-transfers measured[6]). At low energy, vibrational energy has just the opposite role of KE, which is quite unexpected on the basis of statistical models in which all energy has more or less the same role and in which all forms of energy would decrease the complex lifetime and presumably the rate constant. The vibrational enhancement onset at an energy near the bending mode and the large magnitude of the enhancement shown in Fig. 1 also occurs for the linear NO_2^+ and CO_2^+ charge-transfers with NO and for CO_2^+ with O_2. The enhancement is very substantial, a factor of 7 at 0.3 eV in $CO_2^+ + O_2$ for example.[6]

In an attempt to explain these observations an Angular Momentum Coupling Model has been proposed.[6] The AM associated with the degenerate bending mode of linear ions ($\sim \hbar$) is assumed to couple to the AM of the orbiting complex ($\sim 100\ \hbar$), allowing transfer of AM and hence KE from orbiting motion into internal vibrational motion, thereby increasing the lifetime of the complex and hence the reaction rate constants for slow ($k<<k_c$) reactions. The fact that vibrational enhancement also occurs in the direct mode (Fig. 1 and ref. 6) as well as the observation of an enhancement[6] for the non-linear ion H_2O^+ in charge-transfer with NO indicates that other mechanisms are operative as well. The origin of the enhancement in the direct regime is not clear. If the increase of k with KE in this region reflected the increase of the adiabatic parameter $k \sim \exp[\frac{-a\Delta E}{hv}]$ with velocity v, then vibrational energy would not be expected to have a large effect.

The Franck-Condon weighted kinetic energy defects, ΔE, might have variations with vibrational state but they should be small and randomly distributed in sign whereas observations so far always show an increase in charge-transfer rate constant with vibrational energy. The rough equivalence between vibrational and kinetic energy in the direct region suggests some sort of energy barrier or activation energy but it is not clear what this might be for an exothermic charge-transfer reaction.

The utility of vibrationally enhanced reactions as a probe for vibrational excitation (and by inference deexcitation) in various buffer gases has already been mentioned. It is clear that this technique is capable of being made quantitative or at least semi-quantitative by varying the number of ion collisions with the buffer gas through the region where the vibrational excitation probability is ~ 1, as deduced from the variation in a vibrationally sensitive reaction rate such as shown in Fig. 1. This could be done either by varying the drift tube length or the buffer gas density. For example, the reaction of O_2^+ with CH_4, which is greatly enhanced by O_2^+ vibrational excitation, is observed to depend on the buffer gas density (N) at a fixed energy (E/N) near threshold in Ar buffer, indicating that $P_{ex} \sim 10^{-3}$ in this region.[9]

Vibrational Quenching Measurements Using Monitor Ion Method

The use of an added neutral which reacts with an excited state of an ion has been used to measure metastable electronically excited ions in FD tubes, e.g. $O_2^+(a^4\pi_u)$,[10,11] $NO^+(a^3\Sigma)$[11,12] and $O^+(^2D)$ ions.[13,14] The principle is that a suitable monitor gas is introduced into the drift tube after the reaction zone and before the mass spectrometer sampling aperture as a detector for excited state ions. For example, Ar is a suitable detector for $N_2^+(v=1)$ ions[15] since N_2^+ charge-transfer with Ar is endothermic (and slow) for v=0 and exothermic (and fast) for v=1. In the case of O_2^+, Xe serves as a monitor for $O_2^+(v\geq 1)$, SO_2 as a monitor for $O_2^+(v\geq 2)$ and H_2O as a monitor for $O_2^+(v\geq 3)$[16] and for NO^+ CH_3I serves as an ideal monitor for NO^+ $(v\geq 1)$[17] since $NO^+(v=0)$ does not react with CH_3I at energies up to 1 eV so that energy dependences of quenching rates of $NO^+(v)$ can be carried out. Results for vibrational relaxation rates which have been measured are listed in Tables I and II, along with the collision rate constant k_c and $Z \equiv k_c/k_q$.

Table I. Vibrational Quenching Rate Constants (cm^3s^{-1}) at 300 K for O_2^+ (v=1,2) Ref. 16.

Quencher	k_{q1}	k_{q2}	k_c	Z_{10}	Z_{21}
He	<2(-15)	<6(-15)	5.0(-10)	>3(5)	>9(4)
Ne	<1(-14)	<2(-14)	4.2(-10)	>3(4)	>2(4)
Ar	1.0(-12)	2.5(-12)	7.1(-10)	710	280
Kr	1.1(-11)	1.7(-11)	7.6(-10)	69	45
H_2	2.5(-12)	5.0(-12)	1.5(-9)	610	305
D_2	6.5(-13)	2.6(-12)	1.1(-9)	1700	430
N_2	1.9(-12)	5.4(-12)	8.0(-10)	420	150
CO	4.4(-11)	6.5(-10)	9.3(-10)	21	14
CO_2	1.0(-10)	2.0(-10)	8.7(-10)	8.7	4.4
H_2O	1.2(-9)	-	1.4(-9)	1.2	-
CH_4	6.0(-10)	-	1.1(-9)	1.9	-
SO_2	5.7(-10)	-	1.2(-9)	2.1	-
SF_6	1.1(-10)	2.0(-10)	9.7(-10)	8.8	4.9
O_2	3.0(-10)	4.0(-10)	7.4(-10)	2.5	1.9

Table II. Vibrational Quenching Rate Constants (cm^3s^{-1}) at 300 K for N_2^+(v=1) Ref. 15 and NO^+(v>0) Ref. 17.

Ion	Neutral	k_q	k_c	Z
NO^+	NO	5(-10)	7.5(-10)	1.5
	N_2	7(-12)	8.0(-10)	114
	CO_2	4(-11)	8.7(-10)	22
	CH_4	3(-11)	11.4(-10)	38
	NO_2	1.5(-10)	11.8(-10)	8
	He	<1(-13)	5.6(-10)	>5600
	Ar	<1(-12)	7.1(-10)	>710
	Kr	<1(-12)	7.6(-10)	>760
	NH_3	1.5(-9)	1.5(-9)	1
	CO	1.0(-11)	9.3(-10)	93

	C_2H_4	8.7(-10)	1.3(-9)	1.5
	O_2	<1(-12)	7.4(-10)	>740
	H_2	<1(-12)	1.5(-9)	>1500
	D_2	<1(-12)	1.1(-9)	>1100
	Ne	<1(-12)	4.2(-10)	>420
N_2^+	NO	1(-11)*	7.5(-10)	75
	Kr	5(-12)	7.6(-10)	152
	O_2	1.2(-10)	7.4(-10)	7.4
	N_2	5(-10)	8.0(-10)	1.6
	Ne	4.5(-12)	4.2(-10)	93

*increases drastically with KE to $\sim 10^{-10}$ cm^3s^{-1} at 0.1 to 1 eV

Table I includes the $O_2^+(v=1)$ and $O_2^+(v=2)$ quenching rates measured in Boulder and Table II includes the $N_2^+(v)$ and $NO^+(v)$ quenching rates measured in Innsbruck. Kinetic energy dependences were measured for $O_2^+(v)$ quenching by Kr, CO_2 and O_2 and for all the $N_2^+(v)$ and $NO^+(v)$ quenching. Fig. 4 shows the results for $O_2^+(v) + O_2$ which is typical of (O_2^+, NO^+, N_2^+) vibrational quenching by the parent neutral of the ion. The rate constants are large, close to the collision rate constant k_c, and essentially independent of energy. It is clear that vibrational relaxation occurs by resonant charge-transfer and this is probably general when exothermic or near thermoneutral.

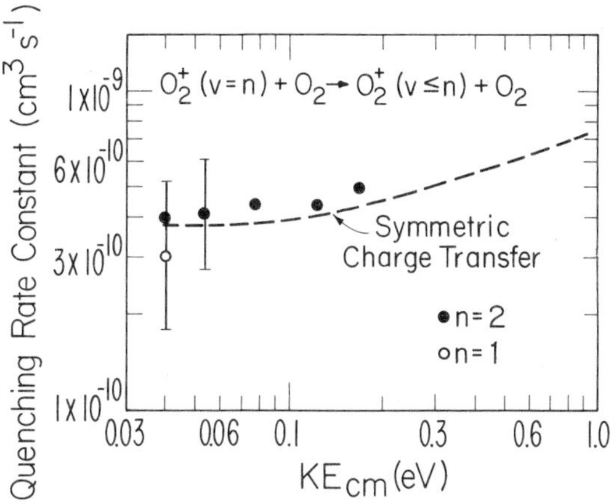

Figure 4. Vibrational relaxation of $O_2^+(v)$ by O_2 as a function of KE.

The $O_2^+ + CO_2$ and Kr quenching rate constants decrease with KE as shown in Fig. 5.

This is a characteristic of the Innsbruck data on $NO^+(v)$ as well, the only exception being $NO^+(v)$ quenching by NO_2 where k_q increases slightly from 1.5×10^{-10} cm^3s^{-1} at 0.06 eV to 2.0×10^{-10} cm^3s^{-1} at 1 eV. Perhaps NO^+ is quenched in this case by chemical reaction, i.e. by O^- abstraction from NO_2.

A strong correlation of k_q with bond energy is found for the O_2^+ and the NO^+ quenching rate constants. This, in conjunction with the negative energy dependences and the large magnitudes generally found for k_q, suggests vibrational relaxation by an intermediate complex mechanism. A decrease in rate constant with energy is a signature that a process is controlled by long range forces. In the complex formed by $O_2^+(v) + X$, the O_2^+ vibrational energy is converted to $O_2^+ - X$ "vibrational" energy, which blows the weak transient (unbound) complex apart.

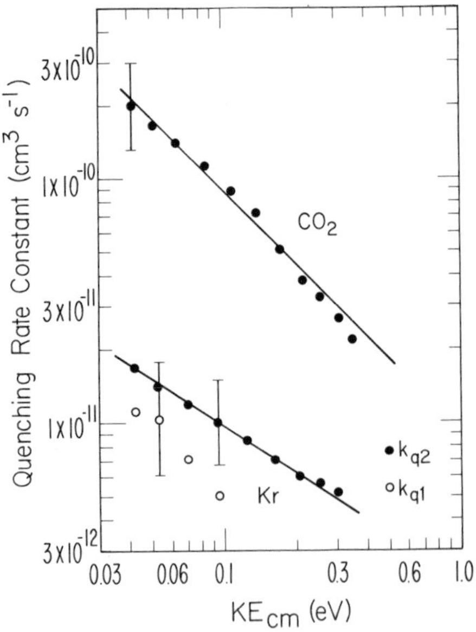

Figure 5. Vibrational relaxation of $O_2^+(v)$ by Kr and CO_2 as a function of KE.

The decrease with KE reflects the shorter lifetime of the transient complex with increased energy content. This immediately suggests an analogy with the widely studied vibrational predissociation of neutral van der Waals molecules or hydrogen bonded clusters, such as $He \cdot I_2$, $(HF)_2$ or $(H_2O)_n$.

A similar correlation between cross-section and interaction energy has been established by Parmenter et al.[18-21] for a variety of collisionally induced state transformations $A^* + M \rightarrow B + M$, $\sigma_M = C_{exp}(\varepsilon_{A^*M}/kT)$, where C is a constant and ε_{A^*M} is the intermolecular well depth between A^* and M. This model has been successfully applied to electronic state deactivation in atoms, intersystem crossings and internal conversions in polyatomics, rotational and vibrational relaxation and predissociation in polyatomics, and in one case vibrational relaxation for a molecular ion $(C_5H_9{}^+)^*$.

The model shown in Fig. 6 has been developed to allow calculations of the vibrational predissociation rate constant. The simple

VIBRATIONAL RELAXATION SCHEME

$$AB^+(\nu) + X \xrightarrow{k_q} AB^+(\nu'<\nu) + X$$

$$AB^+(\nu) + X \underset{k_u = \tau^{-1}}{\overset{k_c}{\rightleftharpoons}} [AB^+(\nu) \cdot X]^* \xrightarrow{k_{vp}} AB^+(\nu'<\nu) + X$$

$$k_q = k_c \frac{k_{vp}}{k_u + k_{vp}} \qquad \begin{array}{l} k_q = k_c\, k_{vp}\, \tau, \text{ if } k_u \gg k_{vp} \\ k_q = k_c \text{ if } k_{vp} \gg k_u \end{array}$$

$$AB^+ + X + M \xrightarrow{k^3} AB^+ \cdot X + M$$

$$[AB^+ \cdot X]^* + M \xrightarrow{k_s} AB^+ \cdot X + M$$

$$k^3 = k_c k_s \tau \qquad k_s \sim k_L = 2\pi e \sqrt{\alpha/\mu}$$

$$\text{for } k_q < k_c \Longrightarrow k_{vp} = k_q k_s / k^3$$

Figure 6. Intermediate complex reaction mechanism for vibrational relaxation.

energy transfer model of three-body association is used to infer the complex lifetime, actually the product of the complex formation rate constant and lifetime which enters the expression for both k_q and k^3. The complex stabilization rate constant k_s is assumed to be Langevin. There is typically 25-50% uncertainty in measured values of k^3, sometimes greater, and in some cases the value of k^3 has been measured with He and at 80 or 200 K rather than 300 K. In these cases k^3 has been empirically adjusted to account for the lower value it would have at a higher temperature and the larger value it would have with a heavier third-body such as N_2, CO_2 or Ar. The third body effect assumed is that a heavy M would be 3.8 times more efficient than He. The temperature correction is taken to be $k^3 \sim T^{-n}$ where

Table III. Vibrational Predissociation Rate Constants

Ion	Neutral	k_q (cm^3s^{-1})	k^3 (cm^6s^{-1})	k_{vp} (10^9s^{-1})
O_2^+ (v=1)	Ar	1.0(-12)	3.5(-31)[a]	2.1
	Kr	1.1(-11)	9.4(-31)[a]	7.9
	H_2	2.5(-12)	5.3(-32)[a]	25
	N_2	1.9(-12)	1.0(-30)	1.4
	CO_2	1.0(-10)	2.6(-29)[b]	2.7
	SO_2	5.7(-10)	5.5(-29)[b]	0.7
	H_2O	1.2(-9)	2.5(-28)	≥3.3 $k_q \sim k_c$
NO^+	Kr	<1(-12)	9.7(-32)[a]	<7
	O_2	<1(-12)	9.0(-32)	<7
	N_2	7.0(-12)	3.0(-31)	17
	CO	1.0(-11)	1.9(-30)	3.7
	CO_2	4.0(-11)	1.4(-29)	2.0
	CH_4	3.0(-11)	<5.8(-29)	>0.4
	NH_3	1.5(-9)	8.8(-28)	≥1.2 $k_q \sim k_c$

a) extrapolated from M = He, 80K to M = N_2, 300K by $3.8(3.75)^{-(\ell/2 + 1)}$

b) extrapolated from M = He, 200K to M = N_2, 300K by $3.8(1.5)^{-(\ell/2 + 1)}$ where ℓ = number of rotational degrees of freedom of reactants, 2 for neutral atoms, 4 for linear neutrals, 5 for non-linear neutrals.

$n = \ell/2 + 1$, ℓ being the number of rotational degrees of freedom of the ion and neutral reactants. These corrections are empirical, based simply on the examination of available data and could easily lead to a factor three or more errors in the extrapolations from 80K. The calculated values in those cases for which both k_q and k^3 data exist are given in Table III. The bunching of most of the values between $10^9 s^{-1}$ and $10^{10} s^{-1}$ is remarkable. It is not a priori clear that the value of k_{vp} is expected to be so nearly constant. In the theory of vibrational predissociation of neutral clusters,[22] k_{vp} depends on the cluster bond strength and the vibrational frequency of the vibrationally excited entity as well as the vibrational frequency of the weak cluster bond. In the case of neutral vibrational predissociation however, one starts with a cold, stable cluster and pumps in a vibrational quantum (or several) and measures the lifetime against predissociation. In the case of vibrational relaxation there is no stabilized cluster at any point, only the transient collison complex with one species vibrationally excited. The dissociation energy in this case is evidently a much less important parameter. There are two exceptionally high values of k_{vp} in Table III. One of these is the $O_2^+(v=1)$-H_2 case, where $k_{vp} \sim 2.5 \times 10^{10} s^{-1}$, a value not well defined since the $O_2^+ \cdot H_2$ complex is so weak ($D < 0.2$ eV) that k^3 has only been measured at 80 K with He buffer. H_2 and H containing compounds have enhanced vibrational relaxation efficiencies for neutral molecules[23] as well and this is usually attributed to the low atomic mass in some way, either to the high H atom velocity or the high rotational energy spacing. We speculate that the O_2^+ collisions rotationally excite H_2, trapping the O_2^+-H_2 pairs for a time long compared to that expected for such a weak electrostatic bond. For o-H_2 (3/4) in J=1 a rotational excitation to J=3 ($\Delta J=2$) requires 0.075 eV which exceeds 3/2kT (0.039 eV) and a p-H_2 (1/4) excitation from J=0 to J=2 takes 0.045 eV > 0.039 eV, both of which would imply trapping if rotational excitation occurs. For D_2 on the other hand the 2/3 o-D_2 in J=0 requires only 0.023 eV < 0.039 for excitation to J=2 and 0.038 $\simeq$ 0.039 for the p-D_2(1/3)J=1 $\rightarrow$ J=3 excitation. If only the p-D_2 is trapped one might roughly expect a ratio of Z for $H_2/D_2 \simeq 3$ which might explain the observed isotopic effect $Z(H_2)/Z(D_2) = 2.8$ for $O_2^+(v=1)$ and 1.9 for $O_2^+(v=2)$.

In the second anomalous case, $NO^+ + N_2$, the possibility for a resonant V-V transfer exists. The difference in vibrational frequencies is only 14 cm^{-1}(exothermic) which is essentially an exact resonance. The anomaly for $NO^+(v) + N_2$ shows up directly in k_q which is 7×10^{-12} cm^3s^{-1}, while $O_2^+(v) + N_2$ has $k_q = 1.9 \times 10^{-12}$ cm^3s^{-1}, in spite of the substantially lower O_2^+ frequency, 472 cm^{-1} less than that of NO^+. An inverse correlation of k_q with the ion vibrational frequency is expected. This is predicted from vibrational predissociation theory[22] or from any vibrational relaxation model and is well established for neutral vibrational relaxation.[23,24] The other 6 common quenchers all have larger k_q's for $O_2^+(v=1)$ than for $NO^+(v=1)$.

There is compelling experimental evidence however that the NO^+ (v) + N_2 vibrational relaxation is <u>not</u> a resonant V→V process. Recent measurements in Birmingham[25] have determined that the reverse resonant vibrational transfer process, $N_2(v=1) + NO^+(v=0) \rightarrow N_2(0) + NO^+(1)$ is very slow $k < 1 \times 10^{-13}$ cm^3s^{-1}. Therefore, from detailed balance, the $NO^+(1) + N_2(0) \rightarrow NO^+(0) + N_2(1)$ rate constant must also be less than $1 \times 10^{-13} cm^3s^{-1}$. Thus the vibrational relaxation of $NO^+(v=1)$ by N_2 produces ground vibrational state NO^+ and N_2. This result was very surprising since it had been speculated that the anomalously fast $NO^+(v)$ quenching by N_2 was due to the resonant V→V process. However, some reflection leads to the conclusion that this was not a reasonable speculation. Theoretical calculations have been carried out for the closely related process, $CO_2(001) + N_2(0) \rightarrow CO_2(000) + N_2(1) + 18$ cm^{-1}, by resonant V→V transfer induced by the long range dipole-quadrupole interaction.[24] The V→V transfer probability is 2×10^{-3} at 300K and goes as T^{-1}. This is in excellent agreement with the experiment (5%). The same theory can be applied to the isoenergetic V→V transfer $NO^+(1) + N_2(0) \rightarrow NO^+(0) + N_2(1) + 14$ cm^{-1} by simply replacing the $CO_2(001)$ Einstein A coefficient, 425 s^{-1}, by that for NO^+, 10.9 s^{-1}, since $P \sim A$. This leads to $P \sim 2 \times 10^{-3} \times 10.9/425 \sim 4 \times 10^{-5}$, much less than the experimental $P \sim 8 \times 10^{-3}$. It is thus clear that the long range dipole-quadrupole interaction is insufficient to account for the vibrational relaxation of $NO^+(v)$. The limit $k_q < 10^{-13}$ cm^3 s^{-1} implies $P < 10^{-4}$ in agreement with the calculation. The rapid quenching of $NO^+(v)$ by N_2 must then be due to vibrational predissociation $NO^+(1)$ +

$N_2(0) \rightarrow NO^+(0) + N_2(0) + 2344\ cm^{-1}$. That vibrational predissociation gives $NO^+(0) + N_2(0)$ rather than the near resonant $NO^+(0) + N_2(1)$ follows from phase space or dynamical arguments. The available phase space in the $NO^+(0) + N_2(1)$ separation coordinate, 14 cm^{-1}, is extremely small, making the process extremely unlikely. From a dynamical point of view, the initial capture of NO^+ by N_2 to form the complex involves rotational excitation of the N_2, and perhaps also the NO^+, which puts the $[NO^+(1).N_2(0)]^*$ complex below the $NO^+(0) + N_2(1)$ dissociation limit in the NO^+-N_2 separation coordinate. Overall energy of course is conserved on the multidimensional $NO^+ + N_2$ potential surface.

The low value of k_q for $NO^+(v)$ with O_2 is consistent with the low three-body association rate constant and can be rationalized by noting that the $NO^+(^1\Sigma)$ and $O_2(^3\Sigma)$ molecules approach on a triplet potential surface which may not be the ground state surface and hence may be repulsive or have only a very shallow well. One might then expect that $O_2(^1\Delta)$ molecules would quench $NO^+(v)$ more readily since the ground state NO_3^+ potential well would be accessible. The low value of k_q for $NO^+(v)$ with O_2 is also implied by the measurements of Smith, et al.[26] in which vibrationally excited NO^+ produced by $N^+ + O_2$ reaction was detected by its infrared emission in the presence of O_2.

Comparisons from Tables I and II give for ion vibrational relaxation by Kr, $k_q(O_2^+) > k_q(N_2^+) > k_q(NO^+)$ as expected from the increasing frequencies $O_2^+ = 1872\ cm^{-1}$, $N_2^+ = 2175\ cm^{-1}$ and $NO^+ = 2344\ cm^{-1}$. Comparing common $O_2^+(v)$ and $NO^+(v)$ quenchers, aside from the N_2 anomaly just discussed, (and omitting parent neutrals of ions because of fast charge-transfer quenching) there are 6 common quenchers, Ar, Kr, CO, CO_2, H_2 and CH_4 and the quenching of $O_2^+(1872\ cm^{-1})$ is faster in every case than NO^+ $(2344\ cm^{-1})$, the ratio rising from > 1 for Ar to 20 for CH_4. Omitting the high ratio for CH_4 the average ratio is $k_q(O_2^+)/k_q(NO^+) > 4$. Interestingly the value k_{vp} predicted from vibrational predissociation theory[22] goes as $\exp[-\omega^{1/2}]$. The ratio of values computed for $NO^+(v) = N_2^+(v) = O_2^+(v) = 1 = 1.6 = 3.5$. The large value of k_q for $O_2^+(v)$ with CH_4, relative to $NO^+(v)$ with CH_4, is a consequence of long-lived complex formation by H^- abstraction to produce $[CH_3^+.HO_2]^*$.[27,28] The large ratio of the $O_2^+(v)$ to $NO^+(v)$ quenching by Kr, > 11, has been shown to be due to a large complex lifetime ratio as determined by the large three-body association rate constant ratios.[29] In this case both the three-body association and the vibrational quenching measurements can be invoked to deduce a specifically chemical O_2^+-Kr interaction in addition to the electrostatic interaction, which would be approximately the same for O_2^+ and NO^+ with Kr.

There are other confirmations of the strong frequency dependence. The ions H_2O^+ and N_2O^+ are both vibrationally excited near their bending frequency thresholds (1379 cm^{-1} and 460 cm^{-1}) in $\sim 10^3$ collisions with Ne, implying a vibrational excitation and hence vibrational deexcitation rate constant $\sim 10^{-12} cm^3 s^{-1}$ or larger.[6] This vastly exceeds the upper limit for $O_2^+(v)$ quenching by Ne, $k_q < 10^{-14}\ cm^3 s^{-1}$.

The increase of k_q with v as observed for $O_2^+(v)$ is observed in van der Waals I_2.He vibrational predissociation[30] where the dependence $P \sim v^2$ has been explained theoretically.[22] The more nearly linear v dependence for $O_2^+(v)$ quenching is a result of the very harmonic nature of O_2^+ vibrations in the bottom of the well, the non-linear v dependence results from vibrational anharmonicity.

The vibrational predissociation lifetime, $\sim 10^{-9}$-10^{-10}s is very long compared with lifetimes of weak ion-molecule collision transient complexes, i.e. those associated with neutrals of low dipole moment and low polarizability. Therefore vibrational relaxation is slow in the sense of occurring only in a small fraction of collisions, $Z \sim 10^2 \rightarrow 10^3$. A minimum collision time is $\sim 10^{-8}$ cm/10^5cm $s^{-1} \sim 10^{-13}$s so that if k_{vp} were as large as $10^9 s^{-1}$ and the model were valid for extremely weak collisions, one would expect $P > 10^{-13} s \times 10^9 s^{-1} = 10^{-4}$ in all cases. Since the relaxation of O_2^+ by He is much less probable than this, $P < 3 \times 10^{-6}$, the model clearly breaks down for such weakly interacting systems. The O_2^+-He potential well depth is ~ 0.026 eV.[31] Indeed the model must breakdown for sufficiently shallow potential wells. Neutral molecules, with weak van der

Waals attractive forces are normally vibrationally relaxed by the short-range repulsive interaction, which leads to small deactivation rate constants with positive temperature dependences. O_2^+ is vibrationally excited in He collisions at high energy and therefore would be deexcited in high energy collisions. The expected general pattern of vibrational relaxation of molecular ions over an extended energy range is a decreasing k_q with energy as shown in Figure 5 followed by an eventual increase with energy when the short-range repulsive interaction becomes dominant. At the extreme of strong complexes, i.e. vibrational relaxation by strongly polar molecules or molecules with large polarizabilities, one expects high efficiencies approaching unity, as occurs for H_2O with O_2^+ and for NH_3 with NO^+ for example. This will be especially true for lower vibrational frequencies. Note that NO^+ has the highest vibrational frequency of any non-hydrogenic molecular ion! Since $k_q = k_{vp}k^3/k_s$, if we take an average value for $k_s \sim 7 \times 10^{-10}$ cm^3s^{-1} and assume $k_{vp} \sim 2 \times 10^9 s^{-1}$ then $k_q \sim 3 \times 10^{+18}$ k^3. For an ion which has a three-body association rate with a neutral $\sim 3 \times 10^{-28}$ cm^3s^{-1}, the expected value of k_q would be $\geq 10^{-9}$ cm^3s^{-1}, the collision rate constant, or the quenching probability would approach unity. There are expected to be many such cases.

In the case of weak complexes the vibrational deexcitation probability is small, $P \sim 10^{-3}$ for $O_2^+(v) + N_2$ for example. This has implications for complex formation, or for vibrational excitation in low energy ion-molecule collisions. If the exothermic vibrational de-excitation probability for $O_2^+(v=1)$ in collisions with N_2 has a low probability, the endothermic vibrational excitation probability must be even less. Thus the possibility of exciting O_2^+ vibrationally in low energy collisions, even where the collision energy at impact due to the ion-induced dipole force exceeds the vibrational quantum energy, must be very low. Therefore the conversion of translational to vibrational energy leading to complex formation is quite inefficient. Presumably, the trapping which occurs is due to rotational excitation induced by the collision of the ion with the anisotropically polarizable molecule. The lifetime of an O_2^+-N_2 collision at 300 K calculated from the three-body association rate constant (Fig. 6) is only $\sim 2 \times 10^{-12}$

sec, which is only the time required for an orbiting path $\sim 20\text{Å}$, so that not much trapping is implied. Rotational excitation only, without vibrational excitation, is consistent with recent theoretical models of Bates[32] and Herbst[33] for the three-body association of ions and molecules. On the other hand, when the well is deep and k_q large, vibrational excitation could also be efficient and play a role in determining complex lifetimes.

Conclusions

In this chapter we have attempted to outline a new approach to studies of vibrational excitation and deexcitation of molecular ions and the effect of molecular ion vibrational excitation on reaction rate constants now available using ion swarm techniques. It can be reasonably anticipated that this new area of study will lead to new insights into molecular interaction processes, as has occurred with neutral molecule vibrational relaxation and reaction studies over a period of many decades. The application of such data to various natural systems as the weak plasmas of planetary atmospheres, interstellar molecular clouds, discharges and combustion systems also represents an exciting challenge.

Acknowledgements

The author gratefully acknowledges the assistance of L.A. Eddy in the preparation of this manuscript. The list of colleagues in the Aeronomy Laboratory ion chemistry program over the years to whom I am indebted for stimulation and education has by now become a very long one. In particular the pioneering efforts of F.C. Fehsenfeld and A.L. Schmeltekopf and subsequently D.L. Albritton made the program go. The specific science of this chapter was contributed to in an essential way by M. Durup-Ferguson, H. Böhringer and W. Lindinger during sabbatical stays in the Aeronomy Laboratory and by David Fahey. The subsequent collaboration with Professor Lindinger and his colleagues and students at Innsbruck has been very stimulating and rewarding and contributes much that is in this review. Defense Nuclear Agency has played a key role in support of the ion chemistry program over many years.

References

1. McFarland, M., Albritton, D.L., Fehsenfeld, F.C., Ferguson, E.E., Schmeltekopf, A.L.: J. Chem. Phys. 59, 6610, 6620, 6629 (1973).
2. Adams, N.G., Smith, D.: Int. J. Mass Spectrom. Ion Phys. 21, 349 (1976); J. Phys. B 9, 1439 (1976).
3. Lin, S.L., Bardsley, J.N.: J. Chem. Phys. 66, 435, (1977).
4. Viehland, L.A., Mason, E.A.: J. Chem. Phys. 66, 422 (1977).
5. Viehland, L.A., Lin, S.L., Mason, E.A.: Chem. Phys. 54, 341 (1981).
6. Durup-Ferguson, M., Böhringer, H., Fahey, D.W., Ferguson, E.E.: J. Chem. Phys. 79, 265 (1983).
7. Albritton, D.L., Dotan, I., Streit, G.E., Fahey, D.W., Fehsenfeld, F.C., Ferguson, E.E.: J. Chem. Phys. 78, 6614 (1983).
8. Ferguson, E.E., Bohme, D.K., Fehsenfeld, F.C., Dunkin, D.B.: J. Chem. Phys. 50, 5039, (1969).
9. Dotan, I., Fehsenfeld, F.C., Albritton, D.L.: J. Chem. Phys. 68, 5665 (1978).
10. Lindinger, W., Albritton, D.L., McFarland, M., Fehsenfeld, F.C., Schmeltekopf, A.L., Ferguson, E.E.: J. Chem. Phys. 62, 410 (1975).
11. Glosik, J., Rakshit, A.B., Twiddy, N.D., Adams, N.G., Smith, D.: J. Phys. B 11, 3305 (1978).
12. Dotan, I., Fehsenfeld, F.C., Albritton, D.L.: J. Chem. Phys. 75, 3280 (1979).
13. Rowe, B.R., Fahey, D.W., Fehsenfeld, F.C., Albritton, D.L.:

J. Chem. Phys. 73 194 (1980).

14. Johnsen, R., Biondi, M.A.: J. Chem. Phys. 73, 190 (1980).
15. Dobler, W., Ramler, H., Villinger, H., Howorka, F., Lindinger, W.: Chem. Phys. Lett. 97, 553 (1983); Dobler, W., Howorka, F., Lindinger, W.: Plasma Chem. and Plasma Processing 2, 353 (1982).
16. Böhringer, H., Durup-Ferguson, M., Ferguson, E.E., Fahey, D.W.: Planet. Space Sci. 3, 483 (1983); Böhringer, H., Durup-Ferguson, M., Fahey, D.W., Fehsenfeld, F.C., Ferguson, E.E.: J. Chem. Phys. 79, 4201 (1983).
17. Dobler, W., Federer, W., Howorka, F., Lindinger, W., Durup-Ferguson, M., Ferguson, E.E.: J. Chem. Phys. 79, 1543 (1983); J. Chem. Phys. to be submitted.
18. Lin, H.M., Seaver, M., Tang, K.Y., Knight, A.E.W., Parmenter, C.S.: J. Chem. Phys. 70, 5442 (1979).
19. Parmenter, C.S., and Seaver, M.: Chem. Phys. Lett. 67, 279 (1979).
20. Parmenter, C.S., and Seaver, M.: Chem. Phys. 53, 333 (1980).
21. Parmenter, C.S., and Seaver, M.: J. Chem. Phys. 70, 5448 (1979).
22. Beswick, J.A., Jortner, J.: Adv. Chem. Phys. 47, 363 (1981).
23. Lambert, J.D.: Vibrational and Rotational Relaxation in Gases, Clarendon Press, Oxford, 1977.
24. Yardley, J.T.: Introduction to Molecular Energy Transfer, Acad. Press, N.Y., 1980.
25. Ferguson, E.E., Adams, N.G., Smith, D., Alge, E.: J. Chem. Phys. to be submitted.
26. Smith, M.A., Bierbaum, V.M., Leone, S.R.: Chem. Phys. Lett. 94, 398 (1983).
27. Rowe, B.R., Dupeyrat, G., Marquette, J.B., Smith, D., Adams, N.G., Ferguson, E.E.: J. Chem. Phys. 50, Jan. 1984.
28. Durup-Ferguson, M., Böhringer, H., Fahey, D.W., Fehsenfeld, F.C., Ferguson, E.E.: J. Chem. Phys. to be submitted.
29. Ferguson, E.E., Smith, D., Adams, N.G.: Int. J. Mass Spectrom. Ion Processes, in press.
30. Johnson, K.E., Wharton, L., Levy, D.H.: J. Chem. Phys. 69, 2719 (1978).
31. Böhringer, H., Arnold, F., Smith D., Adams, N.G.: Int. J. Mass Spectrom. Ion Phys. 52, 25 (1983).
32. Bates, D.R.: J. Phys B 12, 4135 (1979); J. Chem. Phys.

73, 1000 (1980).
33. Herbst, E.: J. Chem. Phys. 70, 2201 (1979); 72 5284 (1980).

Kinetic and Internal Energy Effects on Ion-Neutral-Reactions

W. Lindinger

Institut für Experimentalphysik, Atom- und Molekülphysik, Universität Innsbruck, A-6020 Innsbruck, Austria

Introduction

The preceding chapter by Ferguson presents a summary of recent drift tube results on vibrational effects on charge transfer reactions of molecular ions as well as on the quenching of the vibrational excitation of molecular ions in nonreactive collisions with neutrals.

Here we discuss two additional groups of ion neutral interactions and their energy dependences for which drift tube experiments have provided a wealth of data and fast progress in understanding over the past years:

(A) Charge transfer processes of atomic ions.

(B) Reactive collisions of molecular ions in selected vibrational states.

The investigation of many of these reactions became possible due to a recent experimental development[1,2] (Fig. 1) involving the modification of a selected-ion -(flow)- drift tube[3-5] by adding an octupole ion storage section[6] between the ion source and the injection quadrupole. Ions, X^+, originating from the source can interact with neutrals R added in trace quantities to the octupole section, and the product ions Y^+ of this interaction are then analyzed by the adjacent quadrupole mass spectrometer and introduced into the drift region by means of a Venturi type inlet.[7,8] As the ions Y^+ do not undergo any further collisions with R, they reach the drift region (containing He

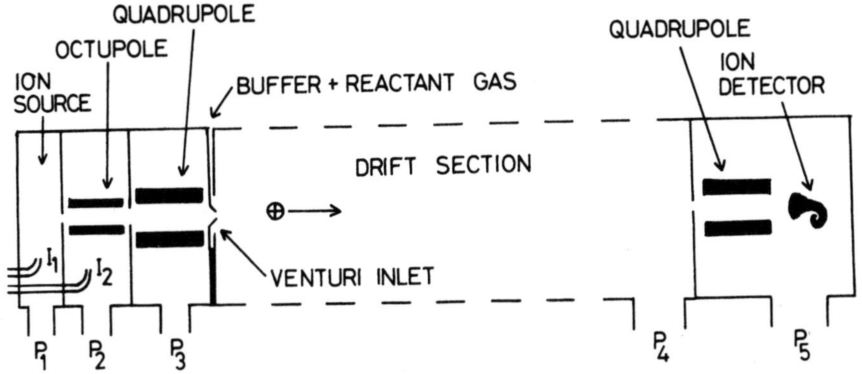

Fig. 1. Schematic representation of a Selected-Ion-(Flow)-Drift-Tube (SI(F)DT) including an octupole section for ion preparation. $P_{1..5}$ denote pumps (Ref.1).

as a buffer, which usually does not change the ionic excitations, as long as E/N in the drift region remains low) in their nascient internal vibrational state distributions, which can be analyzed by adding proper monitor gases to the drift region. At the same time further reactions of these excited product ions Y^+ can be investigated as a function of E/N and thus as a function of the relative kinetic energy KE_{cm} between the reactants.

Besides their academic interest, data of this kind are of importance for plasma modelling, where the further reactions of excited ions, produced in a preceding reactive step, need to be taken into account, before they can reach equilibrium between their internal state distribution and the plasma temperature.

Charge Transfer Reactions of Atomic Ions

Fast charge transfer reactions (with k being close to the limiting values k_c, where k_c indicates the Langevin - or ADO-limiting values[9,10]) hardly ever show any significant energy dependences[11,12] (Fig. 2) and despite of the existence of favourable Franck Condon factors they frequently do not proceed via resonant charge transfer at thermal energies, as has been shown e.g. for the reactions of $He^+ + O_2$[13], $Ar^+ + H_2O$[14] and $O^+ (^2D) + O_2$ (and N_2)[15,16]. Slow charge transfer

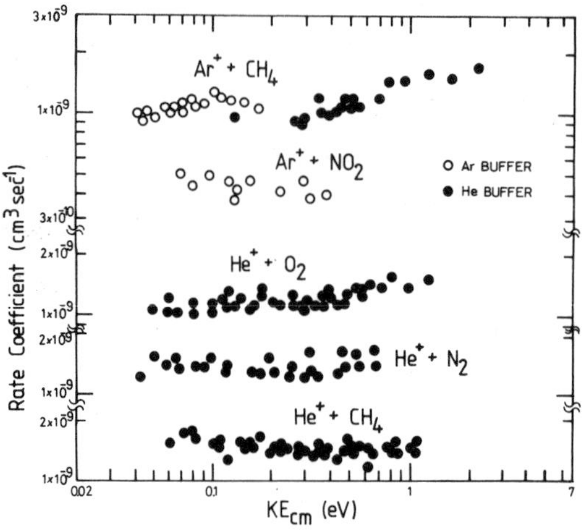

Fig. 2. Rate coefficients, k, as a function of KE_{cm} for the charge transfer processes of He^+ and Ar^+ with various neutrals (Refs. 11,12).

reactions however show pronounced variations as a function of KE_{cm} from which behavior informations on the reaction mechanism can be inferred.

By far the most intensively studied charge transfer process is the one between Ar^+ and N_2[17-20]. The rate coefficient for this process increases from a thermal value $k = 1 \times 10^{-11}$ cm^3 sec^{-1} to ~5×10^{-10} cm^3 sec^{-1} at $KE_{cm} = 2$ eV, (Fig. 3) a behavior indicating the slightly endoergic path of the process (Equ. 1a), leading to the product ion N_2^+ (X,v=1) rather than N_2^+ (X,v=o),

$$Ar^+ \ (^2P_{3/2}) + N_2 \xrightarrow{k_{1a}} N_2^+(X,v=1) + Ar - 0.09 \text{ eV} \qquad (1a)$$

That indeed vibrationally excited N_2^+ results as the main product from this reaction at thermal energies (and above) has been proven by using the reverse process

$$N_2^+(X,v=1) + Ar \rightarrow Ar^+ + N_2, \qquad (2)$$

as a test reaction in drift experiments[17], but also by means of other than swarm techniques, such as the Threshold Elec-

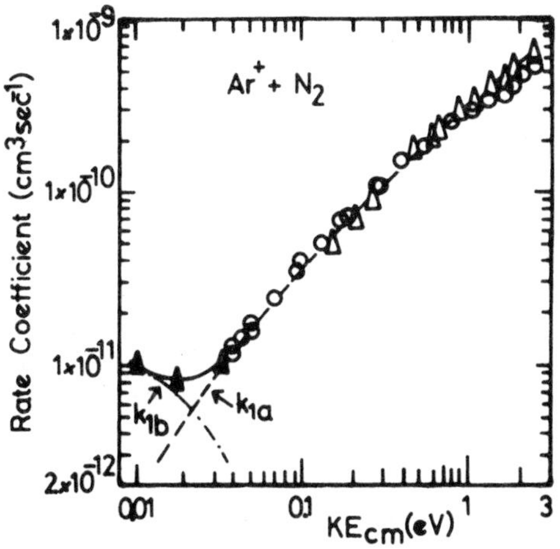

Fig. 3. Rate coefficients, k, as a function of KE_{cm} for the charge transfer between Ar^+ and N_2. Open triangles and circles denote data from Refs. 17 and 29 respectively obtained in drift experiments, and solid triangles denote data from Ref. 18, obtained in a temperature variable SIFT.

tron-Secondary Ion Coincidence (TESICO) technique[21] and Threshold Photoelectron -Photoion Coincidence using pulsed synchrotron radiation.[19] From the Newton-diagrams of a low energy beam experiment on reaction (1) performed at 1 eV collision energy it not only was evident, that the product N_2^+ ions are vibrationally excited, but also that unreactively scattered Ar^+ as well as the product ion N_2^+ originated from an orbiting complex with common origins in the center of mass velocity vector.[20,22] These data suggest that even at $KE_{cm} \simeq 1$ eV complex formation occurs at a Langevin rate, falling apart as $Ar^+ + N_2$ or as $N_2^+(v) + Ar$ at a rate of 2 : 1, just as one would expect from the ratio $k_L:k_1 \simeq 3:1$ at $KE_{cm} = 1$ eV.

Measurements of Smith and Adams using a temperature variable Selected Ion Flow Tube (SIFT) found (Fig. 3) that k_1 does not decrease towards lower temperatures as would be expected from the endoergicity in (1a), but increases instead, after passing through a minimum at 140°K thus showing a competition between channel (1a) and the exoergic channel (1b),

$$Ar^+(^2P_{3/2}) + N_2 \overset{k_{1b}}{\rightarrow} N_2^+ (X,v=o) + Ar + 0.18\ eV \qquad (1b)$$

as indicated in Fig. 3.

A consistent explanation for the $Ar^+ \rightleftarrows N_2^+(v)$ system was recently provided by Govers[23] and coworkers[19] on the basis of transitions between vibronic potential curves according to the Bauer-Fisher-Gilmore model[24] and using the Landau-Zener formalism for the calculation of the transition probabilities. For the interaction between Ar^+ and N_2 at low (thermal) energies only the lowest lying potential curves need to be taken into account for explaining the mechanism of the charge transfer. Depending on the angle Ø (N-N ... Ar angle) the potential curves (for the attractive part) for Ar^+-N_2, lie between the extreme positions indicated by the solid lines in Fig. 4a and b respectively.[19] The average dipole

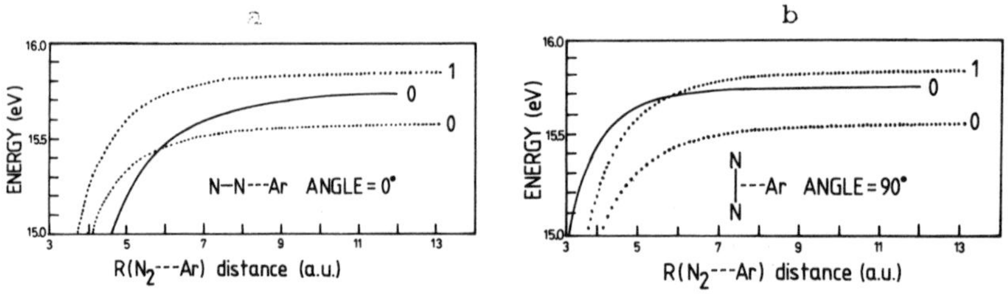

Fig. 4. Attractive parts of the diabatic potentials pertaining to $N_2^+(X,v'' = 0,1) + Ar \rightarrow N_2\ (X,v = o) + Ar^+ (^2P_{3/2})$, as a function of $R(N_2 \ldots Ar)$ for a fixed angle Ø = 0° (Fig. 4a), and Ø = 90° (Fig. 4b), from Ref. 19.

orientation comes close to case 4a, (the polarizability of N_2 is α = 2.38 Å for Ø = 0° and α = 1.45 $Å^3$ for Ø = 90°, the average polarizability α = 1.73 Å, much closer to the case of Ø = 90° rather than Ø = 0°) where no crossing between the 0 - 0 curves exists at favorable N_2 - Ar distances, so that a transfer only can proceed, when the interaction energy is high enough to allow for a 0 - 1 transition i.e. for the passage into the slightly endoergic channel resulting in N_2^+

(X,v=1). Even for cases close to the situation of 4b, charge transfer into N_2^+ (X,v=o) does not occur at appreciable rates, as the calculated transition probabilities 0 - 0 are very small.[19] This picture explains very well the increase of k_1 from its thermal value (which is $k_1 = 1 \times 10^{-11}$ cm^3 sec^{-1} $\simeq k_L \cdot e^{-\Delta E/kT}$; with ΔE being the endoergicity of 0.09 eV of reaction (1a)) towards higher relative kinetic energies between the reactants in connection with the observed production of N_2^+(X, v≠ o) ions.

Even the observed increase of the channel (1b) with decreasing temperature and thus the minimum in the overall rate coefficient $k_1 = k_{1a} + k_{1b}$ is consistent with the model of Govers et al.[19] Any interaction between Ar^+ and N_2 at angles $\emptyset \neq 0°$ leads to a torque on the N_2 (due to the anisotropy in its polarizability) and thus induces rotation in the N_2 molecule. Whenever the initial relative kinetic energy between Ar^+ and N_2 is small enough to be comparable to a quantum of rotation, the Ar^+ and N_2 can become "trapped" in an $(ArN_2)^{+*}$ complex, the lifetime of which is long enough to allow for the charge transfer from Ar^+ to N_2 to take place. The probability for this "trapping" increases of course with decreasing KE_{cm}, thus leading to the increase of k_{1b} with $T \to 0°$. This mechanism of energy "absorption" by rotational excitation of either the neutral or the ionic reactant by conversion of relative kinetic energy into rotational energy has been considered by Ferguson et al.[25] to explain association and isotope exhange reactions, where long complex lifetimes are required. The same idea is also included in the statistical models of Bates[26,27] and Herbst[28] on complex lifetimes.

Similarly as in the Ar^+-N_2 case, several other charge transfer reactions have low transition probabilities at low KE_{cm}, but possess high probabilities for transitions via endogergic channels and thus are strongly promoted by increasing KE_{cm}. Ar^+ has a small rate coefficient $k = 5 \times 10^{-11}$ cm^3 sec^{-1} for thermal charge transfer to O_2,

$$Ar^+ + O_2 \xrightarrow{k_3} O_2^+ + Ar \qquad (3)$$

k_3 decreasing to a value of 2×10^{-11} cm^3 sec^{-1} at KE_{cm} (Ar^+ - O_2) = 0.4 eV but then increasing strongly with further increasing KE_{cm} (Fig.5).[29] 0.35 eV higher than the Ar^+ recombination energy lies the ($a^4\Pi_u$) state of O_2^+ and positive proof was made[30] by using consecutive reactions of O_2^+ produced via charge transfer from Ar^+ at various KE_{cm}, that indeed O_2^+ ($^4\Pi_u$) is the dominant product ion for process (3) at energies at and above 1 eV.

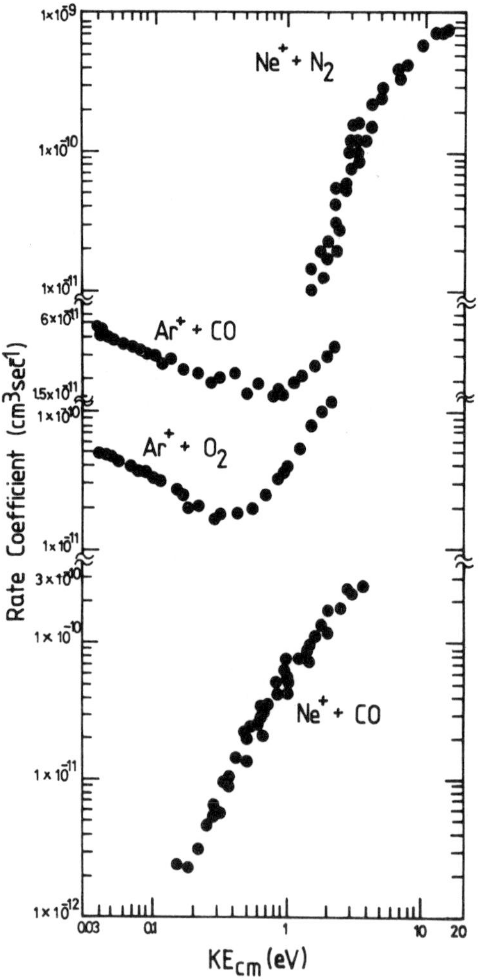

Fig. 5. Rate coefficients, k, as a function of KE_{cm} for the charge transfer reactions of Ar^+ with CO and O_2 and of Ne^+ with N_2 and CO.

Practically the same situation is given for the system Ar^+ + CO,

$$Ar^+ + CO \xrightarrow{k_4} CO^+ + Ar \qquad (4)$$

where the strong increase[29] of k_4 at $KE_{cm} > 1$ eV (Fig. 5) seems to reflect the promotion of the channel

$$Ar^+ + CO \rightarrow CO^+ (A) + Ar + 0.9 \text{ eV} \qquad (5)$$

which is endoergic by 0.9 eV.

Both the reactions of Ne^+ with N_2[31,32] and CO[31] respectively are immeasurably slow at room temperature, but show strongly increasing values k with increasing KE_{cm}[33-35] (Fig. 5). This behavior reflects in both cases the strong promotion of endoergic charge transfer paths by kinetic energy.

For N_2, low transition probabilities seem to exist for all the exoergic product channels, i.e. into N_2^+ ($X^2\Sigma_g^+$; $A^2\Pi_u$; $B^2\Sigma_u^+$ and $^4\Sigma_u^+$). The strong increase of k_6 occurs obviously due to strong transition probabilities into the C state N_2^+ ($^2\Sigma_u^+$ (v)) which is endoergic by > 2.0 eV. This is supported by the observed dominance of the dissociative charge transfer channel (6b),

$$Ne^+ + N_2 \rightarrow N_2^+ + Ne \qquad (6a)$$

$$\rightarrow N^+ + N + Ne + 2.7 \text{ eV} \qquad (6b)$$

at $KE_{cm} \geq 3$ eV, as the $N_2^+(C\Sigma_u^+)$ state is known to be highly predissociative for $v \geq 3$. Also the kinetic energy loss of Ne^+ in the charge transfer with N_2 as observed in a low energy beam experiment infers population of the N_2^+(C) state.[36]

Similarly in the reaction of Ne^+ with CO a dramatic increase of k_7 with KE_{cm} is connected with a change in the product ion from predominately CO^+ below 1 eV KE_{cm} to C^+ above 1 eV KE_{cm}.[2] This behavior is in favourable agreement with the energetics of the channel (7b),

$$Ne^+ + CO \rightarrow CO^+ + Ne \quad (7a)$$
$$\rightarrow C^+ + O + Ne - 1.0\ eV \quad (7b)$$

The energy dependences of the total rate coefficients together with the product ion distributions or the internal excitation of the product ions clearly reflect the energetics involved in the processes discussed above. We want to stress the point however, that not in each case such clear correlations are observed, e.g. in the interactions of Ne^+ with CO_2 and O_2,[35] where the observed energy dependences of the respective rate coefficients cannot be related to changes in the reaction paths.

Reactive Collissions of Molecular Ions in Selected Vibrational States

Molecular ions drifting in heavy buffer gases at moderate and higher E/N become readily vibrationally excited, and even Helium has proven to be able to enhance the vibrational temperature of ions drifting in it at elevated E/N.[37,38] The influence of this "internal heating" on charge transfer reactions is discussed in detail in the above chapter by Ferguson[39]. A recently developed theory by Viehland et al.[40,41] results in the following relation between the internal vibrational temperature of the drifting ions, T_i, and the ion drift velocity, v_d,

$$\frac{3}{2} k_B T_i = \frac{3}{2} k_B T_g + \frac{1}{2} M_b v_d^2 (1-b) \quad (8)$$

where T_g and M_b denote the temperature and the mass of the buffer gas respectively and b is a correction term usually considerably smaller than unity. This has been shown recently for N_4^+ drifting in He and N_2, and for O_2^+ as well as N_2^+ drifting in a Helium buffer[37,38]. For Helium the second term in equ. (8) is small at low E/N and thus ions do not significantly become internally excited. At the same time Helium is quite inefficient in quenching of the internal excitation of the same ions at low E/N, so that ions created in or injected into Helium keep their internal state distribution unchanged in thermal collisions with this buffer gas (radiative decay is not prevented, of course). This has been exploited in a variety of experiments, where ions were produced (e.g. in an octupole) in specific states or state distributions by selective reactions, before their entering a Helium buffered drift section, where the reactions of these excited ions with neutrals could then be investigated. The principle involved in that kind of investigation is easily understood[42,43] by means of Fig. 6. Let us assume that a

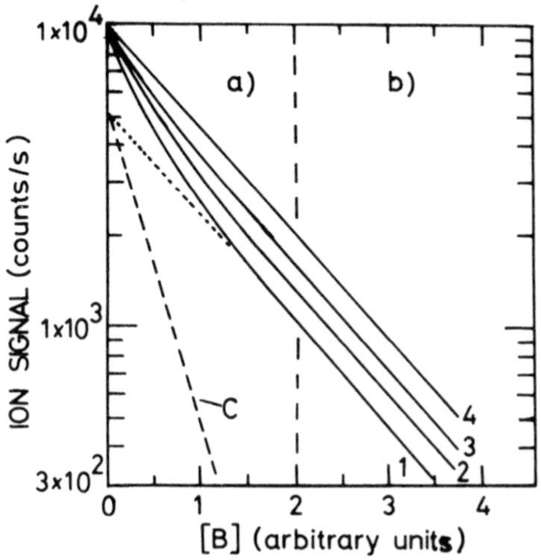

Fig. 6. Declines of the ion signal, when an initial fraction of 50% ground state A^+ and 50% excited A^{+*} is present primarily, with $k_{q1} = 0$ and $0 < k_{q2} < k_{q3} < k_{q4}$. k_o is obtained from the declines in region b) and k_r from the decline of C, which is the difference between ... and — (line 1).

mixture of excited and nonexcited ions A^{+*} and A^+ (at a ratio of 1:1) is introduced into a drift section and reactant gas B is added to allow for reactive and quenching processes (9a) and (9b) respectively,

$$A^{+*} + B \xrightarrow{k_r} \text{Products other than } A^+ \qquad (9a)$$

$$\xrightarrow{k_q} A^+ + B \qquad (9b)$$

and for losses of ground state A^+ according to

$$A^+ + B \xrightarrow{k_o} \text{Products} \qquad (10)$$

With $k_r \sim 3\, k_o$, declines of the ion signals are observed as indicated by the solid lines in Fig. 6. Line 1 corresponds to a case, where the quenching is zero (k_{q1} = 0) and lines 2 through 4 correspond to cases, where $0 < k_{q2} < k_{q3} < k_{q4}$. In cases, where the ratio $[A^{+*}]/[A^+]$ is small and k_q comparable or larger than k_r or k_o, differences in the rate coefficients k_r and k_o hardly can be recognized, as the declines of the total ion signal will be close to case of line 4 in Fig. 6. Thus, in the first attempt using only regular ion sources, differences between k_r and k_o for the reactions of $N_2^+(X,v=1)$ and N_2^+ (X,v=o), with neutrals X(X = O_2, NO, Kr, Ar)[43-45]

$$N_2^+ (X, v=1) + X \xrightarrow{k_r} \text{Products other than } N_2^+ \qquad (11a)$$

$$\xrightarrow{k_q} N_2^+ (X,v=o) + X \qquad (11b)$$

$$N_2^+ (X,v=o) + X \xrightarrow{k_o} \text{Products} \qquad (12)$$

only were observed for X = Kr, Ar, where at thermal energies k_r is larger than k_o by about one or several orders of magnitudes respectively, but in cases where k_r and k_o only differs by about a factor of two, differences were not recognizable. Due to the use of an octupole system (as indicated and discussed above in connection with Fig.1) a ratio $[N_2^+(X,v=1)] : [N_2^+(X,v=o)] \geq 4$ has been reached by allowing for production of $N_2^+(X,v=1)$ via the charge transfer process (1a)[42]. With this high ratio of excited to nonexcited N_2^+, k_r and k_o for reactions with O_2 could be obtained separately though only differing by about a factor of two and despite $k_q > k_r > k_o$. The obtained data are shown in Fig. 7.

The differences in the charge transfer rate coefficients k_r and k_o most likely are explicable on the basis of transitions between vibronic potential curves in a similar fashion as shown by Govers et al.[19] for the $Ar^+ - N_2^+$ case, but so far no quantitative estimates have been made on the $N_2^+ - O_2$ system for thermal energies.

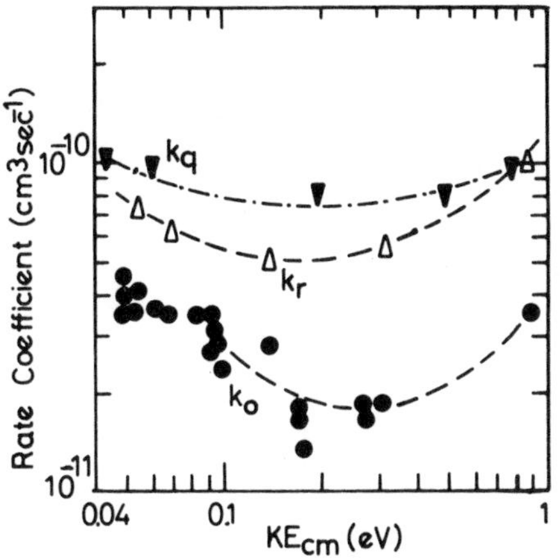

Fig. 7. Rate coefficients k_r, k_q and k_o for the reactions of N_2^+ (X,v=1) and N_2^+(X,v=o) with O_2 (see equ. 11 and 12).

A completely different situation is given in types of processes which are best represented by the reaction between O_2^+ and CH_4. While in the above case (N_2^+- O_2) the addition of a quantum of vibration or a comparable amount of kinetic energy does not provide access to new reaction paths, here the product ion pattern is changed dramatically by either adding kinetic or internal energy to the reactants.[46]

$O_2^+ + CH_4$	$\rightarrow CH_3O_2^+ + H$	+ 111	kcal mole^{-1}	(13a)
	$\rightarrow H_3O^+ + CO + H$	+ 93	-"-	(13b)
	$\rightarrow H_2O^+ + H_2CO$	+ 53	-"-	(13c)
	$\rightarrow HCO^+ + H_2O + H$	+ 69	-"-	(13d)
	$\rightarrow H_2CO^+ + H_2O$	+ 94	-"-	(13e)
	$\rightarrow CH_3O^+ + OH$	+ 79	-"-	(13f)
	$\rightarrow CH_3^+ + HO_2$	- 4	-"-	(13g)
	$\rightarrow CH_4^+ + O_2$	- 14	-"-	(13h)

This reaction has been studied extensively in many swarm type experiments[46-51] and the most detailed investigation on both, its dependence on KE_{cm} and on the O_2^+ vibrational excitation was done recently by Durup-Ferguson et al.[46], who found O_2^+(X, v=1) to be quenched in thermal collision with CH_4 in about half of the orbiting collisions,

$$O_2^+(X, v=1) + CH_4 \xrightarrow{k_q} O_2^+ (X, v=o) + CH_4 \quad (14)$$

with $k_q = 6 \times 10^{-10}$ sec^{-1}. $O_2^+(X, v=2)$ reacts as follows

$$O_2^+ (X, v=2) + CH_4 \rightarrow CH_3^+ + HO_2 \quad (15a)$$

$$\rightarrow CH_3O_2^+ + H \quad (15b)$$

while $O_2^+(X, v=3)$ only seems to undergo simple charge transfer

$$O_2^+ (X, v=3) + CH_4 \rightarrow CH_4^+ + O_2 \quad (16)$$

The dominant product ion in the reaction of O_2^+ (X, v=o) with CH_4 at thermal and subthermal energies is $CH_3O_2^+$, being protonated formic acid, $HC(OH)_2^+$, as could be demonstrated by isotopic exchange reactions[52] and by investigating the collisional breakup of this ion[53]. At elevated KE_{cm}, the slightly endoergic products CH_3^+ and CH_4^+ become dominant[46] (Fig. 8),

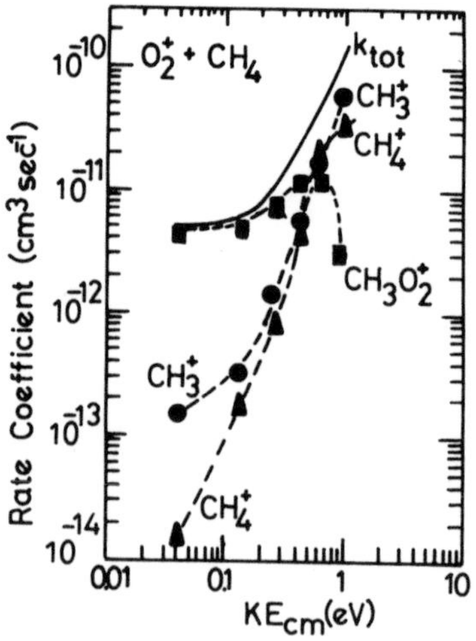

Fig. 8. Total rate coefficient, k_{tot}, as well as product ion distribution for the reaction of O_2^+ with CH_4 as a function of KE_{cm} (Ref. 46).

while the exogeric reaction paths (13b) through (13f) are not observed. Channel (13a) requires the breaking and formation of several chemical bonds, thus the total rate coefficient is small at room temperature ($k \simeq 6 \times 10^{-12}$ cm^3 sec^{-1}) as the complex lifetime, τ, is too short to allow for these rearrangements to occur with high efficiency. Towards $T \rightarrow 0°K$, and thus with increasing τ, the rate coefficient increases[60] towards its Langevin value (1.1×10^{-9} cm^3 sec^{-1}), which makes this reaction extremly important for interstellar molecular synthesis and may well represent the main source for formic acid in interstellar clouds[54] ($CH_3O_2^+$ most likely recombines with electrons to form HCOOH and H).

On the basis of all the experimental information available, Durup-Ferguson et al.[46,51] were able to deduce a detailed picture of the mechanism of reaction 13. In this model the long lifetime of the complex $(O_2^+ \cdot CH_4)^*$ of about 10^{-9} sec at 300° is believed to be due to an initial H^- abstraction from CH_4 to form $CH_3^+ .. HO_2$, which reduces the O-O bond from 6.7 eV in O_2 to 2.7 eV in HO_2 thus facilitating the breaking of this bond and (after one CH bond is broaken and another one made) formation of $CH_3O_2^+$ + H in a successive exothermic path.

An important application of reactions, which are very sensitive to the state of excitation of the reactant ions is their use as monitors in the investigations of quenching processes, as discussed above but they also represent a useful tool for investigating the internal state distribution of ions, resulting as products from various ion molecule reactions, thus giving information on the energy partitioning involved.

A typical example is the vibrational state distribution of N_2H^+, (NN-H^+-bond), which is strongly different whether this ion is produced in one or the other of the two reactions,[55]

$$N_2^+(X, v=o) + H_2 \rightarrow N_2H^+(v \geq 2) + H + \leq 27 \text{ kcal mole}^{-1} \quad 40\% \quad (17a)$$

$$\rightarrow N_2H^+(v=1) + H + 36 \quad \text{-"-} \quad 37\% \quad (17b)$$

$$\rightarrow N_2H^+(v=o) + H + 46 \quad \text{-"-} \quad 23\% \quad (17c)$$

and

H_3^+ (more than 90% in ground state)

$$+ N_2 \rightarrow N_2H^+(v=1) + H \; + \; 3 \text{ kcal mole}^{-1} \quad 25\% \qquad (18a)$$

$$\rightarrow N_2H^+(v=o) + H \; + \; 12 \quad \text{-"-} \quad 75\% \qquad (18b)$$

These state distributions were obtained by using the endothermic proton transfer processes

$$N_2H^+ + Kr \rightarrow KrH^+ + N_2 \; - \; 12 \text{ kcal mole}^{-1} \qquad (19)$$

and

$$N_2H^+ + Ar \rightarrow ArH^+ + N_2 \; - \; (15 \text{ to } 25) \text{ kcal mole}^{-1} \qquad (20)$$

as monitor reactions. When Kr is added to N_2H^+ produced via reaction (17), a decline of the N_2H^+ ion signal is observed at $KE_{cm} \simeq 0.53$ eV as shown in Fig. 9a[57]. The two breaks in the measured ion signal indicate the presence of three different N_2H^+ species, one, $N_2H^{+'}$, reacting at a very slow rate ($k_a = 1.3 \times 10^{-11}$ cm^3 sec^{-1}) and thus being N_2H^+ in the ground state and a second one, $N_2H^{+''}$ reacting at an intermediate rate of $k_b = 2.7 \times 10^{-10}$ cm^3 sec^{-1}, thus indicating a slightly endoergic reaction. This is consistent with the assumption, that $N_2H^{+''} \equiv N_2H^+$ (X,v=1). With the value of the fundamental frequency of HNN^+ being 3233,95 cm^{-1},[56] and the difference in the proton affinities of N_2 and Kr $\Delta PA \sim 12$ kcal/mole[58], we conclude that the proton transfer from N_2H^+(X,v=1) to Kr is slightly endoergic, in agreement with the rate coefficient k_b deduced from Fig. 9. Any N_2H^+ excited to $v \geq 2$ is expected to transfer its proton to Kr at a Langevin rate, as is consistent with the observed fast rate constant $k_c = 1.1 \times 10^{-9}$ cm^3 sec^{-1} attributed to the species $N_2H^{+'''} \equiv N_2H^+$ (X, $v \geq 2$). The energy dependences of the rate coefficients k_a, k_b and k_c, shown in Fig. 10[57] fully support this interpretation. When traces of Ar is added to the Helium buffer prior to Kr addition, the fast reacting fraction $N_2H^{+'''}$ has vanished (Fig. 9b), indicating that N_2H^+ (X, $v \geq 2$) makes fast proton transfer to Ar. This yields a limit for the proton affinity PA(Ar) $\geq$ 94 kcal/mole.

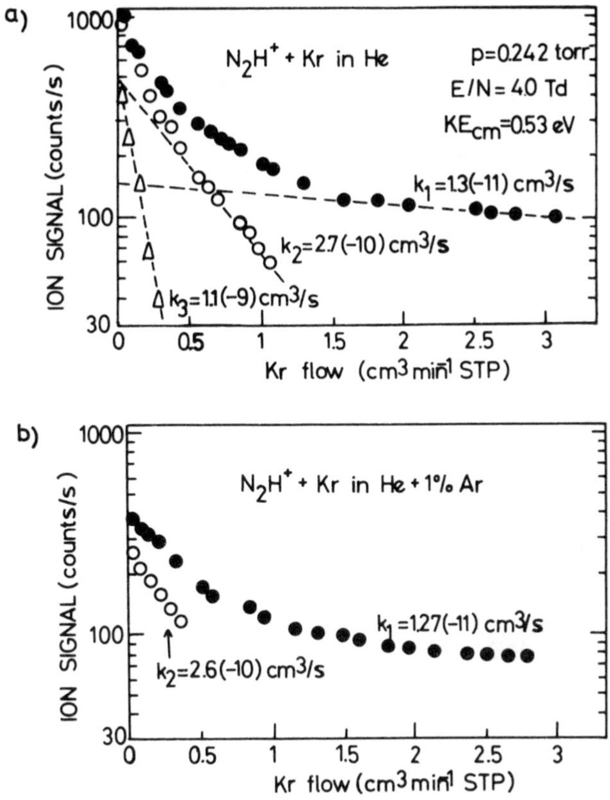

Fig. 9. Count rates of N_2H^+ ions (produced via reaction (17) as a function of the Kr reactant gas flow (a: pure Helium buffer gas, b: Helium with Argon addition).

Together with the known difference between the proton affinities of Ar and H_2 (0.4 eV $\leq \Delta PA \leq 0.6$ eV)[59], a best estimate for the proton affinity of Ar is PA(Ar) = 93 $\pm$ 2 kcal/mole.

Conclusion

Due to recent experimental developments a wide new field of information on the kinetic and internal energy dependences of ion molecule reactions is now accessible. The cases discussed in this chapter provide new insight into the reaction mechanisms involved and yield new thermodynamic information.

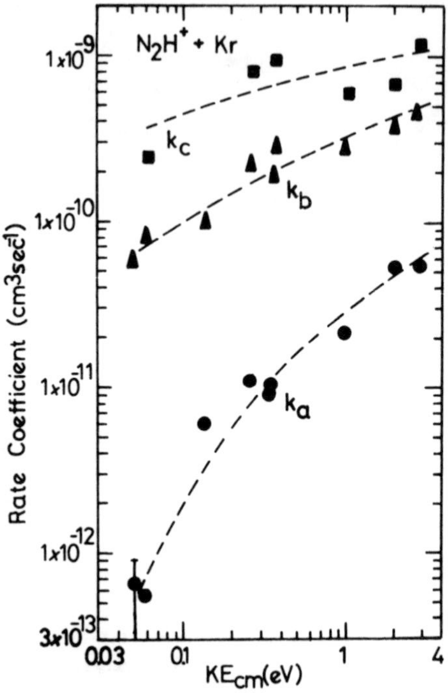

Fig. 10. Rate coefficients k_a, k_b and k_c for the proton transfer from N_2H^+ (X,v=o), N_2H^+ (X, v=1) and N_2H^+ (X, $v \geq 2$) to Kr respectively as dependent on KE_{cm} (Ref. 57).

Acknowledgements

Much of the work mentioned in this chapter was initiated directly or indirectly by Prof. Eldon Ferguson (Boulder) and Prof. David Smith (Birmingham). Most encouraging discussions and collaborations with them have helped our Innsbruck group to overcome many obstacles over the past years. I am especially indebted to my coworkers, in particular to Dr. Hannes Villinger, for their efforts to do their best even under extremely unfavorable conditions. The assistance of M. Heigl in the preparation of this manuscript is greatly appreciated.

References

1. Villinger, H., Futrell, J.H., Saxer, A., Richter, R. and Lindinger, W., J. Chem. Phys. 80, (March 1984)
2. Lindinger, W. and Smith, D., in "Reactions of small transient species", (Eds. A. Fontijn and M.A.A. Clyne), Acad. Press (1983) Chapter 7.
3. Howorka, F., Dotan, I., Fehsenfeld, F.C. and Albritton, D.L., J. Chem. Phys, 73, 758 (1980).
4. Dobler, W., Howorka, F., and Lindinger, W., Plasma Chemistry and Plasma Processing, 2, 353 (1982)
5. Johnsen, R., in: "Electron and Ion Swarms"(Ed. Christophrou L.C.) Pergamon, N.Y. (1981)
6. Teloy, E. and Gerlich, D., Chem. Phys., 4, 417 (1974)
7. Adams, N.G. and Smith, D., J. Phys. B9, 1439 (1976)
8. Adams, N.G. and Smith, D., Int. J. Mass Spectrom. Ion Phys., 21, 349 (1976)
9. Gioumousis, G. and Stevenson, D.P., J. chem. Phys., 29, 294 (1958)
10. Su, T. and Bowers, M.T., Int. J. Mass Spectrom. Ion Phys. 12, 347 (1973)
11. Lindinger, W., Albritton, D.L. and Fehsenfeld, F.C., J. Chem. Phys., 62, 4957 (1975); Lindinger, W., Alge, E., Störi, H., Varney, R.N, Helm, H., Holzmann, P. and Pahl, M., Int. J. Mass Spectrom. Ion Phys. 30, 251 (1979)
12. Störi, H., Alge, E., Villinger, H., Egger, F. and Lindinger, W., Int. J. Mass Spectrom. Ion Phys., 30, 263 (1979)
13. Mauclaire, G., Derai, R., Fenistein, S., Marx, R. and Johnsen, R., J. Chem. Phys. 70, 4023 (1979)
14. Derai, R., Fenistein, S., Gerard-Ain, M., Govers, T.R., Marx, R., Mauclaire, G., Profous, C.Z. and Sourisseau, C, Chem. Phys. 44, 65 (1979)
15. Johnsen, R. and Biondi, M.A., Geophys. Res. Lett. 7, 401 (1980)
16. Rowe, B.R., Fahey, D.W., Fehsenfeld, F.C. and Albritton, D.L., J. Chem. Phys., 73, 194 (1980)
17. Lindinger, W., Howorka, F., Lukac, P., Kuhn, S., Villinger, H., Alge, E. and Ramler, H., Phys. Rev. A23, 2319 (1981)

18. Smith, D. and Adams, N.G., Phys. Rev. A23, 2327 (1981)
19. Govers, T.R., Guyon, P.M., Baer, T., Cole, K., Fröhlich, H. and Lavollee, M.; Chem. Phys. (in press, 1984)
20. Fröhlich, B., Trafton, W., Rockwood, A., Howard, S. and Futrell, J.H., J. Chem. Phys. (in press, 1984)
21. Kato, T., Tanaka, K. and Koyano, I., J. Chem. Phys. 77, 834 (1982)
22. Wen Hu, D., Friedrich, B., Rockwood, A., Howard, St., Lindinger, W. and Futrell, J.H., 4th Symp. Atomic and Surface Physics (Howorka, F. et al., eds.) Contr. pp. 106, Inst. f. Atomphys., Univ. Innsbruck, (1984)
23. Govers, T.R., 4th Symp. Atomic and Surface Phys. (Howorka, F. et al., eds.) Contrib., pp. 96, Inst. f. Atomphys. Univ. Innsbruck (1984)
24. Bauer, E., Fisher, E.R. and Gilmore, F.R., J. Chem. Phys., 51, 4173 (1969)
25. Ferguson, E.E., Smith, D. and Adams, N.G., J. Chem. Phys. in press
26. Bates, D.R., J. Phys. B12, 4135 (1979)
27. Bates, D.R., J. Chem. Phys., 73, 1000 (1983)
28. Herbst, E., J. Chem. Phys. 70. 220 (1979); ibid. 72, 5284 (1980)
29. Dotan, I. and Lindinger, W., J. Chem. Phys. 76, 4972 (1982)
30. Lindinger, W., Villinger, H. and Howorka, F., Int. J. Mass Spectrom. Ion Phys., 41, 89 (1981)
31. Laudenslager, J.B., Huntress, W.T. and Bowers, M.T., J. Chem. Phys., 61, 4600 (1974)
32. Rakshit, A.B. and Twiddy, N.D., Chem. Phys. Lett., 60, 400 (1979)
33. Lindinger, W., Alge, E., Ramler, H., and Peska, K., 3rd Symp. Atomic and Surface Phys. (Lindinger, W. et al., eds.), Contrib., pp. 274, Inst. f. Atomphys. Univ. Innsbruck (1980)
34. Jones, T.T.C., Villinger, H., Lister, D.G., Tichy, M., Birkinshaw, K. and Twiddy, N.D., J. Phys. B, 14, 2719 (1981)
35. Villinger, H., Futrell, J.H., Richter, R., Saxer, A., Niccolini, St. and Lindinger, W., Int. J. Mass Spectrom. Ion Phys., 47, 175 (1983)

36. Futrell, J.H., private communication
37. Lindinger, W., Villinger, H., Saxer, A., Ramler, H. and Richter, R., 3rd Swarm Seminar (Lindinger, W. et al.,eds), Proceedings pp 153, Inst. Experimentalphysik, Univ. Innsbruck (1983)
38. Federer, W., Ramler, H., Villinger, H., Barbier, L., Wagner, H.E. and Lindinger, W., 4th Symp. Atomic and Surface Physics (Howorka, F. et al., eds), Contributions pp 135, Inst. f. Atomphys., Univ. Innsbruck (1984)
39. Ferguson, E.E., preceding chapter in this book
40. Viehland, L.A. and Mason, E.A., Ann. Phys. 110, 287 (1978)
41. Viehland, L.A., Lin, S.L. and Mason, E.A., Chem. Phys. 54, 341 (1981)
42. Villinger, H., Ramler, H., Saxer, A., Richter, R., Tosi, P. and Lindinger, W., XVI Int. Conf. on Phenomena in Ionized Gases (Bötticher, W., et al., eds), Proceedings pp. 564, Inst. Theoret. Physics, Univ. Düsseldorf (1983)
43. Dobler, W., Ramler, H., Villinger, H., Howorka, F., and Lindinger, W., Chem. Phys. Lett., 97, 553 (1983)
44. Alge, E. and Lindinger, W., J. Geophys. Res. 86, 871 (1981)
45. Dobler, W., Villinger, H., Howorka, F., and Lindinger, W., Int. J. Mass Spectrom. Ion Phys., 47, 171 (1983)
46. Durup-Ferguson, M., Böhringer, H., Fahey, D.W., Fehsenfeld, F.C. and Ferguson, E.E., J. Chem. Phys., in press
47. Dotan, I., Fehsenfeld, F.C. and Albritton, D.L., J. Chem. Phys. 68, 5665 (1978)
48. Albritton, D.L., in "Kinetics of Ion-Molecule Reactions" (P. Ausloos, ed.) Plenum Press, N.Y. (1979) p. 119
49. Alge, E., Villinger, H., and Lindinger, W., Plasma Chem. and Plasma Processings, 1, 65 (1981)
50. Alge, E., Villinger, H. Peska, K., Ramler, H., and Lindinger, W., Journal de Physique, C7, 83 (1979)
51. Durup-Ferguson, M., Böhringer, H.and Ferguson, E.E., 3rd Int. Swarm Seminar (Lindinger, W. et al., eds) Proceedings pp 118, Inst. f. Experimentalphysik, Univ. Innsbruck (1983)
52. Villinger, H., Richter, R., and Lindinger, W., Int. J. Mass Spectrom. Ion Phys., 51, 25 (1983)

53. Villinger, H., Saxer, A., Richter, R. and Lindinger, W., Chem. Phys. Lett. 96, 513 (1983)

54. Villinger, H., Saxer, A., Ferguson, E.E., Bryant, H.C. and Lindinger, W., 3rd Int. Swarm Seminar (Lindinger, W. et al. eds) Proceedings, pp 127, Inst. f. Experimentalphysik, Universität Innsbruck (1983)

55. Glosik, J., Villinger, H., Saxer, A., Richter, R., Futrell, J., and Lindinger, W., 4th Symp. Atomic and Surface Phys. (Howorka, F. et al., eds) Contrib. pp 132, Inst. Atomphysik, Universität Innsbruck (1984)

56. Gudeman, C.S., Begemann, M.H., Pfeifer, J. and saykally, R.J., J. Chem. Phys., 78, 5837 (1983)

57. Villinger, H., Futrell, J.H., Saxer, A., Richter, R., Lindinger, W., J. Chem. Phys. (März 1984)

58. Walder, R. and Franklin, J.L., Int. J. Mass Spectrom. Ion Phys., 36, 85 (1980)

59. Villinger, H., Futrell, J.H., Howorka, F., Duric, N. and Lindinger, W., J. Chem. Phys., 76, 3529 (1982)

60. Rowe, B.R., Dupeyrat, G. and Marquette, J.B., 3rd Int. Swarm Seminar (Lindinger, W. et al. eds) Proceedings, pp 122, Inst. Experimentalphysik, Univ. Innsbruck (1983)

Cluster Ion Association Reactions: Thermochemistry and Relationship to Kinetics

A. W. Castleman, Jr., and R. G. Keesee

Department of Chemistry, The Pennsylvania State University, University Park, PA 16802, U.S.A.

Introduction

Interest in the formation and properties of cluster ions has dramatically increased during the last decade. From a fundamental point of view, this is due in large measure to recognition of the value of work on them in bridging the gap between the gas and condensed phase. Another important aspect is that data on the bonding of molecules to ions serves to provide a direct measure of the depth of the potential well of interaction between the ion and the collection of neutral molecules. Work on clusters is also beginning to shed light on nucleation phenomena, the formation of surfaces, and in some cases insight into the physical basis for catalysis. Finally, cluster ion research is useful in unraveling certain problems pertaining to such areas as reaction rate theory and energy transfer.

In terms of applications, work in this area has a bearing on a broad range of fields including atmospheric sciences, combustion research, radiation chemistry, biochemistry, electrochemistry, analytical chemistry, fission product behavior in nuclear reactors, as well as the fields of corrosion science, fusion, and magnetohydrodynamics. The entire research area has been the subject of a recent extensive review (1) to which the interested reader is referred.

Although still in a developing stage, the most advanced aspect of the subject is that pertaining to the thermochemical properties of cluster ions.

Cluster ion thermochemistry has been discussed in several early reviews including three general ones by Kebarle (2-4) covering the period up through 1976. Several others devoted largely to the authors' own works, but with some attention to the general field include Kebarle (5,6) and Castleman and coworkers (7-10). Other general reviews (11-18) also contain some information related to this topic. The available thermodynamic data has grown many fold in the last five years, but there is presently no single tabulation of these. Recently, we have compiled all known thermodynamic data on the bonding of ligands to ions in the gas phase and these data will be published elsewhere (19).

Thermodynamics of Cluster Reactions

Cluster formation can be represented by the general association reaction:

$$I\cdot(n-1)L + L + M = I\cdot nL + M \qquad (1)$$

Here, I designates a positive or negative ion, L the clustering neutral (ligand), and M the third-body necessary for collisional stabilization of the complex. Taking the standard state to be 1 atm, and making the usual assumptions (20) concerning ideal gas behavior and the proportionality of the chemical activity of an ion cluster to its measured intensity, the equilibrium constant $K_{n-1,n}$ for the nth clustering step is given by

$$\ln K_{n-1,n} = \ln \frac{C_n}{C_{n-1}P_L} = -\frac{\Delta G^\circ_{n-1,n}}{RT} = -\frac{\Delta H^\circ_{n-1,n}}{RT} + \frac{\Delta S^\circ_{n-1,n}}{R} \qquad (2)$$

Here, C_{n-1} and C_n represent the respective measured ion intensities, P_L the pressure (in atm) of the clustering species L, $\Delta G^\circ_{n-1,n}$, $\Delta H^\circ_{n-1,n}$, and $\Delta S^\circ_{n-1,n}$ the standard Gibbs free energy, enthalpy, and entropy changes, respectively, R the gas-law constant, and T absolute temperature. By measuring the equilibirum constant $K_{n-1,n}$ as a function of temperature, the enthalpy and entropy change for each sequential association reaction can be obtained from the slope and intercept of the van't Hoff plot ($\ell n\ K_{n-1,n}$ vs. $1/T$).

Thermodynamic information also can be obtained by studying switching or exchange reactions of the form

$$I \cdot nL + L' = I \cdot (n-1)L \cdot L' + L \qquad (3)$$

The thermodynamic quantities for the association of L' onto $I \cdot (n-1)L$ are the sum of those for reaction (1) plus reaction (3).

Experimental techniques that employ van't Hoff plots lead to enthalpy changes derived from slopes which are representable as straight lines over moderate temperature ranges. In actuality, the enthalpy change is a weak function of temperature due to the difference in heat capacity, ΔC_p, between products and reactants.

$$\Delta H_{T_2} = \Delta H_{T_1} + \int_{T_1}^{T_2} \Delta C_p(T)dT \qquad (4)$$

The various experimental techniques measure and report various related values: the enthalpy change ΔH_T^o of association, the bond dissocation energy D_o ($= -\Delta H_0^o$), or the potential well depth D_e ($= D_o + \frac{1}{2} \sum_i h\nu_i$).

In almost all situations of interest to the field of cluster ions, the electronic contribution to the heat capacity is negligible. Important contributions to the heat capacity, then, are those arising from translation, rotation and vibration. At temperatures above a few tens of degrees Kelvin, rotation is usually fully activated and it is the quantitative evaluation of the vibrational contribution which is difficult to make because it requires a knowledge of the vibrational frequencies of the cluster. Since ion-neutral bonds are relatively weak, the frequencies associated with these are typically low. Therefore, they are particularly important in calculating ΔC_p in the temperature range 100-600 K over which most association reaction thermochemical data are derived.

The vibrational contribution to the heat capacity is usually calculated from the well known relationship

$$C_v(vib) = R \sum_{i=1}^{3N-6} \left(x_i^2 \frac{e^{x_i}}{(e^{x_i} - 1)^2}\right) \quad (5)$$

where $x_i = h\nu_i/kT$ with h Planck's constant, k Boltzmann's constant, T the absolute temperature, and ν_i the frequency of the ith vibrational mode. A few investigators (e.g., Conway and coworkers (21-24) and Castleman and coworkers (10,25)) have considered this problem in detail. For example, in the case of Cl^- associated with water (10,25), the measured enthalpy change ΔH^o_{470} is -14.9 kcal/mole. Using the calculated vibrational frequencies of Kistenmacher et al. (26), ΔH^o_{298} and ΔH^o_0 (= $-D_o$) were calculated to be -14.9 and -14.2 kcal/mole, respectively. In the case of NO^+ with N_2 (21), $-\Delta H^o_{210}$ were measured to be 5.2 kcal/mole, while $-\Delta H^o_{298}$, D_o, and D_e were estimated to be around 5.0, 5.0 and 5.7 kcal/mole, respectively. So, although technically incorrect, the common practice in the literature to discuss measured enthalpy changes in terms of "bond energies", especially for comparison with theoretical results, does not appear to be a critical failing.

The van't Hoff plots also enable a determination of the entropy for each individual association reaction. The method involves an extrapolation to determine the intercept of the plot as $T\to\infty$. Rigorously the entropy is also dependent on temperature, although only weakly so. Again based on the calculated frequencies and structure of $Cl^-\cdot H_2O$ (26), the entropy change at 470°K for the association of water onto Cl^- is calculated to be -19.1 cal/K-mole compared to the experimentally determined value of -19.7. At 298°K, the entropy change is computed to be -18.9 cal/K-mole. An alternative approach sometimes used is one which depends on measuring the $\Delta G°$ value at one temperature and using a calculated value of $\Delta S°$ to obtain $\Delta H°$. This technique is most often employed when data are available at only one temperature, and requires an estimation of $\Delta S°$. This can be accomplished through use of the Sackur-Tetrode equation (27) with knowledge of both the structure, to determine moments of inertia, as well as the vibrational frequencies where there are both rotational and vibrational contributions. For an example of the application of simple statistical mechanical computational techniques for estimating $\Delta S°$ contributions, the reader is referred to Dzidic and Kebarle (28) and Castleman et al. (29).

The contribution due to translation is always large compared to the values for rotation and vibration and the electronic effects are generally negligible. However, the vibrational contribution can sometimes be rather significant due to the low frequency modes present in ion clusters. Generally they are not known experimentally, but in a few cases are available on the basis of molecular orbital calculations. Since the calculated corrections are generally rather uncertain, it is evident that the preferred method for deducing $\Delta H°$ values for sequential clustering reactions is to derive these from van't Hoff plots made over a fairly wide range of temperatures rather than using estimated entropy values and a measured $\Delta G°$ at one temperature.

Experimental Techniques

The Knudsen cell technique (30) was apparently the method which provided one of the first direct measurements of a thermodynamic quantity for the formation of a cluster ion ($K^+ \cdot H_2O$) which has stood the test of time. Other early observations of ion clusters were obtained in ion sources operated in the neighborhood of 10^{-4} torr, but equilibrium conditions were generally not attainable with the few collisions taking place and thermodynamic parameters could usually not be measured with confidence. Field (31), Melton and Rudolf (32). and Wexler and Marshall (33) were successful in observing reactions which required a third-body for stabilization by using essentially conventional mass spectrometric ion sources, but equipped with small ion exit slits and improved pumping. However, it was generally impossible to ensure that complete thermalization of the ions and the attainment of equilibrium with respect to clustering had occurred.

The advent of high pressure mass spectrometry (HPMS), originally developed by Kebarle and coworkers (34), has been particularly valuable in quantitatively determining the stability of ion clusters. In this technique, ions effuse from a high pressure source (typically a few torr) through a small aperture into a mass filter where the distribution of ion clusters is determined. Ionization may be initiated by various methods including radioactive sources, heated filaments, and electric discharges. The pressure of the ion source is maintained sufficiently high such that ions reside in a region of well-defined temperature for a time adequate to ensure the attainment of equilibria among the various ion cluster species of interest, but the pressure must be low enough to avoid additional clustering via adiabatic expansion as the gas exits the sampling orifice.

Other variations of the theme include low field drift tubes with sampling mass spectrometer (DTMS) and pulsed ionization sources as in pulsed high pressure mass spectrometry (PHPMS) or stationary afterglow-mass spectrometry (SAMS). In pulsed ion sources, the kinetics (with corrections for diffusional losses) and approach to equilibrium with increasing residence time of the ions in the high pressure source can be directly monitored. Thermodynamic data can be obtained at lower source pressures in the pulsed mode compared to continuous ionization modes since the collection of data can be delayed for some time after the pulse, thus avoiding those ions which exit the source with insufficient residence time.

The flowing afterglow technique (FA) developed by Ferguson, Fehsenfeld, and Schmeltekopf (35) and other related flow reactors such as the selected ion flow tube (SIFT) (36) have provided a wealth of data on general ion-molecule reactions (37) and in the process several ion clusters have also been studied.

In the flowing afterglow apparatus, the ionization with, for instance a microwave discharge or electron gun, occurs upstream directly in the carrier gas. The flow tube is generally about 1 meter long and 8 cm in diameter. Flow velocities are on the order of 10^2 m sec^{-1} and tube pressures are typically around 1 torr. While most of the gas is pumped away, a small fraction is sampled through an orifice where the ions are mass identified and counted. Reactant gases are added into the flow, so kinetic data and the approach to equilibrium can be determined by varying the position of the reactant injection, the flow rate of reactant into the tube, or the bulk flow velocity. In comparison to PHPMS, the flow tube technique affords more versatility in making kinetic measurements and identifying mechanisms, whereas high pressure mass spectrometry is more amenable to temperature control and enables measurements at higher pressures.

All the experimental techniques thus far described involve extraction of ions from a relatively high pressure into the high vacuum region of a mass spectrometer. In these methods, draw-out potentials must be kept small to avoid cluster fragmentation. Additionally, Conway and Janik (24) pointed out that measurements made on larger clusters may be slightly influenced by unimolecular decomposition of the cluster ions following their exit from the high pressure region. They specifically made estimates on the $O_2^+ \cdot nO_2$ cluster system. Sunner and Kebarle (38) have also considered this problem in the $K^+ \cdot nH_2O$ system.

Ion cyclotron resonance (ICR) experiments are typically performed at pressures of 10^{-5} torr or less, so three-body association reactions are not likely to achieve equilibrium during typical ion trapping time (on the order of a second). Consequently, ICR data on ion clusters has been restricted to measuring the free energy change of switching reactions where the initial ion-molecule complex is formed by an elimination reaction such as (39)

$$Li^+ + (CH_3)_2CHCl \rightarrow Li^+(CH_3CH{=}CH_2) + HCl \qquad (6)$$

If a switching reaction involves an ion-molecule complex whose ΔG° of association is known by some other technique, then an absolute scale can be affixed to the ICR data. Enthalpy changes are estimated by calculating the entropy changes of the switching reactions based on the translational and rotational contributions (40,41). The latter requires some assumption about he structure of the complex, but the result is not usually sensitive to the assumed structure. Also the vibrational contribution to the entropy change of the switching reaction is commonly assumed to be negligible. Some systems for which relative values are available, but an absolute scale is lacking, are $\eta^5C_5H_5Ni^+$ (42), Al^+ (43), Mn^+ (44), Cu^+ (45) Ni^+ (46), $FeBr^+$ (47), and Co^+ (48), largely with organic ligands, and alkoxide ions with alcohols (49).

Photofragmentation (PF) and collison induced dissociation (CID) involve measurement of the energy thresholds of dissociation of ions and ion clusters in beams. Photoionization (PI) and electron impact ionization (EI) thresholds for clusters in neutral beams also have been used to derive bond energies D_o for ion clusters. The bond energies can be derived from measurements of appearance potentials if it is assumed that adiabatic values are obtained from the measurements and if the bonding of the neutral precursor is known or can be adequately estimated. The bonding of ammonia to NH_4^+ has been derived from the photoionization of ammonia clusters (50), where similar measurements have been determined by Stephan et al. (51) using electron impact ionization. The values differ significantly from those derived by high pressure mass spectrometric techniques in the cases where there is "internal" reaction following ionization.

Other methods which have produced information on the bonding in ion-molecule association complexes include inversion of ionic mobility data (M) in rare gases which lead to potential well-depths, D_e, scattering experiments (S), emission spectroscopy (ES), reactive energy thresholds (RET), and various drift tube experiments (DT). Arnold and coworkers (52,53) have made rough estimates of thermodynamic quantities of several cluster ions found in the stratosphere based on balloon measurements of relative ion densities along with estimates of atmospheric temperature and appropriate neutral concentrations. McDaniel and Vallee (54) measured halide-hydrogen halide bond energies by measuring the heat of absorption of HX into a crystal MX and assuming that this quantity was identical to the gas phase process $X^- + HX \rightarrow HX_2^-$ where M^+ was chosen to minimize the lattice energy of the crystal.

Consideration of Enthalpy Values and Bonding

One major impetus for studies of ion clustering is to quantify the potential of interaction between ions and neutral molecules, i.e., the well depth. It is of particular interest to elucidate the forces for larger degrees of clustering which provide important data that bridge a gap between the gaseous and the condensed phase. As discussed later, the bonding in the first step is often significantly different from that in the sequential steps, and systems that have one pattern of ordering at small sizes may display different behaviors at larger degrees of clustering. The most extensive data that enable a semi-quantitative interpretation is that of ligand attachment to the spherically symmetric ions, namely the positive alkali metals and the negative halide ions.

Comparing the relative bond strengths for a variety of ligands about a given positive ion is very instructive in elucidating the role of the ligand. In this context, the most extensive data are available for sodium and secondly for potassium. Data for a wide range of molecules clustered to sodium are available, including ones with large permanent dipole moments and polarizabilities to those having low permanent dipole moments but relatively large quadrupole moments. Comparing the data for the trends in bonding given in Table I shows that they are in general accord with expectations based on simple electrostatic considerations.

Using simple electrostatic considerations, Spears (56,57) has been able to account for the hydration of simple ions including Na^+. Diercksen and Kraemer (58) and Clementi and coworkers (26,59,60)

Table I. Properties of Small Molecules

Ligand	Dipole Mqment D_z (Debye)	Polarizability $\alpha_z, \alpha_y, \alpha_x$ ($Å^3$)	Quadrupole moment Q_{zz}, Q_{yy}, Q_{xx} (10^{-26} esu·cm^2)	$\Delta H_{0,1}$ (exp) (kcal/mole) for attachment to Na^+ (cf. Table II)	BE_{calc}[d] (kcal/ mole)	axes 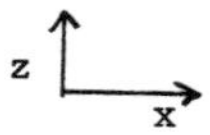
H_2O	1.85[a]	1.452,1.651,1.226[b]	-.26,-5.0,5.26[c]	-24		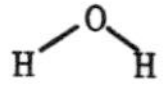
SO_2	1.63[e]	2.653,4,273,4.173[e]	2.6,8.0,-10.6[f]	-18.9	-15.8	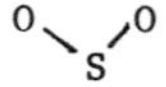
CO	.1098[g]	2.601,1.624,1.624[h]	-5.0,2.5,2.5[i]	-12.6	-11.6	
N_2	0[a]	2.39,1.46,1.46[j]	-2.41,1.2,1.2[k]	-8.0	-7.7	
CO_2	0[a]	4.05,1.95,1.95[ℓ]	-8.64,4.32,4.32[m]	-15.9	-11.1	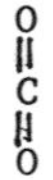
CH_4	0[a]	2.56[n]	0	-7.2	-2.75	
NH_3	1.47[a]	2.388,2.1,2.1[o]	-4.64,2.32,2.32[p]	-29.1	-31.4	
HCl	1.084[a]	2.727,2.477,2.477[q]	7.6,-3.8,-3.8[i]	-12.2	-11.4	

NOTE: All lettered references are found in Ref. 55

performed ab initio SCF-MO calculations on the alkali ion hydrates and alao have shown that the bonding is almost purely electrostatic. Using semi-quantitative calculations, Castleman (61) found that there was slightly more transfer of charge in the case of ammonia compared to water, but still found that the bonding is essentially electrostatic.

The ordering of the ligand molecules with respect to the enthalpy change involved in forming the first cluster with Na^+ follows the order DME(dimethoxyethane) $>NH_3>H_2O>SO_2>CO_2>CO\gtrsim HCl>N_2\gtrsim CH_4$ (see Table II). Additional considerations (62) suggest that DME forms a bidentate bond with Na^+. Comparing the stronger bond for ammonia compared to water in light of the electrostatic properties given in Table I reveals the importance of the ion-induced dipole interaction in governing bond strength. Although ammonia has the smaller dipole moment, its much larger polarizability enhances its bonding at small cluster sizes. The role of the quadrupole moment is also seen by comparing the relative bonding of SO_2 with NH_3. Both have approximately comparable dipole moments and polarizabilities, but the quadrupole moment of SO_2 leads to a repulsive interaction compared to the attractive one of NH_3, consistent with the smaller bonding strength of SO_2. Molecules without permanent dipoles have comparatively low bond energies as seen for N_2 and CH_4; the relatively stronger bonding of CO_2 compared to N_2 and CH_4 is accountable in terms of the large polarizability of CO_2.

Another interesting finding is evident by comparing the binding of various ligands to K^+ and Li^+ (20,40). The trend in bond strengths correlates inversely with differences in ionization potential between the respective ligand and the neutral metal.

Since there is so little electron transfer in these systems, alkali ion-molecule interactions can be considered in terms of a Lewis acid-base interaction (4). However, the Lewis acid-base concept is not completely general. Rather convincing evidence that the bonding of molecules to certain ions does involve some covalent bonding comes from a number of experimental measurements which show that the bonding strengths exceed

Table II. Stepwise heats of association, $-\Delta H_{n-1,n}$ in kcal/mole, for Na^+ with various neutral species L.

	(n-1,n)						
L	(0,1)	(1,2)	(2,3)	(3,4)	(4,5)	(5,6)	Ref.
$(CH_3OCH_2)_2$	47.2	35.1	23.2				62
CH_3CN	----	24.4	20.6	14.9	12.7		63
NH_3	29.1	22.9	17.1	14.7	10.7	9.7	20
H_2O	24.0	19.8	15.8	13.8	12.3	10.7	28
HNO_3	20.6						64
SO_2	18.9	16.6	14.3	12.3			64,62
CO_2	15.9	11.0	9.7	8.4			65
CO	12.6	7.5					62
O_3	12.5						66
HCl	12.2						64
N_2	8.0	5.3					64
CH_4	7.2						62

those expected on the basis of simple electrostatic considerations. This is evident in considering the bonding of H_2O to Pb^+ (67) as well as water to Sr^+ and Bi^+ (68,69) and water and ammonia to Ag^+ and Cu^+ (70). The stabilities of the cluster ions N_4^+, O_4^+, and $(CO)_2^+$ are all much greater than expected for ion induced dipole interactions in the case of the first two, and ion-dipole interaction for the last species. It is suggested that bonding must arise due to the sharing of an electron by the two molecules (71,24,4).

As in the case of the cations, for anions the most extensive set of measurements are also for hydration, but considerable data are available for the clustering of HCl, SO_2, MeCN, MeOH, and HCOOH to Cl^- (72-76). The same general tendency of the ΔH values to approach the ΔH of condensation of the individual ligands at large cluster sizes is seen for these as with H_2O. Interestingly, however, data for MeCN bound to Cl^-, and water to I^-, show that it is possible for the absolute values of the heat of association to fall below that of heat of condensation at intermediate cluster sizes. This is understandable in the case of systems where the ion-ligand bond is comparatively weak and more than one binding site is available between ligands in the condensed phase or in the case where the initial orientation of ligands about the ion significantly hinders the development of secondary solvation shells.

A rather interesting trend (73) in bond energies as shown in Table III is seen for the case of the first ligand attachment to a variety of ions. The trend for systems including SO_2, water, and CO_2 shows the inequality to be as follows:

$$OH^- > F^-,\ O^- > O_2^- > NO_2^- > Cl^- > NO_3^- > CO_3^- > SO_4^- \approx SO_3^- \approx Br^- > I^- \qquad (7)$$

The above inequality appears to parallel the order of gas phase basicity of the negative ions where the strongest bases exhibit the largest bond dissociation energies.

Table III. Dissociation energies (kcal/mole) of ion-neutral complexes.

Ion	SO_2	H_2O	CO_2
OH^-	---	25[a]	56[b]
O^-	>60[c]	<30[d]	52[e]
F^-	43.8[f]	23.3[g]	>18[h]
O_2^-	---	18.4[a]	18.5[i]
$O_2^-\cdot H_2O$	---	17.2[a]	~14[c]
NO_2^-	25.9[j]	15.2[j]	9.3[j]
Cl^-	21.8[j]	14.9[j]	8.0[j]
NO_3^-	18.2[k]	14.6[j]	---
CO_3^-	>14.2[c]	14.1[j]	7.1[j]
SO_4^-	~14[c]	~12.5[c]	---
HSO_4^-	13.7[l]	11.9[m]	---
SO_3^-	13.3[j]	~13[c]	6.5[j]
I^-	12.9[j]	11.1[j]	5.6[j]

[a]ref. 77; [b]from ab initio calculations in ref. 78; [c]from free energy changes in ref. 79 assuming ΔS = −25 and 0 cal/K-mole for addition and switching reactions, respectively; [d]ref. 79; [e]ref. 80; [f]ref. 41; [g]ref. 81; [h]ref. 82; [i]ref. 83; [j]ref. 73; [k]ref. 84; [l]ref. 85; [m]ref. 86

Another interesting comparison is that of the relative bond dissociation energies for a given ion with different ligands such as SO_2, H_2O, and CO_2. The hydrogen, sulfur, and carbon are the centers which are attracted to a negative charge. Water has a slightly larger dipole moment than sulfur dioxide but the quadrupole moment of water is repulsive when the dipole is attractive to a negative ion. Alternately, the quadrupole moment of SO_2 is attractive to a negative ion and carbon dioxide has no dipole but does have a significant quadrupole. Considering only charge-dipole or charge-multipole interactions, the bond strength for a ligand attached to a given ion would be expected to be in the order of $H_2O \approx SO_2$, but both being greater than CO_2.

For a weakly basic or large ion like I^-, the order $SO_2 \gtrsim H_2O > CO_2$ is observed. However, as the ions become smaller or more basic, SO_2 bonds relatively more strongly than water. With O_2^- the enthalpy change for addition of CO_2 becomes comparable to that of H_2O. Finally, for a small ion like O^-, the order $SO_2 > CO_2 > H_2O$ actually occurs. Interestingly, the mean polarizabilities for SO_2, CO_2, and H_2O follow the same order as their relative bond energies to small ions. Polarization energies are known to be relatively more important for smaller ions due to the ability of the neutral to closely approach the ion, and thereby become more influenced by the ionic electric field with the attendant result of a larger induced dipole. Qualitatively, consideration of the polarizabilities partially explains the increased bonding strength of SO_2 and CO_2 over that of water in clustering to smaller ions.

In some cases, the bonds are strong enough to be considered as actual chemical ones instead of merely weak electrostatic effects. In other words, the bond may be of a covalent nature which is equivalent to stating that significant charge transfer occurs between the original ion and the clustering neutral. An example is $OH^- \cdot CO_2$ which is more properly considered as the bicarbonate ion. Likewise, $Cl^- \cdot SO_2$ has a rather strong bond energy for the first cluster addition, but a very weak one for the second addition. The first cluster addition may be thought of as forming a "molecule" over which the negative charge becomes relatively widely dispersed resulting in the significant difference energetically between the first SO_2 addition and the second (73).

In terms of charge transfer, the electron affinity of the clustering neutral is a relevant factor to consider in assessing the relative bonding trends. SO_2 has an electron affinity of 1.0 eV (87) and forms a stable gas-phase negative ion, whereas the negative ion of water has not been observed (88). CO_2 has been detected in high energy processes in which the linear neutral can be bent to form the ion although it is short lived and auto-detaches (89). Nevertheless, $(CO_2)_2^-$ is apparently stable (90). Collective effects are evidently important in the formation and stabilization of certain negative ion complexes. This point is further realized by recent observations of Haberland et al. (91), who have reported observation of the hydrated electron when eleven or more water molecules are clustered together.

An examination of the bonding of neutral molecules onto ions has particular value in elucidating the properties which govern stability. An interesting result is found by comparing the association of ammonia, water, sulfur dioxide, and carbon dioxide to both Na^+ and Cl^-. The ion-dipole interaction is the most important electrostatically attractive force between an ion and a neutral molecule. Both sodium and chloride ions have closed electronic configurations and spherical symmetry, although the ionic radius of Na^+ is considerably smaller. Consequently, with other factors being equal, a neutral molecule would be expected to bind more strongly to the smaller Na^+. However, covalent bonding, charge transfer, and higher electrostatic moments are also important.

Based on experimentally determined step-wise heats of association as given in Table IV, it is evident (92) that only sulfur dioxide and hydrogen chloride bind more strongly to Cl^- than to Na^+ for the first association step. Ammonia

Table IV. Comparison of the stepwise heats of association, $-\Delta H_{n-1,n}$, in kcal/mole, for reaction (1) of Na^+ and Cl^-.

L	I	(0,1)	(1,2)	(2,3)	(3,4)	(4,5)	(5,6)
H_2O	Cl^- [a]	14.9	12.6	11.5	10.9	---	---
	Na^+ [b]	24.0	19.8	15.8	13.8	12.3	10.7
SO_2	Cl^- [a]	21.8	12.3	10.0	8.6	---	---
	Na^+	18.9[c]	16.6[d]	14.3[d]	---	---	---
NH_3	Cl^- [e]	7.8	---	---	---	---	---
	Na^+ [f]	29.1	22.9	17.1	14.7	10.7	9.7
CO_2	Cl^- [a]	8.0	---	---	---	---	---
	Na^+ [d]	15.9	11.0	9.7	---	---	---
HCl	Cl^- [g]	23.7	15.2	11.7	10.3	---	---
	Na^+ [c]	12.2	---	---	---	---	---
Xe	Cl^- [h]	3.1	---	---	---	---	---
	Na^+ [i]	9.5	---	---	---	---	---

[a]ref. 73

[b]ref. 28

[c]ref. 64

[d]ref. 62

[e]ref. 85

[f]ref. 20

[g]ref. 72

[h]ref. 93 (well-depth D_e)

[i]ref. 94 (well-depth D_e)

exhibits the largest difference between $Na^+$and Cl^-, with its binding being much weaker to the negative ion. The relative stability of the ion-molecule complexes for ammonia, water, and SO_2 are in reverse order for the two ions. In terms of magnitude, after the first ligand addition, the enthalpy change for the clustering of SO_2 onto Cl^- is much smaller than for Na^+. Many of these trends are in agreement with expectations from electrostatic considerations. For instance water, sulfur dioxide, and ammonia have similar dipole moments. However, when the dipoles are aligned in the electrostatic field of the ion, the small quadrupole moments of water and the considerably larger one of ammonia are repulsive to a negative charge and attractive to a positive one. In the case of sulfur dioxide, the situation is reversed. Thus, the ion-quadrupole interaction is consistent with the different ordering of the first bond energies for water, ammonia, and sulfur dioxide between the positive Na^+ and the negative Cl^-.

Entropies

Entropy values find use in elucidating the structure of cluster ions through comparison with calculations. All other things being equal, large negative values are indicative of structural ordering and in the case of sequential reactions, restricted motion of the ligands due to crowding in a given "shell". In terms of applications, these data contribute to other problems such as contributing to further understanding of ion induced nucleation (8). A consideration (29) of the experimental entropy values in terms of the usual classical equation for nucleation (95) shows that the actual trends are not in agreement with predictions. The problem is that the simple classical model assumes a random liquid-like disordered structure, while in actual fact the entropies pass through a maximum negative value before becoming increasingly positive (toward the limiting value for the case of a liquid). In the region between, the trends are greatly influenced by cluster structure that is not accounted for by the simple models.

Ligand-ligand steric effects and crowding of the solvation shells at very small degrees of clustering greatly influence entropy values.

Entropy values are also of use on considering the kinetics of clustering reactions. Chang and Golden (96) have made a careful re-analysis of the kinetics for ion molecule association reactions using a model similar to RRKM theory as an extension of the general formalisms of Troe (97) which enable a treatment of the so-called fall-off region in association kinetics. They showed that for reactions at the low pressure limit, the requisite information is the density of states of the association complex, which is related to entropy. Detailed calculations were made for the low pressure and fall-off regime of these association reactions, showing very good accord with experimental measurements. Their considerations demonstrate one of the major problems which has been inherent in study of ion molecule association reactions, namely where measurements of the temperature dependence of the reactions have usually been made at the fixed pressure rather than number density. When the measurements are not at the true limiting low pressure limit, erroneous determinations of the temperature coefficient can easily result.

According to Chang and Golden, the preferred method is to first compute the low pressure limit rate constant in the unimolecular direction for a complex using an appropriate theory (98,99) and a simple harmonic oscillator model followed by a few correction factors which account for anharmonicity, energy dependence of the density of states, overall rotation, and internal rotation. Thereby, the low pressure decomposition rate constant is readily calculated and in cases where thermochemical data are available, the forward one can be computed directly from the equilibrium constant of the reaction. The authors conclude that uncertainties in enthalpies for ion-molecule complexes has a relatively small influence on the calculated rate constants compared to their sensitivity to entropic properties.

Unfortunately, insufficient attention has been focused on methods for deriving better entropy values and attention to this problem is clearly warranted.

Ion Solvation

A number of authors have utilized thermochemical data to interpret trends and expectations concerning ions in liquids. The interested reader is referred to a few relevant references (5,11,16).

One of the first attempts to predict solvation energies in the condensed state was based on the Born relationship which involves a change in energy to create a cavity in a dielectric medium. Although this approach has been of moderate success in general terms, it is obviously too simple for quantitative treatment of systems which display individual trends of bonding to a variety of ligands. Nevertheless, a very simple extension of this model by Lee et al. (100) has enabled a successful treatment by employing some microscopic surface tension concepts to correct the general solvation energies predicted on the basis of a Born-like equation. The results of stepwise gas-phase association enthalpies, summed from the first to the nth ligand and divided into the accepted value for the liquid phase single ion heat of solvation display a rather rapid convergence to a single value at cluster sizes in the neighborhood of six ligands. This correlation, which has also been demonstrated for the case of ammonia clustered to both simple and complex cations suggests the possiblity that gas-phase energies may be directly utilized to predict solvation energies in the condensed phase. Further evidence for the relationship of cluster data to ionic solvation comes from observations of discontinuities in otherwise smooth trends in ΔH with size (20,70,74,101,102) which indicate the possible formation of solvation shells.

The importance of measuring intrinsic basicities and acidities of ligands clustered to ions in order to understand the origin of solution phase ordering has been suggested by a number of investigators including Beauchamp (13), Bartmess and McIver (103), Aue and Bowers (12), and Kebarle

(5) among others. Recent studies by Taft and coworkers (11,104) have addressed the importance of understanding the solvation by a single molecule of a ligand in the gas phase, for comparison with the effects of molecular structure and solvation by bulk solvent in the liquid phase. The importance of accounting for interaction among the ligands and solvent molecules was discussed in a recent article by Arnett et al. (105). Castleman (7) has considered this effect for the ion OH^- solvated in water and clustered with CO_2 to form HCO_3^-.

In order to elucidate the nature of some S_N2 type reactions in the solution phase, Bohme et al. (106) have exploited a similar technique to investigate a few analoguos reactions. In particular, they investigated reactions of the type

$$B^-\cdot S_n + AH = A^-\cdot S_m + (n-m)S + BH$$

and

$$B^-\cdot S_n + CH_3Br = Br^-\cdot S_m + (n-m)S + CH_3B$$

Additionally, they studied the influence of stepwise hydration on the kinetics of proton transfer reactions from the hydronium ion to a variety of polar molecules.

Kebarle and coworkers (107,108) have reported a relationship between the basicity of A^- and the hydrogen bond in the monohydrate $A^-\cdot H_2O$ complex which showed that the strength of the hydrogen bond increases with the gas-phase basicity of A^-. Similarly, it was found that the hydrogen bond in $BH^+\cdot H_2O$ increases with the gas-phase acidity of BH^+. These relationships for the bonding of a water molecule in terms of the basicity of A^- or the acidity of BH^+, provided an explanation on a one molecule solvation basis for the attenuation mechanisms observed in the solution phase. Thus, it was suggested that a substitutent that increases the acidity of AH decreases the basicity of A^- and therefore decreases the hydrogen bonding in $A^-\cdot H_2O$. As with the cation clusters there is much to be learned about solutions of organic constituents.

For certain organic systems, Davidson et al. (107) have also made a comparison of proton transfer free energies between the gaseous and solution phase. They concluded that the large attenuation due to the substituent effect in solution must result from an effect of the substituent on solvation that partially cancels the effect on the isolated molecule. For instance, an electron withdrawing substituent-like CN, which increases the intrinsic acidity of a material like phenol, is expected to unfavorably affect the solvation of cyanophenoxide ion and thus reduce the acidity increase of cyanophenol in aqueous solution. Cluster ion research has a great deal to contribute to the field of solvation involving organic constituents, but further discussion is beyond the scope of this review.

Acknowledgments

The support of the Department of the Army, Grant No. DAAG29-82-K-0160, the National Science Foundation, Grant No. ATM-82-4010, and the Department of Energy, Grant No. DE-AC02-82-ER60055 is gratefully acknowledged.

References

1. Mark, T. D., and Castleman, Jr., A. W. (1984) Adv. At. Molec. Phys., 20, in press.

2. Kebarle, P. (1972) in Ions and Ion Pairs, (Szwarc, E., Ed.) Wiley, New York.

3. Kebarle, P. (1972) In Ion Molecule Reactions (Franklin, J. L., Ed.) Plenum, New York, 315-362.

4. Kebarle, P. (1977) Ann. Rev. Phys. Chem., 28, 445-476.

5. Kebarle, P. (1974) Mod. Asp. Electrochem., 9, 1-45.

6. Kebarle, P. (1975) in Interactions Between Ions and Molecules (Ausloos, P., Ed.) Plenum Press, New York, Vol. 6, 459-487.

7. Castleman, A. W., Jr. (1979) in: NATO Advance Study Institute, Kinetics of Ion Molecule Reactions (Ausloos, P., Ed.) Plenum Press, New York, 295-321.

8. Castleman, A. W., Jr. (1979) Advances in Colloid and Interface Science, 10, 73-218.

9. Castleman, A. W., Jr., and Keesee, R. G. (1981) Electron and Ion Swarms. Proceedings, Second International Swarm Seminar, Oak Ridge, TN (Christophorou, L. G., Ed.) Pergamon Press, 189-201.

10. Castleman, A. W., Jr., Holland, P. M., and Keesee, R. G. (1982) Radiat. Phys. Chem., 20, 57-74.

11. Taft, R. W. (1983) Prog. Phys. Org. Chem., 14, 247-350.

12. Aue, D. M., and Bowers, M. T. (1979) Chapter 9 in Gas Phase Ion Chemistry, Vol. II, Academic Press, 1-51.

13. Beauchamp, J. L. (1971) Ann. Rev. Phys. Chem., 22, 527-561.

14. Lias, S. G., and Ausloos, P. (1975) Second ACS/ERDA Research Monograph in Radiation Chemistry, published by the American Chemical Society.

15. Friedman, L., and Reuben, B. G. (1971) in: Advances in Chemical Physics (Prigogine, I., and Rice, S. A., Eds.) Vol. XIX, Wiley, 35-140.

16. Schuster, P., Wolschann, P. and Tortschanoff, K. (1977) in Chemical Relaxation in Molecular Biology (Pecht, I., and Rigler, R., Eds.) Springer Verlag, Vol. 24, 107-190.

17. Wiegand, W. J. (1982) Chapter 3 in: Applied Atomic Collision Physics, Vol. 3 (McDaniel, E. W., and Nighan, W. L., Eds.) Academic Press, 71-96.

18. Franklin, J. L., and Harland, P. W. (1974) Ann. Rev. Phys. Chem., 25, 485-526.

19. Keesee, R. G., and Castleman, Jr., A. W. (1984) J. Phys. Chem. Ref. Data, submitted.

20. Castleman, A. W., Jr., Holland, P. M., Lindsay, D. M., and Peterson, K. I. (1978) J. Am. Chem. Soc., 100, 6039-6045.

21. Turner, D. L., and Conway, D. C. (1976) J. Chem. Phys., 65, 3944-3947.

22. Janik, G. S., and Conway, D. C. (1967) J. Phys. Chem., 71, 823-829.

23. Turner, D. L., and Conway, D. C. (1979) J. Chem. Phys., 71, 1899-1901.

24. Conway, D. C., and Janik, G. S. (1970) J. Chem. Phys., 53, 1859-1866.

25. Keesee, R. G. (1979) Ph.D. Thesis, Univeristy of Colorado.

26. Kistenmacher, H., Popkie, H., and Clementi, E. (1973) J. Chem. Phys., 59, 5842-5848.

27. McQuarrie, D. A. (1976) Statistical Mechanics, Harper and Row.

28. Dzidic, I., and Kebarle, P. (1970) J. Phys. Chem., 74, 1466-1474.

29. Castleman, A. W., Jr., Holland, P. M., and Keesee, R. G. (1978) J. Chem. Phys. 68, 1760-1766.

30. Chupka, W. A. (1959) J. Chem. Phys., 30, 458-461.

31. Field, F. J. (1961) J. Am. Chem. Soc., 83, 1523-1534.

32. Melton, C. E., and Rudolf, P. S. (1960) J. Chem. Phys., 32, 1128-1131.

33. Wexler, S., and Marshall, R. J. (1964) J. Am. Chem. Soc., 86, 781-787.

34. Kebarle, P., and Hogg, A. M. (1965) J. Chem. Phys., 42, 798-799.

35. Ferguson, E. E., Fehsenfeld, F. C., and Schmeltekopf, A. L. (1969) Adv. At. Mol. Phys., 5, 1-56.

36. Smith, D., and Adams, N. G. (1979) Chapter 1 in: Gas Phase Ion Chemistry, Vol. 1 (M. T. Bowers, ed.) Academic Press, pp. 1-44.

37. Albritton, D. L. (1978) Atomic Data Nuclear Tables, 22, 1-101.

38. Sunner, J., and Kebarle, P. (1981) J. Phys. Chem., 85, 327-335.

39. Staley, R. H., and Beauchamp, J. L. (1975) J. Am. Chem. Soc., 97, 5920-5921.

40. Woodin, R. L., and Beauchamp, J. L. (1978) J. Am. Chem. Soc., 100, 501-508.

41. Larson, J. W., and McMahon, T. B. (1983) J. Am. Chem. Soc., 105, 2944-2950.

42. Corderman, R. R. and Beauchamp, J. L. (1976) J. Am. Chem. Soc., 98, 3998-4000.

43. Uppal, J. S., and Staley, R. H. (1982) J. Am. Chem. Soc., 104, 1235-1238.

44. Uppal, J. S., and Staley, R. H. (1982) J. Am. Chem. Soc., 104, 1238-1243.

45. Jones, R. W. and Staley, R. H. (1982) J. Am. Chem. Soc., 104, 2296-2300.

46. Kappes, M. M., and Staley, R. H. (1982) J. Am. Chem. Soc., 104, 1813-1819.

47. Kappes, M. M., and Staley, R. H. (1982) J. Am. Chem. Soc., 104, 1819-1823.

48. Jones, R. W. and Staley, R. H. (1982) J. Phys. Chem., 86, 1387-1392.

49. Bartmess, J. and Caldwell, G. (1984) in: NATO Advanced Study Institute, Ionic Processes in the Gas Phase (M. A. Almoster Ferreira, ed.) Series C, Vol. 118, 346-347; Blair, L. K., Isolani, P. C., and Riveros, J. M. (1973) J. Am. Chem. Soc., 95, 1057-1060.

50. Ceyer, S. T., Tiedemann, P. W., Mahan, B. H., and Lee, Y. T. (1979) J. Chem. Phys., 70, 14-17.

51. Stephan, K., Futrell, J. H., Peterson, K. I., Castleman, A. W., Jr., Wagner, H. E., Djuric, N., and Mark. T. D. (1982) Int. J. Mass Spectrom. Ion Phys., 44, 167-181.

52. Arnold, F., Viggiano, A. A., and Schlager, H. (1982) Nature, 297, 371-376.

53. McCrumb, J. L., and Arnold, F. (1981) Nature 294, 136-139.

54. McDaniel, D. H., and Vallee, R. E. (1963) Inorg. Chem., 2, 996-1001.

55. (a) Handbook of Chemistry and Physics (1974) Chemical Rubber Co., 55th Edition. (b) S. P. Liebmann and J W. Moskowitz (1971) J. Chem. Phys., 54, 3622. (c) J. Verhoeven and A. Dymanus (1970) J. Chem. Phys., 52, 3222. (d) Calculated electrostatic energy, see Ref. 217. (e) D. Patael, D. Margolese, T. R. Dyke (1979) J. Chem. Phys., 70, 2740. (f) J. M. Pochan, R. G. Stone, W. H. Flygare (1969) J. Chem. Phys., 511, 4278. (g) J. S. Muenter (1975) J. Molec. Spect., 55, 490. (h) J. O. Hirschfelder, C. F. Curtis, R. B. Bird, Molecular Theory of Gases and Liquids (Wiley, New York, 1964). (i) D. E. Stogryn and A. P. Stogryn (1966) Mol. Phys., 11, 371. (j) J. Trapy, J. Cl. Lelievre, J. Picard (1974) Phys. Lett., 47A, 9 (1974). (k) M. A. Morrison, P. J. Hay (1979) J. Chem. Phys., 70, 4034. (l) A. Koide and T. Kihara (1974) Chem. Phys., 5, 34. (m) A. E. Barton, A. Chablo, B. J. Howard (1979) Chem. Phys. Lett., 60, 414. (n) A. A. Maryott, F. Buckley (1953) U.S. NBS Circular 537. (o) J. Koch, S. Friberg, T. Larsen, Zahlenwerte und Funktionen, ed. H. A. Landolt, R. Bornstein (1962). (p) S. G. Kukolich (1970) Chem. Phys. Lett., 5, 401. (q) F. A. Gianturco, C. Guidott (1978) J. Phys. B., 11, L385.

56. Spears, K. G. (1972) J. Chem. Phys., 57, 1842-1844.

57. Spears, K. G. (1972) J. Chem. Phys., 57, 1850-1858.

58. Diercksen, G. H. F., and Kraemer, W. P. (1977) in Metal-Ligand Interactions in Inorganic Chemistry and Biochemistry, Part 2, (Pullman, B., and Goldblum, N., Ed.) Riedel, Dordrecht-Holland.

59. Clementi, E., and Popkie, H. (1972) J. Chem. Phys., 57, 1077-1094.

60. Clementi, E. (1976) in Determination of liquid water structure coordination numbers for ions and solvation for biological molecules. Lecture Notes in Chemistry, Springer-Verlag, Berlin, Heidelberg, New York.

61. Castleman, A. W., Jr. (1978) Chem. Phys. Letts., 53, 560-564.

62. Peterson, K. I. (1982) Ph.D. Thesis, University of Colorado, University Microfilms; Castleman, Jr., A. W., Peterson, K. I., Upschulte, B. L., Schelling, F. J. (1983) Int. J. Mass Spect. Ion Phys., 47, 203-206.

63. Davidson, W. R., and Kebarle, P. (1976) J. Am. Chem. Soc., 98, 6125-6133.

64. Perry, R. A., Rowe, B. R., Viggiano, A. A., Albritton, D. L., Ferguson, E. E., and Fehsenfeld, F. C. (1980) J. Geophys. Res., 7, 693-696.

65. Peterson, K. I., Mark, T. D., Keesee, R. G., and Castleman, A. W. (1984) "Thermochemical Properties of Gas-Phase Mixed Clusters: H_2O/CO_2 with Na^+," J. Phys. Chem. (submitted).

66. Rowe, B. R., Viggiano, A. A., Fehsenfeld, F. C., Fahey, D. W., and Ferguson, E. E. (1982) J. Chem. Phys., 76, 742-743.

67. Tang, I. N., and Castleman, A. W. (1972) J. Chem. Phys., 57, 3638-3644.

68. Tang, I. N., Lian, M. S., and Castleman, A. W. (1976) J. Chem. Phys., 65, 4022-4027.

69. Tang, I. N., and Castleman, A. W. (1974) J. Chem. Phys., 60, 3981-3986.

70. Holland, P. M., and Castleman, A. W., Jr. (1982b) J. Chem. Phys. 76, 4195-4205.

71. Teng, H. H., and Conway, D. C. (1973) J. Chem. Phys., 59, 2316-2323.

72. Yamdagni, R., and Kebarle, P. (1974) Can. J. Chem., 52, 2449-2453.

73. Keesee, R. G., Lee, N., and Castleman, A. W., Jr. (1980) J. Chem. Phys., 73, 2195-2202.

74. Yamdagni, R., and Kebarle, P. (1972) J. Am. Chem. Soc., 94, 2940-2943.

75. Yamdagni, R., Payzant, J. D., and Kebarle, P. (1973) Can. J. Chem., 51, 2507-2511.

76. Luczynski, Z., Wlodek, S., and Wincel, H. (1978) Int. J. Mass Spect. Ion Phys., 26, 103-107.

77. Payzant, J. D., Yamdagni, R., and Kebarle, P. (1971) Can. J. Chem. 49, 3308-3314.

78. Jonsson, B., Karlstrom, G., and Wennerstrom, H. (1978) J. Am. Chem. Soc. 100, 1658.

79. Fehsenfeld, F. C., and Ferguson, E. E. (1974) J. Chem. Phys., 61, 3181-3193.

80. Hiller, J. F., and Vestal, M. L. (1980) J. Chem. Phys., 72, 4713-4722.

81. Arshadi, M., Yamdagni, R., and Kebarle, P. (1970) J. Phys. Chem., 74, 1475-1482.

82. Spears, K. G., and Ferguson, E. E. (1972b) J. Chem. Phys., 59, 4174-4183.

83. Pack, J. L., and Phelps, A. V. (1966) J. Chem. Phys., 45, 4316-4329.

84. Wlodek, S., Luczynski, Z., and Wincel, H. (1983) Int'l. J. Mass Spect. Ion Phys., 49, 301-309.

85. Evans, D. H., Keesee, R. G., and Castleman, Jr., A. W., unpublished results.

86. Bohringer, H., Fahey, D. W., and Fehsenfeld, F. C. (1984) Symposium on Atomic and Surface Physics '84 (W. Lindinger, F. Howorka, and T. D. Mark, eds.) Maria Alm/Salzburg, Austria, January 31-February 3, 1984, pp. 210-212.

87. Janousek, B. K., and Brauman, J. I. (1979) in Gas Phase Ion Chemistry, Vol. 2. (Bowers, M. T., Ed.) Academic Press, 53-86.

88. Caledonia, G. E. (1975) Chem. Rev., 75, 333-351.

89. Paulson, J. F. (1970) J. Chem. Phys., 52, 963-964.

90. Klots, C. E., and Compton, R. N. (1977) J. Chem. Phys., 67, 1779-1780.

91. Haberland, H., Oschwald, M., Waltenspiel, R., Winterer, M., and Worsnop, D. R. (1983) Proc. 9th Int. Symp. Molecular Beams, Freiburg, 120-122.

92. Keesee, R. G., and Castleman, Jr., A. W. (1984) in: NATO Advanced Study Institute, Ionic Processes in the Gas Phase (M.A. Almoster Ferreira, ed.) Series C, Vol. 118, 340-342.

93. Thackston, M. G., Eisele, F. L., Pope, W. M., Ellis, H. W., McDaniel, E. W., and Gatland, I. R. (1980) J. Chem. Phys., 73, 3183-3185.

94. Takebe, M. (1983) J. Chem. Phys., 78, 7223-7226 and references therein.

95. Abraham, F. F. (1974) in Homogeneous Nucleation Theory, The Pretransition Theory of Vapor Condensation, Supplement I, Advances in Theoretical Chemistry, Academic Press, New York.

96. Chang, J. S., and Golden, D. M. (1981) J. Am. Chem. Soc., 103, 496-500.

97. Troe, J. (1977) J. Chem. Phys., 66, 4758.

98. Su, T., and Bowers, M. T. (1973) J. Chem. Phys., 58, 3027-3029.

99. Hsieh, E. T.-Y., and Castleman, A. W., Jr. (1981) International Journal of Mass Spectrometry and Ion Physics, 40, 295-329.

100. Lee, N., Keesee, R. G., and Castleman, A. W., Jr. (1980) Journal of Colloid and Interface Science, 75, 555-565.

101. Lau, Y. K., Ikuta, S., and Kebarle, P. (1982) J. Am. Chem. Soc., 104, 1462-1469.

102. Arshadi, M. R., and Futrell, J. H. (1974) J. Phys. Chem., 78, 1482-1486.

103. Bartmess, J. E., and McIver, Jr., R. T. (1979) in: Gas Phase Ion Chemistry, Vol. 2, Academic Press, pp. 88-119.

104. Bromilow, J., Abboud, J. L. M., Lebrilla, C. B., Taft, R. W., Scorrono, G., and Lucchini, V. (1980) J. Am. Chem. Soc., 103, 5448-5453.

105. Arnett, E. M., Jones, F. M., III, Taagepera, M., Henderson, W. G., Beauchamp, J. L., Holtz, D., and Taft, R. W. (1972) J. Am. Chem. Soc., 94, 4724-4726.

106. Bohme, D. K., Mackay, G. I., and Tanner, S. D. (1979) J. Am. Chem. Soc., 101, 3724-3730.

107. Davidson, W. R., Sunner, J., and Kebarle, P. (1979) J. Am. Chem. Soc., 101, 1675-1680.

108. Yamdagni, R., and Kebarle, P. (1971) J. Am. Chem. Soc., 93, 7139-7143.

Studies of Plasma Reaction Processes Using a Flowing-Afterglow/Langmuir Probe Apparatus

D. Smith and N. G. Adams

Department of Space Research, University of Birmingham, Birmingham B15 2TT, England

1. Introduction

As recently as some twenty five years ago even the order-of-magnitude of the thermal energy rate coefficient, k, for any ion-molecule reaction could not be quoted with conviction, and the forms of the temperature variations of rate coefficients were quite unknown. The first reliable data under well-defined conditions, were obtained using stationary afterglow techniques [1], but a truly dramatic increase in the understanding of thermal energy ion-molecule reactions resulted from the development and exploitation of the flowing afterglow technique by E.E. Ferguson, F.C. Fehsenfeld and A.L. Schmeltekopf in 1964 [2,3]. During the first few years, flowing afterglows were operated only at room temperature and the k were determined for a large number of different types of ion-molecule reactions (e.g. charge transfer, proton transfer, etc.). These studies showed that the large majority of exoergic binary (two-body) ion-molecule reactions proceeded rapidly at room temperature, i.e. $k \sim k_c$ (the collisional rate coefficient). For non-polar molecules, k_c is the classical collision (Langevin) rate coefficient [4]; for polar molecules, k_c is somewhat greater [5].

An important application of ion-molecule rate data is to aeronomy [6,7] and indeed this was a major motivation for the development of stationary and flowing afterglows. For this reason, and also because of fundamental interest, the need arose for data on the temperature variation of rate coefficients. The stationary afterglow study of the temperature dependence of k for the reaction,

$$O^+ + O_2 \longrightarrow O_2^+ + O \qquad (1)$$

which indicated that $k \sim T^{-\frac{1}{2}}$ in the thermal energy régime [8], was the prelude to similar flowing afterglow studies of several reactions [9,10], some being studied over the wide temperature range 80-900 K [11]. The general conclusion drawn from these early studies, which has been substantiated by subsequent studies, is that the k for rapid binary ion-molecule reactions (i.e. those for which $k \sim k_c$) are sensibly invariant with temperature in the thermal régime, whereas for slow reactions, k almost invariably changes with temperature [2] although the sense of the variation (increase or decrease) with temperature is not often predictable (except for isotope exchange reactions which are discussed later).

Following this pioneering afterglow work, the Selected Ion Flow Tube (SIFT) was developed [12]. In this case, a major motivation for the SIFT development was the desire to obtain data for use in interstellar ion-chemical models [13-15]. This technique possesses most of the features of the flowing afterglow but has the essential difference that the reaction region is not a plasma medium but rather a flowing gas supporting a relatively low density (swarm) of either positive ions or negative ions (as required). The ions are generated in an ion source which is isolated from the main flow tube by a quadrupole mass filter which facilitates the injection of mass selected ions into the carrier gas. The venturi effect is exploited to minimise the backflow of the carrier gas into the mass spectrometer chamber. The essential features of both the flowing afterglow and the SIFT are described in more detail in a review [16]. Another vital feature of the SIFT is that the source gas from which the primary ions are generated is not present in the flow tube (only in the remote ion source). Therefore studies of the reactions of primary (injected) ionic species with reactant gases which are introduced into the carrier gas are not complicated by the presence of the source gas. This feature is especially important in studies of isotope exchange in ion-molecule reactions (see below). The additional features offered by the SIFT have greatly extended the variety of reactions which can be studied at thermal energies and has allowed product ion distributions to be accurately determined. The recently developed variable-temperature SIFT (VT-SIFT), has allowed these reactions to be studied over the temperature range 80-600 K [16] and much of the data presented in this paper was obtained using the VT-SIFT. Several other important experiments, in which ion-molecule reactions can be studied at very low temperatures, have recently become productive. These include a static drift tube experiment which can be operated down to about 30 K [17], a supersonic jet experiment in which measurements can be made down to about 20 K [18] and an ion trap experiment in which measurements have been made below 10 K [19]. Some data obtained from each of these new experiments will be presented. It is pertinent to mention also the recently-developed, variable-temperature selected ion flow drift tube (VT-SIFDT) in which ion-molecule reactions can be studied at any fixed carrier gas temperature within the available range and over a range of ion-molecule centre-of-mass kinetic energies, KE_{cm}. With the VT-SIFDT, the influence of both true temperature and KE_{cm} on reaction rates and product distributions can be studied in the same experiment.

In this short paper, we discuss some of the most recent results concerning the temperature dependences of positive ion-molecule reactions, highlighting some of the more interesting and unusual results obtained at low temperatures. It should be noted that the techniques used in these positive ion studies can just as well be applied to the study of negative-ion molecule reactions (these are discussed elsewhere [20,21]). Thus we discuss a number of binary reactions which are appreciably exoergic, and some binary reactions which are nearly-thermoneutral and for which equilibrium constants can be determined by measuring forward and reverse rate coefficients. Included in the latter group is a number of isotope exchange reactions. The temperature dependences of the rate coefficients for a number of ternary (three-body) association reactions are also presented and compared with current theoretical descriptions of these reactions. To conclude some of the first data obtained using the VT-SIFDT apparatus will be briefly mentioned and discussed.

2. Binary Reactions

2.1. Exoergic Reactions

As previously mentioned, it is now recognised that the k for fast exoergic binary ion-molecule reactions do not vary appreciably with temperature, whereas the slower reactions have temperature dependent k. Some data illustrating this are shown in Fig. 1. Amongst the slow reactions studied, most exhibit a decreasing k with increasing temperature (see ref. 22 and Fig. 1). The slow $O^+ + O_2$ reaction (1) ($k = 2 \times 10^{-11}$ cm^3 s^{-1} at 300 K) has been studied in several experiments [8,11,23] and it is well established that $k(1) \sim T^{-\frac{1}{2}}$. Because the reaction is slow, it was initially considered to proceed via ion-atom interchange which necessarily requires an intimate encounter between the reactants. The $T^{-\frac{1}{2}}$ dependence was

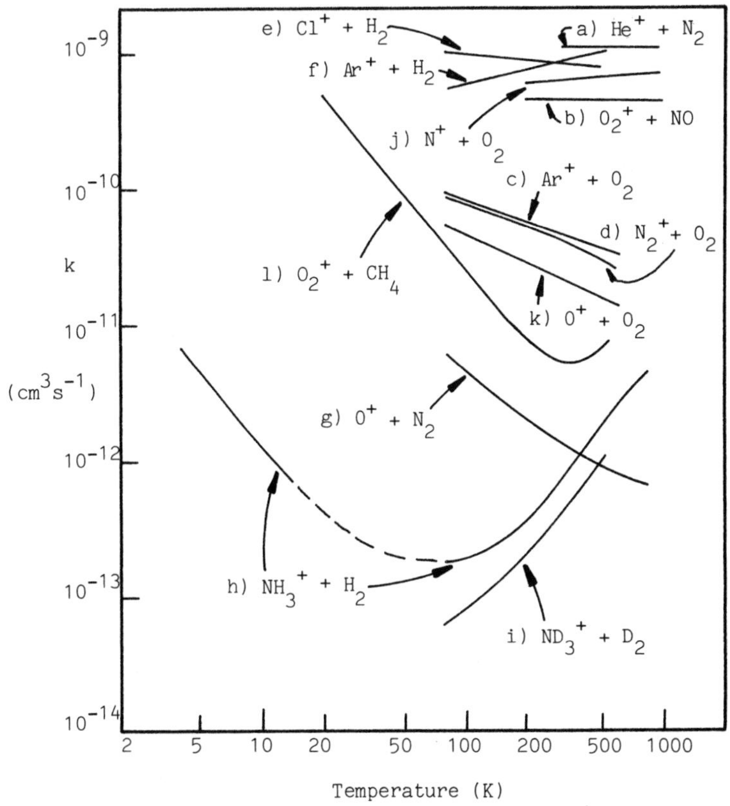

Fig. 1. The variation with temperature of the rate coefficients, k, for several types of exoergic binary ion-molecule reactions, including several charge transfer reactions (a,b,c,d), atom abstraction reactions (e,f,g,h,i) and two reactions (j,k) which proceed via more than one (parallel) mechanisms. Reactions (h) and (l) are discussed in detail in the text (including source references to the data), and source references to the other data are given in the review by Adams and Smith [26] except for reaction (i) which is unpublished VT-SIFT data. Note the deep minima in the k for reactions (h) and (l). Minima in the k also occur for several of the other reactions at elevated ion kinetic energies (see the review by Lindinger and Smith [33]).

correlated with a $(\text{velocity})^{-1}$ dependence which in turn was correlated with the lifetime of the $(O_3^+)^*$ intermediate complex. However isotope labelling experiments [24,25] have shown that ion-atom interchange is only a minor channel and that charge transfer dominates. It is probable therefore that charge transfer occurs within the $(O_3^+)^*$ complex and not, as is often supposed, via a long range electron transfer. The idea that ion-molecule reactions usually proceed via relatively long-lived complexes at thermal energies, even when charge transfer is a major channel, is gaining ascendency as a result of recent work. This idea will be invoked to explain many of the results discussed in this paper. It is interesting that the magnitudes of k for the slow charge transfer reactions $Ar^+ + O_2$ and $N_2^+ + O_2$ are essentially identical and vary as $T^{-\frac{1}{2}}$ (see Fig. 1) and this is perhaps indicative that these reactions, like reaction (1), also proceed via a close encounter and not via long-range electron transfer. Support for this premise is gained from the fact that the Franck-Condon factors are very unfavourable for electron transfer from O_2 into the states of the O_2^+ product ion which are accessible in these reactions, thus greatly inhibiting long-range electron transfer. Complex formation is not, of course, restricted to those reactions which are slow. Included in Fig. 1 are the data for two fast atom abstraction reactions, i.e. $Cl^+ + H_2$ and $Ar^+ + H_2$, in which ion-molecule complexes must be formed. Interestingly, for the former reaction k increases towards k_c with decreasing temperature whereas for the latter reaction k increases towards k_c with increasing temperature! So prediction of the sense of the temperature dependence of k is very hazardous, even for ostensibly simple reactions involving only three atom complexes.

An especially interesting reaction which is thought to be an important step in the production NH_3 in interstellar clouds [13-15] is the atom abstraction reaction

$$NH_3^+ + H_2 \longrightarrow NH_4^+ + H \qquad (2)$$

This reaction is exoergic by about 0.5 eV, yet at room temperature it is very slow ($k(2) \sim 10^{-4} k_c$). Flowing afterglow (and flow drift tube) studies [28] have shown that k(2) increases rapidly with temperature above 300 K (and with increasing KE_{cm}) suggesting that an activation energy barrier inhibits the reaction. This tempted modellers of interstellar ion chemistry to assume that the form of the temperature dependence would be the same below 300 K and that k(2) would therefore be extremely small under the very low temperature conditions of interstellar dense clouds [14]. Thus they expected that it could not be involved in NH_3 synthesis.

However, recent SIFT studies [29] have shown that k(2) does not reduce below 300 K at the rapid rate expected on the basis of the higher temperature data, and indeed it appeared from the SIFT data that k(2) might actually have reached a minimum at 80 K. Subsequent ion trap experiments conducted at very low temperatures [30] have shown that k(2) does indeed increase again dramatically below 80 K as is illustrated in Fig. 1. This suggests that the reaction is mechanistically very different at the high and very low temperatures, perhaps indicating the formation of a long-lived complex at the very low temperatures in which there is sufficient time for the entire reaction surface to be explored, locating regions without activation energy barriers. Somewhat surprisingly, the situation is very different for the fully deuterated analogue of reaction (2) (i.e. $ND_3^+ + D_2$) in that there is no evidence at 80 K for a minimum in k.

A reaction which exhibits rather similar temperature behaviour to reaction (2) is:

$$O_2^+ + CH_4 \longrightarrow CH_3O_2^+ + H \qquad (3)$$

At room temperature k(3) is very small ($\sim 5\times10^{-12}$ cm^3s^{-1}; [31,32]) and flow-drift-tube data suggested that k(3) might be near a minimum value at 300 K [32]. So, following the very recent development of their supersonic jet experiment (CRESU) by Rowe and co-workers (with which reactions can be studied at temperatures as low as 20 K), a co-ordinated CRESU and SIFT study of reaction (3) was carried out, k(3) being measured over the uniquely wide temperature range 20-560 K [27]. The data obtained are represented in Fig. 1 where it can be seen that k(3) is indeed at a minimum around room temperature and increases dramatically to approach to within about 50% of k_c at 20 K! Presumably at some lower temperature, k(3) will become equal to k_c. In the overlapping temperature range of the CRESU and SIFT experiments the data are in excellent agreement. The low temperature data can be fitted to a power law of the form $k(3) = 1.1\times10^{-7}T^{-1.8}$. This, by analogy with the power law behaviour of ternary association reactions (see Section 3), strongly indicates that the $CH_3O_2^+$ product is formed via rearrangement in a long-lived intermediate $(CH_4O_2)^+$ complex, the lifetime of which largely controls the rate of the reaction. It has been suggested [27] that this rearrangement proceeds via H^- abstraction within the complex forming an electrostatically bound $(CH_3^+.HO_2)$ ion which can then undergo concerted molecular rearrangement leading to $CH_3O_2^+$ and a separate H atom. The reaction of O_2^+ with CH_4 has also been studied in a flow drift tube experiment [32] and it is observed that, when KE_{cm} exceeds ~ 0.5 eV, then the endoergic CH_3^+ and CH_4^+ product channels become dominant. This emphasises the important point that, when k versus T plots suggest different reaction mechanisms at different temperatures, vital information concerning the respective mechanisms can be obtained by carefully monitoring the reaction products at each temperature. Such has provided a much better insight into the mechanism of many ion-molecule reactions studied in drift tube experiments [33]. Further reference to reaction (3) is made in Section 4.

It is clear from the above (and the data given in Fig. 1) that it is the slow reactions which exhibit the interesting temperature effects and it is therefore most probable that further studies of such reactions will provide better insight into reaction mechanisms. A list of reactions which are slow at room temperature is given in our recent review [26]. A nice discussion of the temperature dependences of binary ion-molecule reactions has been given recently by Magnera and Kebarle [34].

2.2. Thermoneutral Reactions: Isotope Exchange

The study of near-thermoneutral ion-molecule reactions has traditionally been pursued by determining equilibrium constants, K, in high pressure ion sources [35,36] and in flowing afterglows [37]. Clearly, this can only be accomplished at room temperature when the standard free energy change in the reaction, $\Delta G°$, is small (since $\Delta G° = -RT\ln K$). By this approach, the proton affinity differences for many atoms and molecules have been determined by observing equilibria in binary reactions of the type:

$$XH^+ + Y \underset{k_r}{\overset{k_f}{\rightleftharpoons}} X + YH^+ \qquad (4)$$

Ideally, K and hence $\Delta G°$ is determined as a function of temperature since then both the enthalpy and entropy changes, $\Delta H°$ and $\Delta S°$, can be determined. Often, however, only room temperature equilibrium data are available and then to determine $\Delta H°$ (and proton affinity differences, for example) the $\Delta S°$ for the reaction has to be calculated or approximated [37]. With the advent of the variable temperature SIFT technique [16,26], it has become possible to determine the forward and reverse rate coefficients, k_f and k_r, for reactions such as (4) as a function of temperature. Hence equilibrium constants, $K = k_f/k_r$, can be determined from the ratio of rate coefficients, which thus avoids inaccuracies in K due to diffusion effects in flow tubes and to mass discrimination effects in mass spectrometers. This approach has been especially valuable in the study of isotope exchange in ion-molecule reactions, especially in those reactions in which multiple isotope exchange can occur since, in these cases, it is important to determine branching ratios into the various exchange channels. The determination of $\Delta H°$ and $\Delta S°$ for such reactions from kinetics studies provides insights into the factors which control the course of the reactions e.g. zero-point-energy differences and relative bond strengths, complex lifetimes, statistical factors, etc.

Before discussing isotope exchange reactions, we will consider some recent data on the near-thermoneutral proton transfer reaction:

$$O_2H^+ + H_2 \underset{k_r}{\overset{k_f}{\rightleftharpoons}} H_3^+ + O_2 \qquad (5)$$

There have been several studies at room temperature of k_f and k_r for this reaction (see the summary in ref. 37) and all indicate that at ~300 K, $k(H_3^+ + O_2) > k(O_2H^+ + H_2)$. However, only $\Delta G°_{300}$ could be obtained from the rate coefficient ratio and to determine $\Delta H°_{300}$ (and hence the proton affinity difference $PA(H_2) - PA(O_2)$) a value of $\Delta S°_{300}$ had to be estimated. This has led to some doubt as to whether $PA(H_2)$ or $PA(O_2)$ was the greater. This has recently been positively resolved by determining $k_f(5)$ and $k_r(5)$ at both 300 K and 80 K in a VT-SIFT [38], whence it was found at 80 K, at which temperature $\Delta H°$ is the dominant component of $\Delta G°$ and thus is the controlling influence on K, that $k(O_2H^+ + H_2) > k(H_3^+ + O_2)$. Hence our designation of $k(O_2H^+ + H_2)$ as k_f and $k(H_3^+ + O_2)$ as k_r in reaction (5), i.e., k_f refers to the direction appropriate to a negative $\Delta H°$, a convention we have adopted for all of the isotope exchange reactions discussed below. From this study, it was established that $PA(H_2)$ is greater than $PA(O_2)$ by 0.33 ± 0.04 kcals mole^{-1}. Similarly, for the analogous reaction:

$$O_2D^+ + D_2 \rightleftharpoons D_3^+ + O_2 \qquad (6)$$

The deuteron affinity of D_2 was found to be greater than that of O_2 by 0.35 ± 0.04 kcals mole^{-1}. It is interesting to note that the $\Delta H°$ is so small for reaction (5) that the small corrections, which had to be made to account for the non-equilibrium ratio of ortho-H_2 to para-H_2 used in the SIFT experiment at 80 K and for the non-Boltzmann distribution amongst the rotational states of H_3^+ and H_2 at 80 K, are significant. These corrections have previously been discussed in detail in relation to reaction (11) below. It is also pertinent to note that the $\Delta S°$ for both reactions (5) and (6) closely approximate to $R\ln 1/3$ indicating that the major factor which determines $\Delta S°$ for these reactions is the statistics of separation of the reactants and products from the $(H_3O_2)^+$ and $(D_3O_2)^+$ intermediate complexes. This is also a feature of isotope exchange in ion molecule reactions.

Implicit in the above discussion is the fact that it is necessary to study reactions down to temperatures for which $T \lesssim \Delta H°/R$ if both $\Delta H°$ and $\Delta S°$ are to be obtained. For most isotope exchange reactions, $\Delta H°$ is small and so it is essential to make measurements below room temperature. This we have done during the last few years using the VT-SIFT [38-46]. These studies were primarily motivated by the need for experimental data on the phenomenon of 'isotope fractionation' which had been suggested as a possible process by which the observed enrichment occurs of rare (heavy) isotopes of several elements in interstellar molecules [47,48]. Isotope fractionation was proposed to occur in the reactions:

$$^{13}C^{+} + {}^{12}CO \underset{k_r}{\overset{k_f}{\rightleftharpoons}} {}^{12}C^{+} + {}^{13}CO \qquad (7)$$

essentially because the vibrational zero-point-energy (zpe) of ^{12}CO is greater than that of ^{13}CO. This means that the reaction is endoergic to the left and so, at sufficiently low temperature, $k_r(7)$ should become smaller than $k_f(7)$. VT-SIFT studies showed that $k_f(7)$ and $k_r(7)$ are not significantly different at 500 K, but at lower temperatures $k_f(7)$ became increasingly larger than $k_r(7)$ [39]. A van't Hoff plot of the data (see Fig. 2) provides a value for $\Delta H°_T$ of $-(3.5 \pm 0.5)$ meV which is closely equal to zpe (^{12}CO) - zpe (^{13}CO) as calculated from known vibrational constants [49]. Note that the van't Hoff plot for this reaction also indicates that $\Delta S°_T$ is zero within error, as might be expected on the basis of the equal probabilities of reactants and products separating from the $(^{12}C^{13}CO)^{+}$ complexes. It is also worthy of note that both $k_f(7)$ and $k_r(7)$ increase with decreasing temperature down to 80 K, this being a manifestation of the increasing lifetime of the $(^{12}C^{13}CO)^{+}$ complexes which increases the efficiency of the reaction in both directions. At lower temperatures it is to be expected that $k_f(7)$ will continue to increase towards k_c (the collisional rate coefficient) but that $k_r(7)$ will begin to reduce towards zero. This has been shown to be the case for several H/D exchange reactions (discussed below), for which reducing k_r are readily observable at the lowest temperature (80 K), at which the VT-SIFT experiment can currently be operated, because $\Delta H°$ is usually greater for H/D exchange than for $^{12}C/^{13}C$ exchange. An important empirical rule for ion-molecule isotope exchange reactions, which has been recognized as a result of these studies, is that $(k_f + k_r) \rightarrow k_c$ at low temperatures (i.e. when $T \lesssim \Delta H°/R$). This greatly facilitates the estimation of k_f and k_r for low temperatures and as such is extremely valuable in assessing the degree of isotope fractionation in interstellar clouds [15].

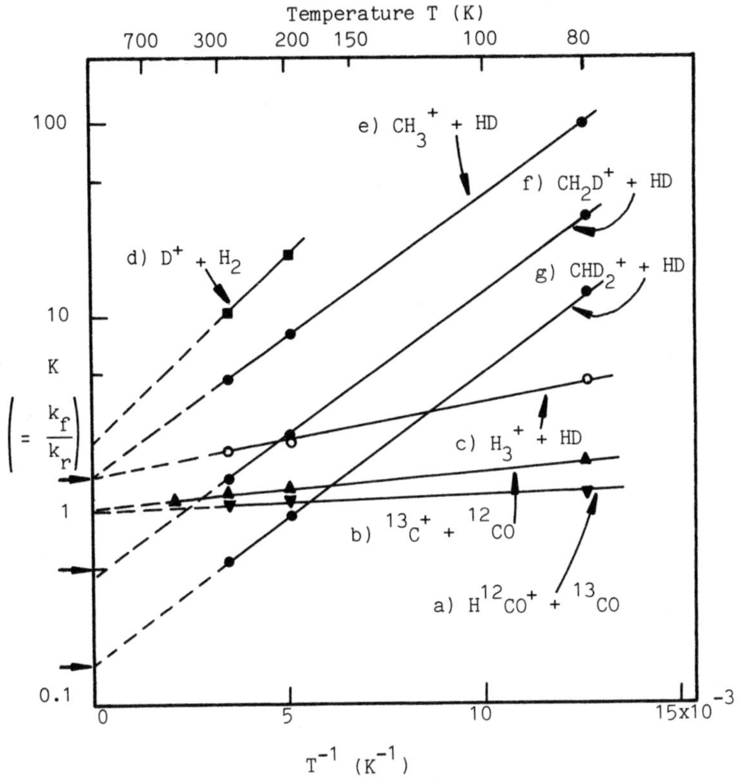

Fig. 2. Van't Hoff plots ($\ln k_f/k_r$ versus T^{-1}) for several isotope exchange reactions. The kinetic data were obtained using a VT-SIFT. Source references are: reactions (a) and (b) [39], (c) [42,45], (d) [40] and (e), (f) and (g) [44]. The ΔH° values derived from the slopes of the lines range from -1 meV for reaction (a) to -50 meV for reaction (d). Note that the ΔH° for reactions (e), (f) and (g) are identical (this has implications regarding bond energies, see text). The ΔS° values, obtained from the intercepts on the K axis, are zero for reactions (a) and (b) (since $\Delta S° = R\ln k_f/k_r$ at high temperature at which for these reactions $k_f/k_r = 1$) and finite for the other reactions. The derived ΔS° for reactions (e), (f) and (g) are close to those predicted based only on the statistical decomposition of the intermediate complexes back to reactants and forward to products (the actual statistical values of k_f/k_r at high temperatures are indicated by the arrows).

The 'isotope exchange' reactions (which are actually proton transfer reactions)

$$H^{12}CO^+ + {}^{13}CO \underset{k_r}{\overset{k_f}{\rightleftharpoons}} H^{13}CO^+ + {}^{12}CO \qquad (8)$$

$$H^{14}N^{14}N^+ + {}^{14}N^{15}N \underset{k_r}{\overset{k_f}{\rightleftharpoons}} H^{14}N^{15}N^+ + {}^{14}N^{14}N \qquad (9)$$

have been shown by VT-SIFT measurements [39,41] to have very small $\Delta H°$ values ($\sim$-0.001 eV, manifest by the very small slope of the van't Hoff plot in Fig. 2), and so even at 80 K, the respective k_f are only a little greater than the k_r. These reactions also considered to be important in low temperature interstellar clouds and will fractionate ^{13}C and ^{15}N into HCO^+ and N_2H^+ respectively [39,41].

For the reaction:

$$D^+ + H_2 \underset{k_r}{\overset{k_f}{\rightleftharpoons}} H^+ + HD \qquad (10)$$

$\Delta H°$ is relatively large ($\sim$-0.04 eV), and so even at room temperature, $k_r(10)$ is about an order-of-magnitude smaller than $k_f(10)$. Of course at lower temperatures, $k_r(10) \ll k_f(10)$, and $k_f(10)$ is expected to approach $k_c(10)$ as is actually observed at temperatures as high as 200 K [40]. Reaction (10) is considered to be the major production process for interstellar HD and since $k_f(10) \gg k_r(10)$ and the concentration ratio $[H_2]/[HD] \gg 1$, then most deuterium in cold interstellar clouds will be in HD [14]. A detailed study of reaction (10) and the analogous $D^+ + HD \rightleftharpoons H^+ + D_2$ reaction has been carried out [40] and it is shown that $\Delta S°$ for these reactions are in approximate accord with expectations based on simple statistics and with more detailed calculations based on partition functions.

Another H/D exchange reaction which has been given a good deal of attention because of its interstellar importance [42,45] is:

$$H_3{}^+ + HD \underset{k_r}{\overset{k_f}{\rightleftharpoons}} H_2D^+ + H_2 \qquad (11)$$

The experimental results for $k_f(11)$ and $k_r(11)$ from a VT-SIFT experiment [42] were as expected in the sense that $k_f(11)$ increased and $k_r(11)$ decreased with decreasing temperature, and a finite entropy change was obtained ($\Delta S° \sim R\ln^3/2$ in accordance with statistical predictions; see Fig. 2). The relatively small value of $\Delta H°$ for this reaction revealed the significant error involved in its determination at low temperatures (< 200 K) when the reactant H_2 has a non-equilibrium ortho-H_2 to para-H_2

ratio, as was the case in the VT-SIFT experiment. Ideally, the kinetics of reactions involving H_2 as a reactant gas should be conducted with equilibriated H_2, especially at very low temperatures. The relatively large spacing between the rotational states of H_2 (and HD) and the so-called "rotational zero-point-energy" of H_3^+ [50] combine to render the $\Delta H°$ for reaction (11) very temperature dependent below about 200 K [45,51]. Experimental evidence for this was obtained from the VT-SIFT study. A detailed study of the reactions of H_3^+, H_2D^+, HD_2^+ and D_3^+ with H_2, HD and D_2 has also been carried out [52].

A detailed study has also been made in the VT-SIFT of the reactions of CH_3^+, CH_2D^+, CHD_2^+ and CD_3^+ with H_2, HD and D_2 [44]. In this series, both binary isotope exchange and ternary association products were observed in some of the reactions. In the exoergic reaction of CH_3^+ with HD the only observed product ion was CH_2D^+. However, for the reaction of CH_2D^+ with H_2 both CH_3^+ and CH_4D^+ were product ions, and thus two fundamentally different reactions occur in parallel i.e. endoergic isotope exchange:

$$CH_2D^+ + H_2 \longrightarrow CH_3^+ + HD \tag{12}$$

and ternary association:

$$CH_2D^+ + H_2 + He \longrightarrow CH_4D^+ + He \tag{13}$$

The endoergic reaction (12) is inhibited and the ternary reaction (13) is enhanced by low temperatures (ternary association is discussed in the next section). However, it is clear from the considerable amount of data on these reactions that the parallel channels (12) and (13) can be viewed as independent reactions and their separate rate coefficients obtained from the measured total rate coefficient and the product branching ratio [44]. Thus the k_f and k_r have been determined for several of the isotope exchange reactions in this $CH_3^+ + H_2$ series and $\Delta H°$ and $\Delta S°$ determined in each case. Van't Hoff plots for three of these reactions are given in Fig. 2. That these plots are parallel indicates that the $\Delta H°$ values are identical for the three reactions and therefore that the difference in the zpe of a C-H and a C-D bond in CH_3^+-like ions, and therefore the difference in the bond strengths, is independent of the degree of deuteration of the CH_3^+ ion. From these kinetic data, an accurate value for the difference in the strength of the C-H and C-D bond in CH_3^+-like ions of 1.56 kcal $mole^{-1}$ has been determined [44]. It is also very clear from the van't Hoff plots for these reactions, that the $\Delta S°$ are very different, but again they are in close accord with expectations based on simple statistics e.g., for reaction (12), $\Delta S° = -R\ln^3/2$.

These isotope exchange studies have clearly demonstrated how fundamental thermodynamic data can be obtained from kinetic studies of isotope exchange reactions and how the lifetimes of the intermediate ion-molecule complexes influence the magnitudes of k_f and k_r relative to k_c. These principles are very clearly operating in the studies of H/D exchange in the reactions of $H_3O^+ + D_2O$, $NH_4^+ + ND_3$ and $CH_5^+ + CD_4$ and their respective 'mirror' reactions (e.g. $D_3O^+ + H_2O$) for which multiple products are possible [43]. The core ion-ligand bond strengths in the complexes formed in these reactions are in the order $H_3O^+.H_2O > NH_4^+.NH_3 > CH_5^+.CH_4$, and therefore τ_d is expected to be greatest for the $H_3O^+.H_2O$ intermediate complexes and least for the $CH_5^+.CH_4$ complexes, an expectation confirmed by ternary association studies [43]. Consistent with this, it is observed experimentally that H/D exchange is facile in the $H_3O^+ + D_2O$ reaction at 300 K and total scrambling of H and D occurs in the intermediate complex, whereas the short τ_d for the ($CH_5^+.CD_4$) complex at 300 K allows for hardly any scrambling of H and D, as is clear from the ion product distribution for the reaction. However, at 80 K, when the τ_d for this reaction is much greater then scrambling can occur, but at this temperature enthalpic effects dominate the product ion distribution for the $CH_5^+ + CD_4$ reaction (and its mirror reaction). From these kinetic studies, fundamental thermodynamic data have been obtained including an estimate of the zpe of CH_5^+ [53]. Just as interesting is the insight which has been obtained into the nature of intermediate ion-molecule complexes formed in these interactions and the motion ("shuttling") of atoms within the complexes. Finally, it is worthy of note that for many of the reactions studied in SIFT experiments, isotope exchange is extremely slow (or immeasurable) and much can be learned about the nature of the ion-molecule interactions from these 'negative' results. A detailed discussion of this and of other isotope exchange reactions is given in a recent review [46].

3. Ternary Association Reactions

It is well established experimentally that the rate coefficients, k, for ternary (3-body) ion-molecule association reactions increase with decreasing temperature, this being consistent with the earliest theoretical predictions. A large number of reactions have been studied, ranging from those involving only three-atom complexes (e.g. $Ar^+ + 2Ar \rightarrow Ar_2^+ + Ar$, $C^+ + H_2 + He \rightarrow CH_2^+ + He$, $O_2^+ + Kr + He \rightarrow O_2^+.Kr + He$) to those involving polyatomic complexes (e.g. $CH_3^+ + H_2O + He \rightarrow CH_3^+.H_2O + He$, $CH_3NH_3^+ + CH_3NH_2 + CH_4 \rightarrow (CH_3NH_2)_2H^+ + CH_4$). Detailed lists are given in several reviews [22,26,54]. The magnitudes of k for different reactions at a given temperature vary over many orders-of-magnitude, the k for reactions involving weakly-bonded complexes generally being small and vice-versa for reactions in which strongly-bonded complexes are formed. This is in general accordance with RRK theory which predicts an inverse temperature dependence for k (i.e. $k \sim T^{-n}$). However, the experimentally obtained n values are generally much smaller than RRK predictions. Lists of k values and their temperature dependences, and an appraisal of RRK ideas as applied to ion-molecule association are given in the review by Meot-Ner [22].

Recent improvements in experimental techniques, notably the advent of ion injection into variable-temperature flow tubes and drift tubes [16,33], have provided opportunities to study more accurately temperature dependences of ternary reactions [55,56]. These developments have proceeded in parallel with developments in the theory of ion-molecule association [57-61] and so more meaningful comparisons between experiment and theory can now be made.

Recent theories due to Bates [57,60] and Herbst [58,59] predict that $k \sim T^{-(\frac{\ell}{2}+\delta)}$, where ℓ is equal to the total number of rotational degrees of freedom in the separated ion-neutral reactants and δ is a parameter which accounts for the temperature variation of the collisional stabilization efficiency factor, f, in excited intermediate complex/third body collisions. Thus, for example, the k for a reaction between a polyatomic ion and a diatomic molecule (i.e. for ℓ = 5), is predicted to vary as $T^{-(5/2+\delta)}$. A more sophisticated treatment of ion-molecule association has been given by Bass et al. [61,62].

Recent data for ternary association reactions of CH_3^+ ions with several diatomic molecules including H_2, N_2, O_2 and CO [55] are in close accord with this prediction; mean values of n over the approximate temperature range 80-600 K have been obtained, varying from 2.3 (for H_2 and CO) to 2.8 (for O_2). These data suggest that δ is relatively small (ranging from -0.2 to +0.3) and thus by implication that f is only a weakly varying function of temperature for these reactions. A summary of the most recent data is given in a recent review [26].

Of course, it is entirely possible that δ is also a slowly varying function of temperature and that k is not described precisely by a simple T^{-n} law. Unfortunately, experimental data are not usually sufficiently precise to determine whether this is so or not and therefore a single power law is generally assumed and a mean value of n obtained over the explored temperature range (as indicated above for the CH_3^+ reactions). However, recent data relating to the much-studied association of N_2^+ with N_2

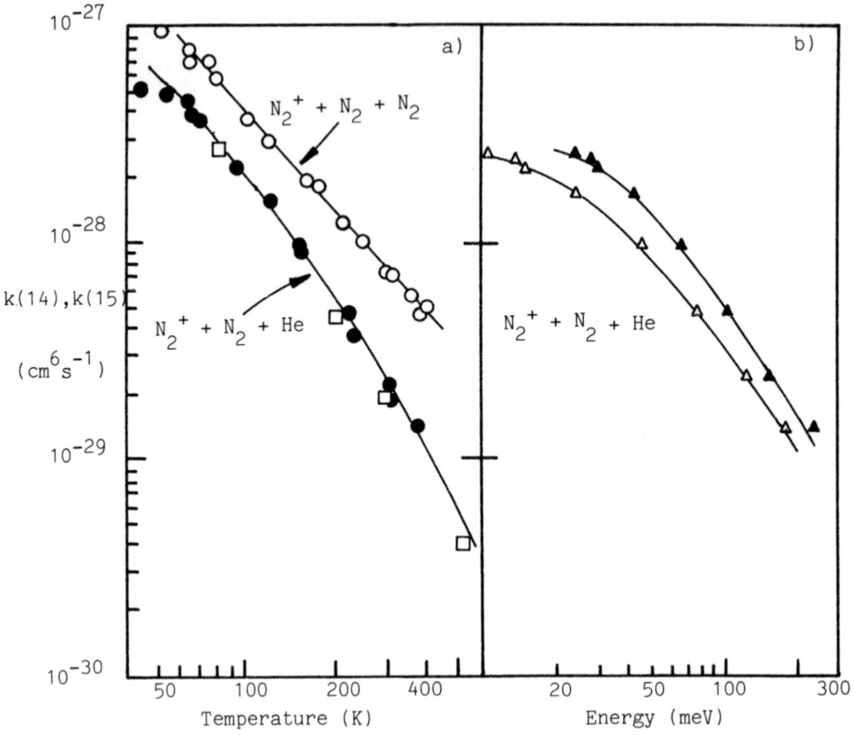

Fig. 3. a) Log-log plots of the ternary rate coefficients, k, versus temperature for the helium and nitrogen stabilised association reactions indicated. Source references: □ are SIFT data [55], ● and ○ are drift tube data [56]. The differences in the magnitudes of k for the He and N_2 stabilised are ascribed to different collisional stabilisation efficiency factors, f, for the reactions. The curvature of the lower plot and its convergence with the upper plot at low temperatures is considered to be due to an increasing f factor with decreasing temperature for the He stabilisation process.

b) Log-log plots of k versus KE_{cm} (Δ) and E_T (▲) for the He stabilised reaction indicated. The data were obtained in the VT-SIFDT at a fixed gas temperature of 80 K [63]. E_T is the total energy of the N_2^+ and N_2 reactants which includes KE_{cm}, the rotational energy of N_2 at 80 K and the rotational energy of the N_2^+ attained in collisions with the He carrier gas (assuming translational/rotational equilibrium). For a discussion of these data see text and ref. 63.

indicate that δ may indeed be a function of temperature for the helium-stabilised reaction:

$$N_2^+ + N_2 + He \rightarrow N_4^+ + He \quad (14)$$

Data obtained for this reaction from a co-ordinated SIFT and static drift tube study [56] are given in Fig. 3 together with data for the corresponding N_2-stabilised reaction:

$$N_2^+ + N_2 + N_2 \rightarrow N_4^+ + N_2 \quad (15)$$

Previously [56], these data have been interpreted simply in terms of a $T^{-2.3\pm0.2}$ dependence for k(14) and a $T^{-1.7\pm0.2}$ dependence for k(15) and it has been suggested that the greater magnitude of k(15), at all temperatures within the available range, was simply due to a larger f factor for the N_2-stabilised reaction. Further, it was suggested [26] that stabilisation of the excited $(N_4^+)^*$ complexes occurred via the switching of N_2 molecules in reaction (15) and via superelastic collisions of helium atoms in reaction (14). This suggestion has been put to the test in a recent VT-SIFDT experiment [63] in which N_4^+ ions, initially in the ground state, were internally excited in collisions with helium atoms in a VT-SIFDT and their reaction studied with isotopically-labelled N_2. The switching reaction:

$$^{14}N_2\,^{14}N_2^+ + {}^{15}N_2 \rightarrow {}^{14}N_2\,^{15}N_2^+ + {}^{14}N_2 \quad (16)$$

became very efficient ($k(16) \rightarrow k_c$) as the $^{14}N_2\,^{14}N_2^+$ was excited towards the dissociation limit. From this observation, it was concluded that the highly excited N_4^+ formed via N_2^+/N_2 association will rapidly switch with N_2, i.e. that f=1 for the stabilisation, and that this switching is indeed the stabilisation mechanism involved in reaction (15). Since reactions which occur at the collisional rate (i.e. when $k = k_c$) are invariably insensitive to temperature, then δ will be near zero and therefore departures from power law dependence for k(15) would not be expected and the power law dependence would be entirely due to the temperature dependence of the lifetime of the complex, τ_d. Indeed the ln k vs ln T data for reaction (15) is a good approximation to a straight line (see Fig. 3). However, since n = 1.7±0.2 (i.e. somewhat less than $\ell/2 = 2$) for this reaction, one has to conclude that all of the rotational modes available for activation in the intermediate N_4^+ complex may not be fully active in prolonging τ_d.

The above reasoning suggests that the f factor for the helium-stabilised reaction (14) is temperature dependent and that the corresponding δ value appropriate to the temperature range explored lies between 0.2 and 1.0 (obtained by considering the limits of n for reactions (14) and (15)).

Actually, it can be argued that for reaction (14), δ itself should be temperature dependent and will approach zero at low temperatures when f approaches unity (at 80 K, f$\sim$0.7 for reaction (14) assuming, as before, that f$\sim$1 for reaction (15); see Fig. 3). If δ is temperature dependent for helium stabilisation then curvature of the log-log plot should result, and some curvature is indeed discernible in comparison with the plot for corresponding nitrogen-stabilised reaction.

So the most reliable data on ternary association reactions involving simple (non-cluster) ions indicate that the index n of the temperature dependence approximates to the appropriate $(\frac{\ell}{2}+\delta)$ predicted by the Bates/Herbst theory and that δ is small compared to $\ell/2$. However, some experimental evidence is available which indicates that n is appreciably greater than $\ell/2$ for association reactions involving cluster negative ions (Viggiano and Paulson, priv. comm. and see ref. 21). This is tentatively attributed to the excitation of "floppy" core ion/ligand bonds in the cluster ions during the association phase of the interaction, in addition to excitation of the normal rotational modes. Indeed, some evidence has been obtained which indicates that n is related to the degree of clustering (i.e. the number of ligands) in the reactant cluster negative ions [21].

Obviously, it is important in determining ternary association rate coefficients to ensure that ternary kinetics are actually obeyed at the bath gas pressures available to the various experiments i.e. to ensure that the experiment is being performed in the low pressure régime [64]. The large majority of ternary association reactions exhibit pure ternary behaviour for normal working pressures ($\sim$0.2 to 1.0 Torr) of flowing afterglow and SIFT experiments at room temperature and above. However, the rapid increase in the rates of these reactions with decreasing temperature can result in departures from ternary kinetics even for the lowest pressures available to the experiments, i.e. the effective binary rate coefficient, k_{eff}, is no longer directly proportional to pressure (or bath gas number density) and the ternary reaction appears to have some binary character. In the limit, k_{eff} becomes independent of bath gas pressure and is then usually an appreciable fraction of the k_c for the reactant ion-reactant molecule collisions. This occurs at 80 K for the relatively fast reaction:

$$CH_3^+ + CO + He \longrightarrow CH_3^+.CO + He \qquad (17)$$

under typical SIFT pressure conditions [65], and $k_{eff}(17)$ is close to $k_c \sim 1 \times 10^{-9}\ cm^3s^{-1}$, indicating that the stabilisation efficiency factor, f, for the reaction is close to unity. Under these circumstances, the reactions is said, loosely, to exhibit 'saturated binary kinetics' and then it is not possible to determine the ternary rate coefficient to the usual accuracy, although an approximate value can be obtained using appropriate techniques to analyse the k_{eff} versus pressure data [64]. 'Saturation' has been observed in room temperature SIFT studies [66] of the reaction:

$$CH_3^+ + HCN + He \longrightarrow CH_3^+.HCN + He \quad (18)$$

At around 0.5 Torr, the k_{eff} was measured to be 2×10^{-9} cm^3s^{-1} with no discernible pressure variation (the ADO value [5] for k_c for the primary interaction is 3.4×10^{-9} cm^3s^{-1}). Recently, we have studied this reaction in the VT-SIFT at 300 K and 580 K. At 300 K, k_{eff} is very close to the value given above, but a small but discernible increase in k_{eff} with increasing pressure occurs and this allowed us to estimate the ternary rate coefficient for (18) to be $\sim 5 \times 10^{-25}$ cm^6s^{-1}, an enormous value! At 580 K, normal ternary kinetics are obeyed in the VT-SIFT experiment (i.e. the 'low pressure' régime is accessible and k_{eff} is proportional to helium pressure) and a ternary rate coefficient of 3×10^{-26} cm^6s^{-1} is obtained, still a very large value which indicates a relatively large τ_d for the $(CH_3HCN)^+$ complex even at this high temperature. A detailed study of the CH_3^+ + HCN reaction has been carried out at low pressures using ICR techniques [64,67], and binary association has been observed (presumably a manifestation of radiative association). A detailed appraisal of the kinetics of this reaction has also been carried out [64]. Further brief reference to reaction (18) is given below in the VT-SIFDT section.

An association channel is sometimes observed in parallel with rapid binary fragmentation channels in 'high pressure' SIFT experiments, even when the binary channels are appreciably exoergic. A good example is the reaction of CH_3^+ with NH_3 at 300 K:

$$CH_3^+ + NH_3 \longrightarrow CH_2NH_2^+(70\%),\ NH_4^+(10\%),\ CH_3NH_3^+(20\%) \quad (19)$$

Several similar reactions have been identified [68], but a detailed study of the rate coefficients and product distributions for such reactions as a function of temperature and pressure has yet to be carried out. Such studies would surely be very profitable. As was mentioned in Section 2.2, a ternary association channel is observed in some of the reactions with deuterated reactants in the $CH_3^+ + H_2$ series, but only when D/H exchange is endoergic. Indeed, ternary association is enhanced by endoergic D/H exchange within the CH_5^+-like complexes. We have termed this interesting phenomenon the "isotope refrigerator effect" [44].

The reaction of C_3H^+ with H_2 is also interesting. At 80 K the only observed product of the reaction is the association ion $C_3H_3^+$, and saturated binary kinetics are observed over the pressure range available in the VT-SIFT. At 300 K, a small (~15%) binary product $C_3H_2^+$ is observed in addition to the major $C_3H_3^+$ product. At 550 K, the $C_3H_2^+$ has increased in importance (~60%) and the association channel has become unsaturated. We now understand this unusual behaviour to be due to the fact that the reaction

$$C_3H^+ + H_2 \longrightarrow C_3H_2^+ + H \qquad (20)$$

is only slightly endoergic and therefore its rate increases with increasing temperature. Conversely the rate of the ternary reaction channel (producing $C_3H_3^+$) decreases, as expected, with increasing temperature. This study of reaction (20) has allowed us to define more closely the heat of formation of $C_3H_2^+$. We have also considered the significance of this $C_3H^+ + H_2$ reaction to interstellar chemistry [70].

4. Initial Studies using the VT-SIFDT

Data relating to both binary and ternary reactions are now being obtained using this apparatus. A major objective is to gain an understanding of the influence of energy partition in the reactants on the course of the various types of ion-molecule reactions. Only a brief report is pertinent now because, although a good deal of data have been obtained at various gas temperatures and over wide ranges of ion-molecule centre-of-mass kinetic energies, KE_{cm}, sufficient to produce large changes in the rate coefficients for some reactions, detailed analysis of much of the data is incomplete.

Our further studies [71] of the binary $O_2^+ + CH_4$ reaction (3) (discussed also in Section 2.1) well illustrate the potential of the VT-SIFDT. The rate coefficient, k, for the reaction has been determined as a function of KE_{cm} at several fixed gas temperatures, T. The data obtained mirror the truly thermal data (see Fig. 1). An increasing KE_{cm} results in a decrease in k for T = 80 K but an increase in k for T = 420 K. At an intermediate T = 200 K, increasing KE_{cm} first results in a decrease in k to a minimum value followed by an increase. These data demonstrate the qualitative equivalence of KE_{cm} and thermal energy (temperature), and preliminary analysis of the data indicates that (using the k as an indicator) there is within error a fixed constant of proportionality between the increase in KE_{cm} (or strictly speaking the increase in total energy) and the increase in T which are required to produce the same change in k, independent of the fixed T at which the KE_{cm} variation of k was obtained.

These data provide an insight into the number of intramolecular modes in the intermediate $(CH_4O_2)^+$ complexes (actually about 8) which are involved in temporary energy storage and which therefore control the lifetime of the complexes against decomposition to the $CH_3O_2^+$ + H products [71]. This is just the kind of insight which it was envisaged would be obtained by exploiting this exciting new experiment.

The VT-SIFDT has important applications to studies of the onset of endoergic binary reactions as KE_{cm} is increased at various T. In this respect, the apparatus is particularly suitable for the study of the ion chemistry of interstellar shocks in which reactant rotational energies are generally low and the kinetic energies can be much greater [72]. We have studied the endoergic reaction $C^+ + H_2 \longrightarrow CH^+ + H$ which is thought to be a very important reaction in interstellar shocks. A rapid increase in the k is observed as KE_{cm} increases towards a value of ~ 0.4 eV (the endoergicity of the reaction). Similarly, studies of the slightly endoergic binary reaction (20), the k of which increases with increasing T (see Section 3), have shown that the k increases for only small increase in KE_{cm}. These and other similar data are being used to improve ion-chemical modelling of interstellar clouds [70].

Following our previous detailed, truly thermal, studies of the ternary association reactions of CH_3^+ with several molecular gases as a function of temperature referred to in Section 3, we have extended these CH_3^+ studies to include measurements of the ternary rate coefficients as a function of KE_{cm} in helium carrier gas. These measurements were made at T = 80 K for H_2, N_2, O_2 and CO, at T = 300 K for CO and HCN and at T = 580 K for HCN reactants. Also, the $N_2^+ + N_2$ + He reaction (14) discussed in detail in Section 3 has been studied at 80 K as a function of KE_{cm} [63]. Predictably in all reactions, the ternary rate coefficients decreased with increasing KE_{cm}, but it is the form of the decrease which is interesting. Whilst detailed analysis remains to be carried out, it is clear that the decrease of k with KE_{cm} is significantly slower than its decrease with T. For example, for the CH_3^+ + diatomic molecule reactions, $k \sim T^{-2.5}$ (as discussed in Section 3) whereas $k \sim KE_{cm}^{-1.5}$. This intriguing result remains to be explained but is indicative of a reduction in the number of internal modes which are involved in the association reaction under drift tube conditions. Similar trends are also apparent in the $N_2^+ + N_2$ (see Fig. 3) and CH_3^+ + HCN data. As previously noted, the CH_3^+ + HCN reaction (18) is pressure saturated at T = 300 K (and KE_{cm} = 0), and so at this temperature an increase in KE_{cm} first moved the reaction from the saturated 'high pressure' to the unsaturated 'low pressure' régime and then k reduced in a predictable manner consistent with the analogous data mentioned above for other reactions. Analysis of the KE_{cm} data for the HCN reaction provides an approximate value for its true ternary rate coefficient at 300 K thus circumventing the 'saturation' problem.

5. Summary

Studies of the temperature dependences of the rate coefficients, k, for a wide variety of ion-molecule reactions have now been made, sufficient to be able to predict with confidence that the k for ternary association reactions will invariably decrease with increasing temperature, a phenomenon well understood theoretically. The situation is not so simple regarding exoergic binary reactions, unpredictable minima often occurring in k versus T plots. However, it seems likely that the k will increase (towards k_c) at sufficiently low temperatures for the majority of exoergic binary reactions which have small k ($< k_c$) near room temperature. This is presumably a manifestation of the importance of intermediate ion-molecule complexes in determining the rates of the reactions. In the case of isotope exchange reactions, the sense of the temperature dependence of the forward and reverse rate coefficients can now be reasonably well predicted as a result of the VT-SIFT work, and there is some hope that the kinetics of (as well as equilibria in) these reactions might even be describable in terms of thermodynamic quantities.

The VT-SIFDT is just beginning to be exploited and clearly has enormous potential for furthering the understanding of the influence of energy partition on the rates and mechanisms of ion-molecule reactions. It is clear that parallel studies of the influence of temperature and energy on rate coefficients and ion product distributions will greatly advance the understanding of molecular physics and chemistry.

References

1. Sayers, J., Smith, D.: Disc. Faraday Soc. 37, 167 (1964).
2. Ferguson, E.E., Fehsenfeld, F.C., Dunkin, D.B., Schmeltekopf, A.L., Schiff, H.I.: Planet. Space Sci. 12, 1169 (1964).
3. Ferguson, E.E., Fehsenfeld, F.C., Schmeltekopf, A.L.: Adv. Atom. Mol. Phys. 5, 1 (1969).
4. Gioumousis, G., Stevenson, D.P.: J. Chem. Phys. 29, 294 (1958).
5. Su, T., Bowers, M.T.: In: Gas Phase Ion Chemistry, Vol. 1 (Bowers, M.T. Ed.), p. 84, New York: Academic Press, 1979.

6. Ferguson, E.E., Fehsenfeld, F.C., Albritton, D.L. In: Gas Phase Ion Chemistry, Vol. 1 (Bowers, M.T. Ed.), p. 45, New York: Academic Press.1979.

7. Smith, D., Adams, N.G. In: Topics in Current Chemistry, Vol. 89 (Vepřek, S., Venugopalan, M. Eds.), p. 1, Berlin: Springer-Verlag.1980.

8. Smith, D., Fouracre, R.A.: Planet. Space Sci. 16, 243 (1968).

9. Dunkin, D.B., Fehsenfeld, F.C., Schmeltekopf, A.L., Ferguson, E.E.: J. Chem. Phys. 49, 1365 (1968).

10. Adams, N.G., Bohme, D.K., Dunkin, D.B., Fehsenfeld, F.C.: J. Chem. Phys. 52, 1951 (1970).

11. Lindinger, W., Fehsenfeld, F.C., Schmeltekopf, A.L., Ferguson, E.E.: J. Geophys. Res. 79, 4753 (1974).

12. Adams, N.G., Smith, D.: Int. J. Mass Spectrom. Ion Phys. 21, 349 (1976).

13. Herbst, E., Klemperer, W.: Ap. J. 185, 505 (1973).

14. Dalgarno, A., Black, J.H.: Rept. Prog. Phys. 39, 573 (1976).

15. Smith, D., Adams, N.G.: Int. Revs. Phys. Chem. 1, 271 (1981).

16. Smith, D., Adams, N.G. In: Gas Phase Ion Chemistry, Vol. 1 (Bowers, M.T. Ed.), p. 1,New York: Academic Press.1979.

17. Böhringer, H. Arnold, F.: Int. J. Mass Spectrom. Ion Phys. 49, 61 (1983).

18. Rowe, B.R., Dupeyrat, G., Marquette, J.B.: J. Chem. Phys: in press.

19. Luine, J.A.: PhD Thesis, University of Colorado, Boulder (1981): Luine, J.A., Dunn, G.H.: Phys. Rev. Lett. submitted.

20. DePuy, C.H. In: Ionic Processes in the Gas Phase (Almoster Ferreira, M.A. Ed.),p. 227, Dordrecht: Reidel. 1984.

21. Viggiano, A.A., Paulson, J.F., see their chapter in this book.

22. Meot-Ner, M. In: Gas Phase Ion Chemistry, Vol. 1 (Bowers, M.T. Ed.), p. 197, New York: Academic Press.1979.

23. Chen, A., Johnsen, R., Biondi, M.A.: J. Chem. Phys. 69, 2688 (1978).

24. Fehsenfeld, F.C. Albritton, D.L., Bush, Y.A., Fournier, P.G., Gowers, T.R.,Fournier, J.: J. Chem. Phys. 61, 2150 (1974).

25. Dotan, I.: Chem. Phys. Lett. 75, 509 (1980).

26. Smith, D., Adams, N.G. In: Reactions of Small Transient Species: Kinetics and Energetics (Fontijn,A., Clyne, M.A.A. Eds.), p. 311, New York: Academic Press.1983.

27. Rowe, B.R., Dupeyrat, G., Marquette, J.B., Smith, D., Adams, N.G., Ferguson, E.E.: J. Chem. Phys. (1984) in press.

28. Fehsenfeld, F.C., Lindinger, W., Schmeltekopf, A.L., Albritton, D.L., Ferguson, E.E.: J. Chem. Phys. 62, 2001 (1975).

29. Smith, D., Adams, N.G.: Mon. Not. R. astr. Soc. 197, 377 (1981).

30. Luine, J.A., Dunn, G.H.: 12th Int. Conf. on Physics of Electronic and Atomic Collisions, Gatlinburg, Tennessee, July 1981.

31. Smith, D., Adams, N.G., Miller, T.M.: J. Chem. Phys. 69, 308 (1978).

32. Dotan, I., Fehsenfeld, F.C., Albritton, D.L.: J. Chem. Phys. 68, 5665 (1978).

33. Lindinger, W., Smith, D. In: Reactions of Small Transient Species: Kinetics and Energetics (Fontijn, A., Clyne, M.A.A. Eds.), p. 387, New York: Academic Press 1983.

34. Magnera, T.F., Kebarle, P. In: Ionic Processes in the Gas Phase (Almoster Ferreira, M.A. Ed.), p. 135, Dordrecht: Reidel. 1984.

35. Kebarle, P.: Ann. Rev. Phys. Chem. 28, 445 (1977).

36. Castleman, A.W. Jr. In: Kinetics of Ion Molecule Reactions (Ausloos, P. Ed.), p. 295, New York: Plenum Press. 1975.

37. Bohme, D.K., Mackay, G.I., Schiff, H.J.: J. Chem. Phys. 73, 4976 (1980).

38. Adams, N.G., Smith, D.: Chem. Phys. Lett. (1984) in press.

39. Smith, D., Adams, N.G.: Ap. J. 242, 424 (1980).

40. Henchman, M.J., Adams, N.G., Smith, D.: J. Chem. Phys. 75, 1201 (1981).

41. Adams, N.G., Smith, D.: Ap. J.(Lett.) 247, L123 (1981).

42. Adams, N.G., Smith, D.: Ap. J. 248, 373 (1981).

43. Adams, N.G., Smith, D., Henchman, M.J.: Int. J. Mass Spectrom. Ion Phys. 42, 11 (1982).

44. Smith, D., Adams, N.G., Alge, E.: J. Chem. Phys. 77, 1261 (1982).

45. Smith, D., Adams, N.G., Alge, E.: Ap. J. 263, 123 (1982).

46. Smith, D., Adams, N.G. In: Ionic Processes in the Gas Phase (Almoster Ferreira, M.A. Ed.), p. 41, Dordrecht: Reidel. 1984.

47. Watson, W.D., Anicich, V.G., Huntress, W.T. Jr.: Ap. J. (Letters) 205, L165 (1976).

48. Watson, W.D.: Ann. Rev. Astron. Astrophys. 16, 585 (1978).

49. Huber, K.P., Herzberg, G.: Molecular Spectra and Molecular Structure, IV Constants of Diatomic Molecules, New York: Van Nostrand Reinhold. 1979.

50. Carney, G.D., Porter, R.N.: J. Chem. Phys. 66, 2756 (1977).

51. Herbst, E.: Astron. Astro. 111, 76 (1982).

52. Adams, N.G., Smith, D.: in preparation.

53. Henchman, M.J., Smith, D., Adams, N.G.: Int. J. Mass Spectrom. Ion Phys. 42, 25 (1982).

54. Good, A.: Chem. Revs. 75, 561 (1975).

55. Adams, N.G., Smith, D.: Chem. Phys. Lett. 79, 563 (1981).

56. Böhringer, H, Arnold, F., Smith, D., Adams, N.G.: Int. J. Mass Spectrom. Ion Phys. 52, 25 (1983).

57. Bates, D.R.: J. Phys. B. 12, 4135 (1979).

58. Herbst, E.: J. Chem. Phys. 70, 2201 (1979).

59. Herbst, E.: J. Chem. Phys. 72, 5284 (1980).

60. Bates, D.R.: J. Chem. Phys. 73, 1000 (1980).

61. Bass, L.M., Chesnavich, W.J., Bowers, M.T.: J. Amer. Chem. Soc. 101, 5493 (1979).

62. Bass, L.M., Jennings, K.R.: Int. J. Mass Spectrom. Ion Phys. (1984) in press.

63. Smith, D., Adams, N.G., Alge, E.: Chem. Phys. Lett. (1984) in press.

64. Bass, L.M., Kemper, P.R., Anicich, V.G., Bowers, M.T.: J. Amer. Chem. Soc. 103, 5283 (1981).

65. Adams, N.G., Smith, D., Lister, D.G., Rakshit, A.B., Tichy, M., Twiddy, N.D.: Chem. Phys. Lett. 63, 166 (1979).

66. Schiff, H.I., Bohme, D.K.: Ap. J. 232, 740 (1979).

67. McEwan, M.J., Anicich, V.G., Huntress, W.T. Jr., Kemper, P.R., Bowers, M.T. In: Interstellar Molecules (Andrew, B.H. Ed.), p. 305, Dordrecht: Reidel. 1980.

68. Smith, D., Adams, N.G.: Ap. J. 217, 741 (1977).

69. Smith, D., Adams, N.G.: Ap. J. (Lett.) 220, L87 (1978).

70. Herbst, E., Adams, N.G., Smith, D.: in preparation.

71. Adams, N.G., Smith, D., Ferguson, E.E.: in preparation.

72. Elitzur, M., Watson, W.D.: Ap. J. (Lett.) 222, L141 (1978).

Reactions of Negative Ions

A. A. Viggiano* and J. F. Paulson

Air Force Geophysics Laboratory, Hanscom AFB, MA 01731, U.S.A.
* Air Force Geophysics Scholar

A. Introduction

Traditionally, negative ion-molecule reactions have been much less studied than those of positive ions. This is due to the fact that the most popular type of ion source, electron impact, produces a much greater variety of positive ions than negative ions and usually in greater abundances. Thus, in order to make workable signals of many types of negative ions, ion sources in which ion-molecule reactions take place have to be used.

The flowing afterglow has been ideally suited to the study of negative ion reactions since in it there exists a region where primary ions can be converted easily into secondary ions /1/. More recently, relatively gas tight electron impact sources, in which the pressure can be a few torr, have been used in selected ion flow tubes (SIFTs) to generate a variety of negative ions. In addition, ion beam experiments, /2/, /3/ and ion cyclotron resonance mass spectrometers, /4/, /5/ have been used to study a number of negative ion-molecule reactions. The latter instrument has been used to a great degree to establish a scale of gas phase acidities /6/. In this review we will cover the research on negative ion reactions performed using swarm experiments in the last five years. The emphasis will be on giving a broad overview of the most recent work in order to show the variety of measurements that can be made with these systems. The work presented covers a wide spectrum of results, including studies of vibrational product distributions and temperature dependences of associative detachment reactions, many studies involving atmospheric species, as well as those relating to electron affinity determinations and isotope exchange. The work involving H_3O^- shows that by choosing conditions carefully one can study species that are difficult to produce. Two main areas are left out of this review: the effects of solvation and the reactions of organic anions. The former topic has recently been reviewed by Bohme /7/ and the latter by DePuy and Bierbaum /8/.

B. Associative Detachment

Associative detachment is an important process in controlling the electron density in a variety of natural plasmas, such as the earth's ionosphere and interstellar space, and has been a much studied process for many years. In spite of this, much new information has become available recently. This new information involves the first studies of the temperature dependence of the rate coefficients of these reactions /9/ and of the infrared emissions from the neutral products of the reactions /10/-/14/. These studies have yielded new insights into the reaction mechanisms as well as details of the kinematics involved. In addition, much recent work has been done on the theoretical aspect of these reactions /15/.

The reactions whose temperature dependences were studied fell into two classes, those which were slow and those for which associative detachment was only one of several channels. These reactions could then be expected to have a significant dependence on temperature in either the rate coefficient or branching ratio. Table 1 lists the results of this study. The temperature dependences are the results of least squares calculations to power law dependences.

For the reactions involving only associative detachment, the temperature dependence was found to be $T^{(-0.75\pm0.1)}$ in all three cases. The authors /9/ concluded that this represented mainly the temperature dependence of the lifetime of the collision complex with respect to dissociation back into reactants. This result should be compared with the theories of Bates /16/ and Herbst /17/, which predict a complex lifetime varying as $T^{-0.5}$ for atomic-diatomic systems.

Among these associative detachment reactions, a particularly interesting one is that of S^- with O_2. This is an example of an insertion reaction, in which one of the reactants must insert itself between two atoms already bonded together. The standard model for an insertion reaction was thought to be a two step process /18/ which is written for this reaction as,

$$S^- + O_2 \rightarrow SO^- + O \qquad [1]$$

$$SO^- + O \rightarrow SO_2 + e \qquad [2]$$

where the products of the first step never separate. The criterion for this to be allowed is that the first step is exothermic. In this example, step 1 is endothermic by 8.7 kcal mole^{-1} and therefore might not be expected to occur. An alternate explanation of this process as an addition reaction to form an isomer of SO_2 also fails, as this is endothermic.

Table 1. Rate Coefficients for Associative Detachment Reactions /9/

Reaction	$k(T)(cm^3\ s^{-1})$
$O^- + NO \rightarrow NO_2 + e$	$3.1(-10)*(300/T)^{0.83}$
$S^- + CO \rightarrow COS + e$	$2.3(-10)(300/T)^{0.64}$
$S^- + O_2 \rightarrow SO_2 + e$	$4.6(-11)(300/T)^{0.72}$
$O^- + C_2H_2 \rightarrow C_2H_2O + e$	$1.1(-9)(300/T)^{0.39}$
$O^- + C_2H_2 \rightarrow$ products	$1.94(-9)$
$O^- + C_2H_4 \rightarrow C_2H_4O + e$	$5.7(-10)(300/T)^{0.53}$
$O^- + C_2H_4 \rightarrow$ products	$9.0(-10)(300/T)^{0.43}$

*3.1 (-10) means 3.1×10^{-10}

The authors /9/ then proposed that the reaction could be explained in terms of the insertion model if the kinetic energy gained during the collision was taken into account. In order to overcome the endothermicity, the reactants must come within 1.94Å of each other. This leads to a reaction efficiency of 8.4%, which compares well with the measured value of 6.2%, lending credence to this explanation. The temperature dependence of this reaction can then still be explained by a change in the complex lifetime.

The reactions of O^- with C_2H_2 and C_2H_4 are fast and have the associative detachment channel as a main channel. The overall rate of the reaction with C_2H_2 was found to be independent of temperature, although the branching ratio was found to have a significant temperature dependence. The reaction of O^- with C_2H_4 was found to have a slight temperature dependence, but the branching ratio was found not to change significantly with temperature. The rate coefficient for the associative detachment channel for each of these reactions was found to vary as $T^{(-0.45\pm0.06)}$.

Over the past several years the flowing afterglow has been used to study the chemiluminescence associated with a number of reactions, many of which were associative detachment reactions. Included in the associative detachment reactions are the reactions of O^- with CO /10/ and of the halide negative ions with hydrogen and deuterium atoms /11/-/14/. The most complete study to date is that of Smith and Leone on the reactions of F^- with H and D /14/, and the present discussion will emphasize these results. The reaction of F^- with H is sufficiently exothermic to produce HF with up to 5 quanta of vibrational energy, and the reaction of F^- with D can produce DF with up to 7 vibrational quanta. The nascent vibrational energy product distributions found for these reactions are shown in Table 2. For the H reaction, it was found that the population in each vibrational level increased up to $v = 4$ and then decreased for $v = 5$. The F^- reaction with D also showed a large amount of product vibrational excitation, although more uniformly distributed as a function of v than the H analog. Smith and Leone /14/ were also able to get some information on the rotational energy distribution of the DF product by studying this reaction in an argon buffer, where the rotational energy quenching rate was slower than in helium. They found a large amount of rotational excitation, equal to about 13% of the available energy.

Associative detachment reactions are an extreme example of a reaction in which there is a large difference in the reduced mass of the products and reactants. Smith and Leone /14/ have pointed out that many aspects of the product distributions can be explained by classical kinematics. Due to the low mass of the outgoing electron, essentially all of the incoming orbital angular momentum must end up in the product neutral. This leads to the unique case in which the entrance channel impact parameter maps directly into the rotational quantum number of the diatomic product. The maximum allowed J then depends on the vibrational level in question and the overall energetics of the reaction. As stated above, Smith and Leone /14/ have qualitatively observed this high level of rotational excitation for the F^- reaction with D in an argon buffer. The observed falloff in population at the highest vibrational level in both the H and D reactions is then explained by the fact that the highest rotational levels are not energetically accessible for states with a large amount of vibrational excitation, and consequently collisions with large impact parameters (large orbital angular momentum) cannot form the product neutral in a high vibrational level and still conserve angular momentum.

Table 2. Relative Product Vibrational Populations from the F^- + H, D Reactions /14/

$F^- + H \rightarrow HF(v) + e$		$F^- + D \rightarrow FD(v) + e$	
v	nascent distribution	v	nascent distribution
1	0.00 ± 0.06	1	0.08 - 0.07
2	0.09 ± 0.01	2	0.09 ± 0.01
3	0.21 ± 0.01	3	0.15 ± 0.02
4	0.41 ± 0.02	4	0.11 ± 0.02
5	0.30 ∓ 0.02	5	0.15 ± 0.01
		6	0.24 ± 0.03
		7	0.18 ± 0.02

In contrast to the fact that the high level of rotational excitation can be explained classically, the high degree of vibrational excitation must be explained quantum dynamically. Smith and Leone /14/ argue that the F^- + H(D) reaction can best be explained by the virtual state model rather than the resonant state model because there exists an open s-wave electron detachment exit channel. In the former model, transitions are facilitated due to a breakdown in the Born-Oppenheimer approximation. The increased nuclear velocity associated with higher vibrational levels then aids the Born-Oppenheimer breakdown, and qualitatively one can expect an increase in the transition rate for this state. Model calculations by Gauyacq /15/ support these arguments. At present, however, there is no explanation for the differences in the product vibrational distributions for H and D.

C. Atmospheric Negative Ion Chemistry

In the past several years a large effort has gone into understanding the ion chemistry of the atmosphere, especially the stratosphere. This interest has been fueled by the advent of balloon-borne mass spectrometers that have yielded the first detailed height profiles of both positive and negative ions in the stratosphere /19/, /20/. In order to explain the results, many of which were unexpected, laboratory measurements had to be performed. This section will deal with the most recent laboratory measurements that pertain to negative ions of atmospheric interest.

The first in-situ measurements of stratospheric negative ions revealed the presence of a series of ions that could be best fit as $R^- (HR)_m(HNO_3)_n$ where HR had a mass of 98 $\pm$ 2 amu /21/. Arnold and Henschen /21/ speculated that HR was sulfuric acid. In order to test this hypothesis, Viggiano et al./22/ studied a number of positive and negative ion reactions with H_2SO_4. In order to get sulfuric acid into the gas phase in a controlled manner, they used a furnace in which dry nitrogen was passed through glass wool covered with several drops of concentrated H_2SO_4. The flow conditions were set such that the flow was viscous, and the H_2SO_4 flow was then proportional to the square root of the N_2 flow. In this manner, they were able to measure the relative rates of the reactions but were unable to set absolute values, since the absolute concentration of sulfuric acid in the flow tube was not known. However, they noted that the ratio of the rate coefficients for the fastest reactions was the same as would be expected for the collision rates based on the masses of the respective reactant ions. By then setting the fastest rate equal to the collision rate, the rate coefficients were put on an absolute basis. Since the time of publication of the results, the dipole moment of H_2SO_4 has been measured, and the rate coefficients have been revised accordingly /23/. The revised results are listed in Table 3.

Table 3. Reaction Rate Coefficients for H_2SO_4 Reactions at 343K /23/

Reaction	$k(cm^3\ s^{-1})x10^9$	$\frac{k_{meas}}{k_{theor}}$
1. $O^- + H_2SO_4 \rightarrow HSO_4^- + OH$	4.2	(1)*
2. $Cl^- + H_2SO_4 \rightarrow HSO_4^- + HCl$	2.7	0.9
3. $NO_3^- + H_2SO_4 \rightarrow HSO_4^- + HNO_3$	2.6	1.0
4. $I^- + H_2SO_4 \rightarrow HSO_4^- + HI$	1.9	0.9
5. $NO_3^-(HNO_3) + H_2SO_4 \rightarrow HSO_4^-(HNO_3) + HNO_3$	2.3	1.1
6. $NO_3^-(HNO_3)_2 + H_2SO_4 \rightarrow HSO_4^-(HNO_3)_2 + HNO_3$	1.1	0.6

* defined as 1

The most important results from an atmospheric viewpoint are the reactions of H_2SO_4 with $NO_3^-(HNO_3)_n$ ions. The latter ions had previously been thought to be the terminal negative ions in the stratosphere, and the fast reactions of these ions with sulfuric acid indicated that HR was correctly identified by Arnold and Henschen /21/. In fact, these measurements have provided the only means at present for determining the gaseous sulfuric acid concentration in the stratosphere /24/. The laboratory studies indicated that sulfuric acid should not play a role in the positive ion chemistry of the stratosphere, and this conclusion is supported by more recent in-situ results. In addition to the atmospheric implications, the reaction of I^- with H_2SO_4 puts a lower limit to the electron affinity of HSO_4 at 4.5 eV.

Another important species in the stratosphere is N_2O_5. Two studies were made of the reactions of atmospheric ions with N_2O_5 /25/, /26/. The results of the negative ion reactions in these studies are listed in Table 4. All the reactions were found to be fast to produce NO_3^- except those involving hydrates of NO_3^-. These results imply that N_2O_5 will speed the conversion rate of primary negative ions in the stratosphere to NO_3^- core ions. The important reactions from this point of view are those of CO_3^-, O_2^- and its hydrates, and NO_2^- and its hydrates, all of which are precursors to NO_3^- in the stratosphere. Particularly interesting is the reaction with CO_3^- since in the mesosphere the conversion of CO_3^- to NO_3^- is slow. The rapidity of this reaction as well as that of the reaction of CO_3^- with HNO_3 ensures rapid production of NO_3^- core ions in the stratosphere. The lack of reaction of the hydrates of NO_3^- with N_2O_5 indicates that the barrier in the reaction of N_2O_5 with H_2O is not overcome in the presence of the NO_3^- core. This last result also holds for positive ion hydrates.

Table 4. Rate Coefficients of Negative Ions Reacting with N_2O_5

Reaction	$k(cm^3s^{-1})$	Ref
$F^- + N_2O_5 \rightarrow NO_3^- + FNO_2$	1.1(-9)	25
$Cl^- + N_2O_5 \rightarrow NO_3^- + ClNO_2$	9.4(-10)	25
$Br^- + N_2O_5 \rightarrow NO_3^- + BrNO_2$	5.9(-10)	25
$I^- + N_2O_5 \rightarrow NO_3^- + INO_2$	5.9(-10)	25
$CO_3^- + N_2O_5 \rightarrow NO_3^- + NO_3 + CO_2$	2.8(-10)	25
$\rightarrow NO_3^- + NO_2 + CO_3$		
$NO_2^- + N_2O_5 \rightarrow NO_3^- + 2NO_2$	7.0(-10)	25
$Cl^- + N_2O_5 \rightarrow NO_3^- + ClNO_2$	9.3(-10)	26
$Cl^-(H_2O) + N_2O_5 \rightarrow$ products	8.2(-10)	26
$O_2^- + N_2O_5 \rightarrow$ products	1.1 (-9)	26
$O_2^- (H_2O) + N_2O_5 \rightarrow$ products	1.0(-9)	26
$O_2^- (H_2O)_2 + N_2O_5 \rightarrow$ products	9 (-10)	26
$NO_2^- + N_2O_5 \rightarrow NO_3^- + 2NO_2$	6 (-10)	26
$NO_2^- (H_2O) + N_2O_5 \rightarrow$ products	5(-10)	26
$NO_3^- (H_2O) + N_2O_5 \rightarrow$ products	<1(-11)	26
$NO_3^- (H_2O)_2 + N_2O_5 \rightarrow$ products	<1(-11)	26

Another interesting subset of the N_2O_5 reactions are those with the halide ions. These are all fast and produce NO_3^- and the nitryl halide. The fast reaction of I^- with N_2O_5 has been exploited in order to study the thermal decomposition rate of N_2O_5 /27/. In this study, the flowing afterglow was used in a novel way as a detector for neutral kinetics. Employing this technique of selective chemical ionization, Viggiano et al./27/ studied the thermal decomposition of

N_2O_5 over a pressure range of 10 to 800 torr and a temperature range of 285 to 384K. The results of this study and that by Connell and Johnston /28/ have been combined by Malko and Troe /29/ and compared to the latest theories on unimolecular decomposition.

CH_3CN has been found to play an important role in the positive ion chemistry of the stratosphere. A study was made by Paulson and Dale /30/ using a SIFT in order to see if CH_3CN could also enter into the negative ion chemistry of the stratosphere. The results of this work are shown in Table 5.

Table 5. Rate Coefficients for Reactions of CN Containing Compounds at 297 K /30/

Reaction	$k(cm^3s^{-1})$
$O^- + CH_3CN \rightarrow CH_2CN^- + OH + 0.54$ eV	3.5(-9)
$\rightarrow OH^- + CH_2CN + 0.85$	<5(-12)
$OH^- + CH_3CN \rightarrow CH_2CN^- + H_2O + 0.89$	3.3(-9)
$\rightarrow CN^- + CH_3OH + 1.00$	<5(-12)
$OH^-(H_2O) + CH_3CN \rightarrow CH_2CN^- + 2H_2O - 0.19$	3.1(-9)
$\rightarrow CH_2CN^-(H_2O) + H_2O$	2.6(-9)
$OH^-(H_2O)_2 + CH_3CN \rightarrow$ products	slow
$CH_2CN^-(H_2O) + CH_3CN \rightarrow CH_2CN^-(CH_3CN) + H_2O$	>5(-10)
$CN^- + CH_3CN + He \rightarrow CN^-(CH_3CN) + He$	3.5(-12)*
$Cl^- + CH_3CN + He \rightarrow Cl^-(CH_3CN) + He$	2.5(-12)*
$O_2^- + CH_3CN + He \rightarrow O_2^-(CH_3CN) + He$	3.7(-11)*
$CH_2CN^- + CH_3CN + He \rightarrow CH_2CN^-(CH_3CN) + He$	1.3(-12)*
$NO_2^- + CH_3CN + He \rightarrow NO_2^-(CH_3CN) + He$	8.5(-12)*
$NO_3^- + CH_3CN + He \rightarrow NO_3^-(CH_3CN) + He$	7.8(-12)*
$CO_3^- + CH_3CN + He \rightarrow CO_3^-(CH_3CN) + He$	1.1(-12)*
$CN^- + HNO_3 \rightarrow NO_3^- + HCN$	2.0(-9)
$CH_2CN^- + HNO_3 \rightarrow NO_3^- + CH_3CN$	1.4(-9)
$CH_2CN^- + NO_2 \rightarrow NO_2^- + CH_2CN$	1.0(-9)

*Two-body rate coefficients at a He pressure of 0.4 torr (1.3×10^{16} cm^{-3}).

In the reactions of acetonitrile with both O^- and OH^-, only the proton transfer channel was observed, although hydrogen atom transfer in the reaction with O^- and nucleophilic displacement in the reaction with OH^- are also exothermic. The latter reaction would lead to $CH_3OH + CN^-$. Reaction with $OH^-(H_2O)$ produced both m/e 40 i.e., CH_2CN^- and m/e 58, written here as $CH_2CN^-(H_2O)$. The ratio of these product ions in the limit of zero reactant flow is 1.22. Production of CH_2CN^- is slightly endothermic, based upon current thermochemical information. Production of m/e 58 might be regarded as nucleophilic displacement if, as suggested by Caldwell et al./31/, the m/e 58 ion product is written not as $CH_2CN^-(H_2O)$ but as $CN^-(CH_3OH)$. With further addition of CH_3CN to the flow tube, however, the m/e 58 species reacts further and is replaced with m/e 81, $CH_2CN^-(CH_3CN)$. This suggests that m/e 58 is $CH_2CN^-(H_2O)$ which then undergoes a solvent switching reaction with further addition of CH_3CN. Reaction of $OH^-(H_2O)_2$ with CH_3CN is slow. The other reactions of CH_3CN listed in Table 5 are three-body association processes. The fate of CH_2CN^- and of CN^- in the stratosphere is determined by the rapid reactions of both of these species with HNO_3 and by the reaction of CH_2CN with NO_2. Based upon the rate coefficients for these reactions and upon recent measurements of stratospheric number densities for HNO_3 and NO_2, the lifetimes of CH_2CN^- and CN^- are about 3 seconds at 35 km, compared to ion lifetimes of thousands of seconds, indicating that these should be minor species, as has been observed /19/, /20/.

Two recent studies pertain to the ion chemistry of the mesosphere. The first of these was made to determine the role of chlorine-containing species in this region /32/. The results are listed in Table 6. HCl was found to react rapidly with O^-, O_2^-, NO_2^- and CO_4^- producing Cl^- (or $ClHO_2^-$ in the latter case). This indicates that these reactions are likely to be the source of Cl^- in the 60-80 km region of the atmosphere. The reaction of CO_3^- with HCl was found to be slow, presumably beause the reaction is endothermic. In contrast, ClO^- was predicted not to be an important species in the atmo sphere. This is due to the fact that the rate constant for ClO^- production from the reaction of Cl^- with O_3 has a small upper limit and that ClO^- was found to react rapidly with NO, NO_2, and O_3. The electron affinity of ClO was found to be 1.95 $\pm$ 0.25 eV in the same study.

Table 6. Reactions of Chlorine Containing Species at 300K /32/

Reaction	$k(cm^3\ s^{-1})$
$O^- + CCl_4 \rightarrow ClO^- + CCl_3$	$(1.4 \pm 0.4) \times 10^{-9}$
$O^- + HCl \rightarrow Cl^- + OH$	$(2.0 \pm 0.6) \times 10^{-9}$
$O_2^- + HCl \rightarrow Cl^- + HO_2$	$(1.6 \pm 0.5) \times 10^{-9}$
$NO_2^- + HCl \rightarrow Cl^- + HNO_2$	$(1.4 \pm 0.4) \times 10^{-9}$
$CO_4^- + HCl \rightarrow ClHO_2^- + CO_2$	$(1.2 \pm 0.4) \times 10^{-9}$
$CO_3^- + HCl \rightarrow$ products	$\leq 3 \times 10^{-11}$
$ClO^- + NO \rightarrow NO_2^- + Cl$	$(2.9 \pm 0.9) \times 10^{-11}$
$ClO^- + NO_2 \rightarrow NO_2^- + ClO$	
$\rightarrow Cl^- + NO_3$	$(3.2 \pm 1.6) \times 10^{-10}$
$\rightarrow NO_3^- + Cl$	
$ClO^- + SO_2 \rightarrow Cl^- + SO_3$	$(1.3 \pm 0.4) \times 10^{-9}$
$ClO^- + CO_2 \rightarrow$ products	$\leq 1 \times 10^{-13}$
$ClO^- + O_3 \rightarrow Cl^- + 2O_2$	$(7 \pm 3.5) \times 10^{-11}$
$\rightarrow O_3^- + ClO$	
$Cl^- + O_3 \rightarrow ClO^- + O_2$	$\leq 5 \times 10^{-13}$

The other study pertaining to mesospheric negative ion chemistry involves silicon-containing species. These measurements were made after a rocket-borne mass spectrometer found an ion at mass 76 that could not be identified with any conventional ions. The results of the rocket flight and the laboratory results are reported by Viggiano et al./33/. An important conclusion was that O_3^- and CO_3^- react rapidly with SiO to produce SiO_3^-. The absolute values of the rate coefficients could not be determined because the flow of SiO could not be measured. The other important laboratory result was that SiO_3^- did not react with NO, NO_2, CO, CO_2, O, O_3, or Cl_2, the most important trace species that could be expected to react with SiO_3^- in the mesosphere. The absence of reaction with these neutrals suggests that they are endothermic and places a lower limit of 4.7 eV on the dissociation energy of the bond between O^- and SiO_2. The authors concluded that the mass 76 ion was probably SiO_3^- and went on to speculate that very heavy ions found in the same flight may be ions of the type SiO_3^- $(SiO_2)_m$ $(MgO)_n$ $(FeO)_o$...formed by electron attachment to particles resulting from meteor ablation. This hypothesis is based on the large stability found for SiO_3^-.

Recently, there has been a study by Fahey et al. /34/ that has looked at the largely unexplored area of tropospheric ion chemistry. This research involves the reactions of $O_2^-(H_2O)_n$ with n = 0 to 4. The study was carried out in a variable temperature flowing afterglow modified so that the reactant ions were created in a high pressure region (27 torr) which was separated from the main flow tube by a membrane having an aperture 0.4 cm in diameter. In this manner, the cluster ions could be produced with a minimum amount of H_2O, which minimized the effects of secondary reactions. The technique had the additional advantage of ensuring that the reactant ions were formed in the source region and not in the reaction region. The results of this study are listed in Table 7.

Table 7. Reaction Rate Coefficients for $O_2^-(H_2O)_n$ Reactions /34/

Reaction	$k(cm^3s^{-1})$	Temp.(K)
$O_2^- + O_3 \rightarrow O_3^- + O_2$	7.8(-10)	335
$O_2^-(H_2O) + O_3 \rightarrow O_3^- + O_2 + H_2O$	8.0(-10)	292-335
$O_2^-(H_2O)_2 + O_3 \rightarrow O_3^-(H_2O) + O_2 + H_2O$	7.8(-10)	235-292
$O_2^-(H_2O)_3 + O_3 \rightarrow O_3^-(H_2O)_2 + O_2 + H_2O$	~6.4(-10)	181-235
$O_2^-(H_2O)_4 + O_3^- \rightarrow O_3^-(H_2O)_3 + O_2 + H_2O$	~4.6(-10)	181
$O_2^-(H_2O) + NO \rightarrow O_2^-(NO) + H_2O$	2(-10)	211-300
$O_2^-(H_2O)_2 + NO \rightarrow O_2^-(NO)(H_2O) + H_2O$	1.5(-10)	176-300
$O_2^-(H_2O)_3 + NO \rightarrow O_2^-(NO)(H_2O)_2 + H_2O$	1.5(-10)	176-211
$O_2^-(H_2O)_4 + NO \rightarrow O_2^-(NO)(H_2O)_3 + H_2O$	1.2(-10)	176-184
$O_2^-(H_2O)_5 + NO \rightarrow O_2^-(NO)(H_2O)_4 + H_2O$	1.2(-10)	184
$O_2^- + SO_2 \rightarrow SO_2^- + O_2$	1.9(-9)	303
$O_2^-(H_2O) + SO_2 \rightarrow SO_4^- + H_2O$	1.8(-9)	303-304
$O_2^-(H_2O)_2 + SO_2 \rightarrow$ products	1.7(-9)	207-304
$O_2^-(H_2O)_3 + SO_2 \rightarrow$ products	1.7(-9)	207-210
$O_2^-(H_2O)_4 + SO_2 \rightarrow$ products	1.6(-9)	207
$O_2^-(H_2O) + CO_2 \rightarrow CO_4^- + H_2O$	>5.2(-10)	295
$O_2^-(H_2O)_2 + CO_2 \rightarrow$ products	7(-11)	292
$O_2^-(H_2O)_3 + CO_2 \rightarrow$ products	<1(-12)	187-213
$O_2^-(H_2O)_4 + CO_2 \rightarrow$ products	<1(-12)	187-213

In the cases in which there is an exothermic exchange of ligands (NO, SO_2, or CO_2 for H_2O), the reactions were all fast, even for the higher hydrates. This could be expected since there are no steric barriers for ligand exchange. The identity of the ion formed in the $O_2^-(H_2O)_n$ reaction with NO is probably the peroxy isomer of NO_3^-, although rearrangement into the more stable form could not be ruled out. The most surprising result of this study was that $O_2^-(H_2O)_n$ ions reacted rapidly with O_3 to produce O_3^- with one less water molecule attached. This process might involve either the transfer of an oxygen atom or the simultaneous transfer of both charge and one or more water molecules. Studies using ^{18}O labeling in the reactant ion showed that only the latter process occurred. The speed of these ligand exchange and charge exchange reactions at large degrees of hydration is in contrast to similar studies involving ion-atom exchange. In the latter case reactivity typically decreased significantly with increased hydration /7/. The findings in this study ensure that the primary $O_2^-(H_2O)_n$ ions expected in the earth's troposphere will quickly react to form more evolved ions and that O_2^- hydrates will not be a major species when ambient tropospheric ions are measured.

D. Electron Affinity Determination

One of the major applications of negative ion-molecule reactions is for electron affinity determination. In the last several years, a number of swarm studies have involved the determination of electron affinities. The determination of electron affinities in a swarm experiment can be done in a variety of ways. A direct method is that of bracketing. This involves studying a number of charge exchange reactions with compounds of known electron affinities. The reactions involved are

$$A^- + B \rightarrow B^- + A \qquad [3]$$

and

$$C^- + A \rightarrow A^- + C. \qquad [4]$$

A fast reaction implies the reactant neutral has the higher electron affinity. By studying a number of such reactions for both A^- and A one can then bracket the electron affinity of species A if the electron affinities of the other species are known. A less direct form of bracketing involves studying more complex reactions than charge exchange, and then by knowing the thermochemistry of the other species involved, one can again put limits on the electron affinity.

An alternative to this method is to study an equilibrium process and derive the electron affinity of one of the negative species from the measured thermochemistry. One may study the equilibrium directly or determine the equilibrium constant by measuring both the forward and reverse rate coefficients independently.

In this manner a number of studies in recent years have yielded electron affinity determinations. In some of the studies the determination of an electron affinity (or limit to it) was an added bonus, as in the studies involving sulfuric acid /22/ and ClO/32/ mentioned in the atmospheric section of this paper. In addition, there have been several studies where the determination of the electron affinity was the prime purpose /35/-/38/. The most recent electron affinity determinations in swarm experiments are listed in Table 8 along with the technique. In some cases, such as HSO_4 and UF_6, only lower limits could be placed. Rather tight limits, as low as 0.1 eV, can be obtained in favorable cases, such as SF_4, with careful attention given to the reactions studied.

Table 8. Recent Electron Affinity (EA) Determinations

Molecule	EA (eV)	Method	Reference
SF_6	1.0 ± 0.2	bracketing	37
SF_4	2.35 + 0.1	bracketing	36
UF_6	>3.61	bracketing	35
HO_2	1.16 ± 0.15	equilibrium constant	38
HSO_4	>4.5	bracketing	22
ClO	1.95 ± 0.25	bracketing	32

One particularly interesting aspect of the above studies not related to electron affinity determinations warrants further discussion. It is usually assumed in flowing afterglow studies that the ambipolar diffusion coefficient is a constant independent of degree of reaction. Streit and Newton /35/, /39/ have found that this is not the case for UF_6 reactions. UF_6^- probably has a much smaller free diffusion coefficient than the lighter reactants used in the study. As the reaction proceeds, the ambipolar diffusion coefficient of the reactant ion in the flowing afterglow increases as the reactant ion is replaced by the slower diffusing UF_6^- /39/. Thus as the flow of UF_6 increases, the decline in the primary signal deviates from linearity in the normal plot of the logarithm of the ion signal versus reactant neutral flow. The effect is that, upon addition of small amounts of UF_6, the decline in the primary signal is more than if only reaction occurred because the ambipolar diffusion coefficient is increasing. Upon large additions of UF_6, the ambipolar diffusion coefficients reach constant values, and the decline in the primary is a straight line. This problem does not occur in a SIFT as charged particles of only one sign are present.

E. Isotope Effects Involving Proton Transfer

Proton transfer is one of the most basic types of ion-molecule reactions. Many studies of proton transfer reactions have been used to establish tables of gas phase acidities and proton affinities /6/, /40/. In this section two studies involving the basic nature of proton transfer will be discussed. Both of these involve the use of isotopically labelled species to help elucidate the reaction mechanism.

The first study to be discussed involves the measurements of the rate coefficients for hydrogen-deuterium exchange between HO^- (and DO^-) with a variety of weakly acidic neutrals, MD(and MH) /41/. The results of this study are listed in Table 9. Grabowski et al. /41/ have pointed out that there is only a weak correlation of the H-D exchange rate with the relative acidity of the exchange reagent. The mechanism used to explain these data is

$$DO^- + MH \rightarrow [DO^-(MH)] \rightarrow [(DOH)M^-] \rightarrow [HO^-(MD)] \rightarrow HO^- + MD. \qquad [5]$$

I II III

The rate of the H-D exchange is then determined by several factors. Molecules with either large dipole moments or large polarizabilities will generally form longer lived ion-molecule complexes and, therefore, will have a better chance for reaction. The slowness of the DO^- reaction with ethylene is an example where the low initial bond strength of the complex causes the reaction to be slow, the reaction efficiency being less than 0.02 percent. In contrast, the analogous reaction with ammonia leads to a complex with large initial bond strength and to a reaction efficiency of 18 percent.

Table 9. Rate Coefficients for H-D Exchange Reactions at 299K /41/

Reaction	$k \times 10^{10}$ (cm^3s^{-1})	k/k_c*	Reaction	$k \times 10^{10}$ (cm^3s^{-1})	k/k_c*
$HO^- + D_2$	0.68	0.060	$HO^- + C_6D_6$	7.5	0.38
$DO^- + H_2$	0.38	0.024	$DO^- + C_6H_6$	6.6	0.34
$HO^- + HD$	0.35	0.027	$DO^- + CH_4$	≤ 0.002	<0.0002
$DO^- + HD$	0.15	0.012	$DO^- + C_2H_4$	≤ 0.002	≤ 0.0002
$HO^- + ND_3$	2.7	0.13	$DO^- + CH_3NH_2$	5.7	0.28
$DO^- + NH_3$	3.8	0.18	$DO^- + (CH_3)_2NH$	7.5	0.38
$HO^- + D_2O$	12	0.50	$DO^- + H_2CO$	exchange observed	
$DO^- + H_2O$	18	0.74			

*k_c means the collision rate

Once a long lived complex is formed, the reaction rate is determined by the rate of intramolecular proton transfer. This rate is determined primarily by two factors, the relative basicity of OD^- and M^- and the relative bond strength of HOD to M^- and that of OD^- to MH (the bond strengths of complex I and II, respectively). These two factors then determine the relative energies of the OD^- (MH) and M^- (HOD) complexes which play a major role in determining the rate of the reaction. Faster rates result when OD^- is the stronger base and the M^- (HOD) complex has the stronger bond. Both effects facilitate proton transfer.

Another effect that has an influence on the reaction efficiency is multiple proton transfer within the collision complex. This problem of intramolecular proton transfer is treated in another study from the same laboratory /42/. This study involved the reactions of amide and hydroxyl with SO_2, CO_2, N_2O and CS_2. The results are listed in Table 10. The observation of unlabeled hydroxide products in the reactions involving $H^{18}O^-$ reactants shows that an intramolecular proton transfer is occurring.

Table 10. Reactions of OH^- and NH_2^- /42/

Reaction	k($cm^3\ s^{-1}$)
$NH_2^- + SO_2 \rightarrow$ (66.5%) $SO_2^- + H_2N$	2.94(-9)
$\rightarrow$ (26.0%) $NSO^- + H_2O$	
$\rightarrow$ (7.5%) $HO^- + HNSO$	
$NH_2^- + CO_2 \rightarrow NCO^- + H_2O$	9.29(-10)
$NH_2^- + N_2O \rightarrow$ (72.1%) $N_3^- + H_2O$	2.88(-10)
$\rightarrow$ (27.9%) $HO^- + HN_3$	
$NH_2^- + CS_2 \rightarrow$(54%) $NCS^- + H_2S$	1.8(-9)
$\rightarrow$(46%) $HS^- + HNCS$	
$H^{18}O^- + S^{16}O_2 \rightarrow H^{16}O^- + S^{16}O^{18}O$	1.25(-9)
$H^{18}O + C^{16}O_2 \rightarrow H^{16}O^- + C^{16}O^{18}O$	5.69 (-10)
$D^{18}O + C^{16}O_2 \rightarrow D^{16}O^- + C^{16}O^{18}O$	6.33 (-10)
$H^{18}O^- + N_2{}^{16}O \rightarrow H^{16}O^- + N_2{}^{18}O$	1.16 (-11)
$H^{34}S^- + C^{32}S_2 \rightarrow H^{32}S^- + C^{32}S^{34}S$	1.53 (-11)

Grabowski /42/ has set up a simple statistical model to help elucidate some of the mechanistic detail of the reactions involving $H^{18}O^-$ and $H^{34}S^-$. In this model, once a complex is formed it can either undergo an intramolecular proton transfer or dissociate into products which may or may not be the initial reactants. The model can then be used to predict the reaction efficiency as a function of the ratio of the proton transfer rate to the dissociation rate. The results of this calculation pertaining to the situation where there are three equivalent heavy atoms (e.g., $H^{18}O^- + CO_2$) are shown in Figure 1. The maximum reaction efficiency is 2/3 since two of the heavy atoms are equivalent. Using this graph one finds that on the average 2 to 3 proton transfers occur in the re-

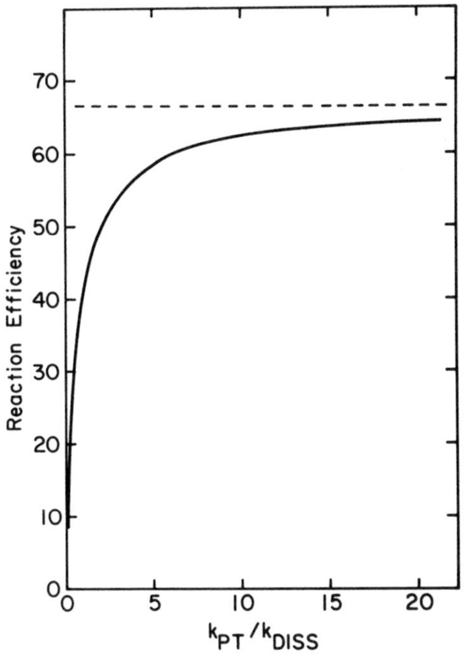

Fig. 1. A plot of the calculated reaction efficiency as a function of the ratio of proton transfer to dissociation rate coefficients for the reactions of $H^{18}O^-$ with SO_2 and CO_2 and the reaction of $H^{34}S^-$ with CS_2. The dashed line is the statistical limit for the reaction efficiency in the event of complete randomization of the proton. Reprinted by permission /42/.

action of $H^{18}O^-$ with CO_2 and 11 to 12 in the reaction with SO_2. The fact that more proton transfers occur in the case of SO_2 is due to the fact that the CO_2 bond energy is less than the SO_2 bond energy. This leads to a reduced reaction efficiency as explained above. By doing experiments in which the hydrogen was replaced by deuterium, the author concluded that the lifetime of the complexes with respect to dissociation, rather than the proton transfer rate, controlled the exchange, since this substitution neither slowed the reaction nor changed the reaction efficiency.

F. Reaction of H_3O^-

The ion H_3O^- has been observed in ion beam experiments /43/ where it is formed in the endothermic reaction

$$OH^-(H_2O) + H_2 \rightarrow H_3O^- + H_2O \qquad [6]$$

and in an ion cyclotron resonance (ICR) apparatus /5/ from the exothermic reaction

$$OH^- + H_2CO \rightarrow H_3O^- + CO \qquad [7]$$

The dissociation energy $D(H^- - H_2O)$ is known to be about 17 ± 3 kcal mol^{-1} from the beam experiments and from molecular orbital calculations /44/. An attempt to observe H_3O^- produced in the reaction between OH^- and H_2CO in a SIFT was unsuccessful /45/. The ions H_3CO^- and $H_3CO_2^-$ were observed, however, and are thought to result from the reactions

$$H_3O^- + H_2CO \rightarrow H_3CO^- + H_2O \qquad [8]$$

$$\rightarrow H_3CO_2^- + H_2$$

which have also been observed in the ICR /5/. When $OD^-(D_2O)$ reacted with H_2CO, no evidence was obtained for the formation of H_2DO^-, and the only ion product was $H_2DCO_2^-$, which corresponds to a solvent switching reaction.

G. Conclusions

In the past several years, a wide range of experimental work has been done on negative ion-molecule reactions. The develeopement of the SIFT apparatus for use with negative ions has greatly expanded the type of experiments that can be performed, as it has for positive ions. As many of the more straight-forward reactions have been studied, experimenters have turned to novel techniques to perform their experiments. Those reported here include the study of vibrational product distributions, use of a membrane ion source region to study large cluster ions, and the study of neutrals difficult to work with, such as H_2SO_4 and N_2O_5. Information has even been obtained on undetected ions such as H_3O^-. All this information has given new insights into the basic mechanisms of ion-molecule reactions as exemplified by the work on proton transfer reactions and associative detachment.

References

1. Ferguson, E.E., Fehsenfeld, F.C., Schmeltekopf, A.L.: Flowing Afterglow Measurements of Ion-Neutral Reactions. Adv. At. Molec. Phys. 5, 1-55 (1969).
2. Paulson, J.F., Dale, F.: Reactions of $OH^{-}\cdot H_2O$ with NO_2. J. Chem. Phys. 77, 4006-4008 (1982).
3. Wu, R.L.C., Tiernan, T.O.: Evidence for Excited States of CO_3^{-}* and NO_3^{-}* from Collisional Dissociation Processes. Planet. Space. Sci. 29, 735-739 (1981).
4. Kleingeld, J.C., Ingemann, S., Jalonen, J.E., Nibbering, N.M.M.: Formation of NH_4^{-} Ion in the Gas Phase. J. Am. Chem. Soc. 105, 2474-2475 (1983).
5. Kleingeld, J.C., Nibbering, N.M.M.: The Long Lived H_3O^{-} Ion in the Gas Phase: Its Formation, Structure, and Reactions. Int. J. Mass Spectrom. Ion Phys. 49, 311-318 (1983).
6. Bartmess, J.E.: Compilation of Gas Phase Anion Thermochemistry. J. Phys. Chem. Ref. Data (in press).
7. Bohme, D.K.: Gas Phase Studies of the Influence of Solvation on Ion Reactivity. In: Nato Advanced Study Institute Volume on Chemistry of Ions in the Gas Phase. Holland. D. Reidel Publishing (in press).
8. DePuy, C.H., Bierbaum, V.M.: Gas Phase Reaction of Organic Ions as Studied by the Flowing Afterglow Technique. Accts. Chem. Res. 14, 146-153 (1981).
9. Viggiano, A.A., Paulson, J.F.: Temperature Dependence of Associative Detachment Reactions. J. Chem. Phys. 79, 2241-2245 (1983).
10. Bierbaum, V.M., Ellison, G.B., Futrell, J.H., Leone, S.R.: Vibrational Chemiluminescence from Ion-Molecule Reactions: $O^{-} + CO \rightarrow CO_2^{*} + e^{-}$. J. Chem. Phys. 67, 2375-2376 (1977).
11. Zwier, T.S., Maricq, M.M., Simpson, C.J.S.M., Bierbaum, V.M., Ellison, G.B., Leone, S.R.: Direct Detection of the Product Vibrational-State Distribution in the Associative Detachment Reaction $Cl^{-} + H \rightarrow HCl(v) + e$. Phys. Rev. Lett. 44, 1050-1053 (1980).
12. Maricq, M.M., Smith, M.A., Simpson, C.J.S.M., Ellison, G.B.: Vibrational Product States from Reactions of CN^{-} with the Hydrogen Halides and Hydrogen Atoms. J. Chem.

Phys. 74, 6154-6170 (1981).

13. Zwier, T.S., Weisshaar, J.C., Leone, S.R.: Nascent Product Vibrational State Distributions of Ion-Molecule Reactions: The $H+F^- \rightarrow HF(v)$ + e Associative Detachment Reaction, J. Chem. Phys. 75, 4885-4892 (1981).
14. Smith, M.A., Leone, S.R.: Product Nascent State Distributions in Thermal Energy Associative Detachment Reactions: F^- + H, D $\rightarrow HF(v), DF(v)$+ e. J. Chem. Phys. 78, 1325-1334 (1983).
15. Gauyacq, J.P.: Associative Detachment in Collisions Between Negative Halogen Ions and Hydrogen Atoms. J. Phys. B. 15, 2721-2739 (1982).
16. Bates, D.R.: Temperature Dependence of Ion-Molecule Association. J. Chem. Phys. 71, 2318-2319 (1979).
17. Herbst, E.: Refined Calculated Ion-Molecule Association Rates. J. Chem. Phys. 72, 5284-5285 (1980).
18. Fehsenfeld, F.C.: Associative Detachment. In: Interactions Between Ions and Molecules. (Ausloos, P., ed.), p. 387-412. New York: Plenum Publishing Corp. 1974.
19. Arnold, F.: Physics and Chemistry of Atmospheric Ions. In: Atmospheric Chemistry. (Goldberg, E.D., ed.) p. 273-300. Berlin: Springer Verlag, 1982 and references therein.
20. Arijs, E.: Positive and Negative Ions in the Stratosphere. Ann. Geophys. 1, 149-160 (1983) and references therein.
21. Arnold, F., Henschen, G.: First Mass Analysis of Stratospheric Negative Ions. Nature. 257, 521-522 (1978).
22. Viggiano, A.A., Perry, R.A., Albritton, D.L., Ferguson, E.E., Fehsenfeld, F.C.: The Role of H_2SO_4 in Stratospheric Negative-Ion Chemistry. J. Geophys. Res. 85, 4551-4555 (1980).
23. Viggiano, A.A., Perry, R.A., Albritton, D.L., Ferguson, E.E., Fehsenfeld, F.C.: Stratospheric Negative Ion Reaction Rates with H_2SO_4 in the Stratosphere. J. Geophys. Res. 87, 7340-7342 (1982).
24. Viggiano, A.A., Arnold, F.: Stratospheric Sulfuric Acid Vapor - New and Updated Results. J. Geophys. Res. 88, 1457-1462 (1983) and references therein.
25. Davidson, J.A., Viggiano, A.A., Howard, C.J., Fehsenfeld,

F.C., Albritton, D.L., Ferguson, E.E.: Rate Constants for the Reactions of O_2^+, NO_2^+, NO^+, H_3O^+, CO_3^-, NO_2^- and Halide Ions with N_2O_5 at 300K. J. Chem. Phys. 68, 2085-2087 (1978).

26. Bohringer, H., Fahey, D.W., Fehsenfeld, F.C., Ferguson, E.E.: The Role of Ion-Molecule Reactions in the Conversion of N_2O_5 to HNO_3 in the Stratosphere. Planet. Space Sci. 31, 185-191 (1983).

27. Viggiano, A.A., Davidson, J.A., Fehsenfeld, F.C., Ferguson, E.E.: Rate Constants for the Collisional Dissociation of N_2O_5 by N_2. J. Chem. Phys. 74, 6113-6125 (1981).

28. Connell, P., Johnston, H.S.: The Thermal Decomposition of N_2O_5 in N_2. Geophys. Res. Lett. 6, 553-556 (1979).

29. Malko, M.W., Troe, J.: Analysis of the Unimolecular Reaction $N_2O_5 + M \rightarrow NO_2 + NO_3 + M$. Int. J. Chem. Kin. 14, 399-416 (1982).

30. Paulson, J.F., Dale, F.: unpublished results.

31. Caldwell, G., Rozeboom, M.D., Kiplinger, J.P., Bartmess, J.E.: Displacement, Proton Transfer or Hydrolysis? Mechanistic Control of Acetonitrile Reactivity by Stepwise Solvation of Reactants. J. Am. Chem. Soc. (in press).

32. Dotan, I., Albritton, D.L., Fehsenfeld, F.C., Streit, G.E. Ferguson, E.E. Rate Constants for the Reactions of O^-, O_2^-, NO_2^-, CO_3^- and CO_4^- with HCl and ClO^- with NO, NO_2, SO_2 and CO_2 at 300K. J. Chem. Phys. 68, 5414-5416 (1978).

33. Viggiano, A.A., Arnold, F., Fahey, D.W., Fehsenfeld, F.C., Ferguson, E.E.: Silicon Negative Ion Chemistry in the Atmosphere - In Situ and Laboratory Measurements. Planet. Space. Sci. 30, 499-509 (1982).

34. Fahey, D.W., Bohringer, H., Fehsenfeld, F.C., Ferguson, E.E.: Reaction Rates Constants for $O_2^-(H_2O)_n$ Ions with n=0 to 4 with O_3, NO, SO_2 and CO_2. J. Chem. Phys. 76, 1799-1805 (1982).

35. Streit, G.E., Newton, T.W.: Negative Ion-Uranium Hexafluoride Charge Transfer Reactions. J. Chem. Phys. 73, 3178-3182 (1980).

36. Streit, G.E., Babcock, L.M.: Negative Ion-Molecule Reactions of SF_4. J. Chem. Phys. 75, 3864-3870 (1981).

37. Streit, G.E.: Negative Ion Chemistry and the Electron Affinity of SF_6. J. Chem. Phys. 77, 826-833 (1982).

38. Bierbaum, V.M., Schmitt, R.J., DePuy, C.H., Mead, R.D., Schulz, P.A., Lineberger, W.C.: Experimental Measurement of the Electron Affinity of the Hydroperoxy Radical. J. Am. Chem. Soc. 103, 6262-6263 (1981).

39. Streit, G.E., Newton, T.W.: The Effect of Chemical Reaction on Diffusive Ion Loss Processes in a Flowing Afterglow. Int. J. Mass Spectrom. Ion Phys. 38, 105-126 (1981).

40. Lias, S.G., Liebman, J.F., Levin, R.D.: An Evaluated Compilation of Gas Phase Basicities and Proton Affinities of Molecules: Heats of Formation of Protonated Molecules. J. Phys. Chem. Ref. Data (in press).

41. Grabowski, J.J., DePuy, C.H., Bierbaum, V.M.: Gas Phase Hydrogen-Deuterium Exchange Reactions of HO^- and DO^- with Weakly Acidic Neutrals. J. Am. Chem. Soc. 105, 2565-2571 (1983).

42. Grabowski, J.J.: Studies of Gas Phase Ion-Molecule Reactions Using a Selected Ion Flow Tube. Doctoral Thesis. University of Colorado (1983).

43. Paulson, J.F., Henchman, M.J.: On the Formation of H_3O^- in an Ion-Molecule Reaction. In: NATO Advanced Study Institute Volume on Chemistry of Ions in the Gas Phase. Holland. D. Reidel Publishing Co. (in press).

44. Squires, R.R.: Ab Initio Studies of the Structures and Energies of Some Anion-Molecule Complexes. In: NATO Advanced Study Institute Volume on Chemistry of Ions in the Gas Phase. Holland. D. Reidel Publishing Co. (in press).

45. Paulson, J.F., Viggiano, A.A.: unpublished results.

Electrons in Dense Gases

L. G. Christophorou[1] and S. R. Hunter[2]

[1] Atomic, Molecular and High Voltage Physics Group, Health and Safety Research Division, Oak Ridge National Laboratory, Oak Ridge, TN 37831, U.S.A.

[2] Department of Physics, The University of Tennessee, Knoxville, TN 37996, U.S.A.

1. Introduction

The study of electron-molecule interactions in dense gases has received considerable attention recently, and a number of workers [1-7] have reviewed and synthesized these advances. It has been pointed out [2] that knowledge of the effects of the density and nature of a gaseous medium on the "isolated molecule"-electron interactions in the intermediate density range between a low-pressure gas and the condensed phase is essential in understanding radiation effects and mechanisms and in linking the wealth of information on electron-molecule interactions in low-pressure gases to that on similar processes in the condensed phase. In this chapter we briefly review recent studies on electron transport and electron attachment to molecules in dense gases (for a discussion of these processes in nonpolar liquids and the linking of gaseous to condensed-phase behavior see [5, 7]).

In low-pressure swarm experiments ($\lesssim$1 atm) electrons are "free" (the electron mean free path is much longer than the electron de Broglie wavelength), and their transport properties are treated within the Boltzmann transport equation; their kinetic energies can be well in excess of thermal depending on the density-reduced applied electric field E/N. In dense gases (pressures $\lesssim$400 atm) and liquids, the electron can interact simultaneously with several molecules. With a few exceptions at high E/N, the electron energies are thermal or epithermal; and the effect of the medium on the electron state, the electron-molecule interaction potential, the intermediate and final products, and the associated reaction energetics and rate constants is often dramatic. Depending on the dense gas (or the liquid) and the temperature, T, excess electrons in such media can be either *quasi-free* (e_f) or *localized* (e_ℓ); of course, during their drift, they can be partly in the quasi-free and partly in the localized state. The interactions of low-energy electrons with molecules in dense media, therefore, depend not only on the molecule itself but also on the electron state and the medium [5, 7].

2. Electron Drift and Diffusion in Dense Gases

In this section we briefly review the changes that have been observed to occur in the electron drift velocity, w, with increasing gas number density, N, and the attempts which have been made to explain these observations.

2.1. Experimental Results

2.1.1. Nonpolar Gases

In the theory of electron motion in gases under the influence of uniform electric fields, it is assumed that the transport coefficients are unaffected by the gas pressure and depend solely on E/N. This assumption implies that (1) the electrons interact with only one molecule in each scattering event, and the presence of neighboring molecules has no influence on the scattering process; and (2) the interaction time between the electron and the gas molecule in a collision is negligible in comparison to the mean free time of the electrons between collisions. Until the late 1960s, this assumption was not questioned as in nearly all the previous experiments that had been performed to obtain transport data in gases, no pressure dependence had been observed in the experimental measurements after the influence of electron diffusion and boundary effects on the transport data had been properly accounted for. This observation is not surprising considering that most electron transport measurements were made at gas pressures below 1 atm; and as has been subsequently observed, most pressure-dependent processes are observable at pressures above 1 atm. The most notable exceptions were the low-temperature w measurements of Lowke [8] in N_2 at 77 K, who observed a small ($\leq 3\%$) unexplainable decrease in w at the lowest E/N values and highest gas pressures ($\lesssim 300$ kPa) in his experiments, and the electron mobility measurements of Levine and Sanders [9] in He at $T \sim 4$ K, who observed a drastic (~ 3 orders of magnitude) decrease in the electron mobility at gas pressures near the critical number density ($\sim 10^{21}$ atoms cm^{-3}) in He. Grünberg [10, 11] subsequently measured w at room temperature in H_2 and N_2 to gas number densities $N \leq 10^{21}$ molecules cm^{-3} and observed a large decrease in w for both H_2 and N_2 with the greatest deviation from the low-pressure behavior occurring at the lowest E/N values.

The density-normalized electron mobility $\mu N(=wE)$ has subsequently been observed to both decrease and increase with increasing N for a number of atomic and molecular gases. In Fig. 1A is shown the dependence of w on N for He [12, 13], and in Fig. 1B similar data are shown for H_2 at 77.6 K [13]. These measurements again indicate that the greatest deviation from the low-pressure mobility occurs at thermal energies for these two gases, the reduction in mobility decreasing in effect at electron energies in excess of thermal. A similar reduction in w with increasing gas pressure

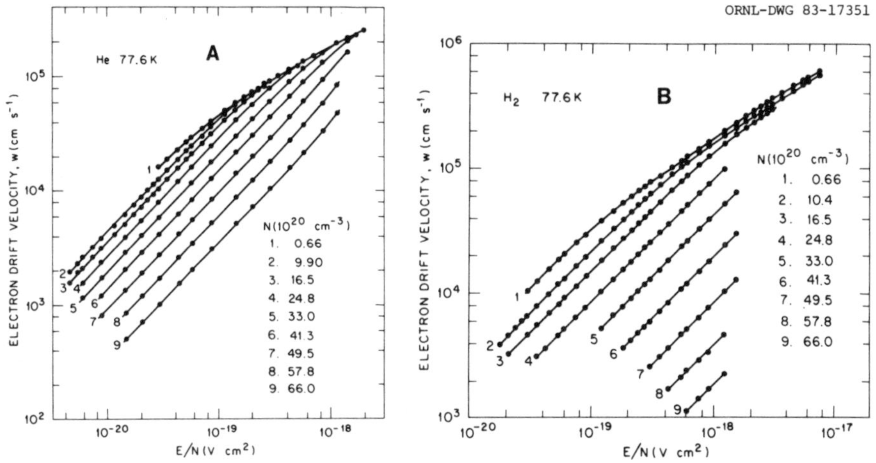

Fig. 1. w versus E/N and N for He (Fig. 1A) and H_2 (Fig. 1B) at 77.6 K [13].

was observed in C_2H_6 and C_3H_8 [14, 15] (Fig. 2) at room temperature. In contrast to the observations in He, H_2, and N_2, ethane was found to exhibit a maximum change in w with increasing N at nonthermal electron energies, while for C_3H_8 the greatest effect occurred at thermal energies. Lehning [16] observed an *increase* in w with increasing N in CH_4 (Fig. 3A) with the maximum change in w occurring at nonthermal electron energies. Bartels [17] also found that w *increased* with increasing N in Ar at room temperature (Fig. 3B), while Robertson [20] observed a small, but significant, *decrease* in w with increasing gas pressure in Ar at 90 K in contrast to the observed increase in w at room temperature. Christophorou [1] has analyzed the drift velocity measurements in the molecular gases and found that the maximum change in w with pressure occurs at thermal energies (~0.038 eV) for H_2 and C_3H_8 but at ~0.06 eV for CH_4 and at ~0.07 eV for C_2H_6.

A considerably larger decrease in w (~3 orders of magnitude) has been observed in CO_2 [21, 22] over a smaller gas number density range ($N \leq 10^{21}$ molecules cm^{-3}) than for the previous measurements. The contrast in the mobility reduction between CO_2 and He and H_2 can be seen in the plot of the relative thermal mobilities for these gases given in Fig. 4A. A large decrease in w has also been recently observed for 1-C_3F_6 [26] (Fig. 5) at even smaller gas number densities ($N \leq 5 \times 10^{18}$ molecules cm^{-3}). The large decrease in mobility in these gases has been attributed [22, 26], at least in part, to temporary negative ion trapping of the electrons by molecular dimers or higher-order clusters.

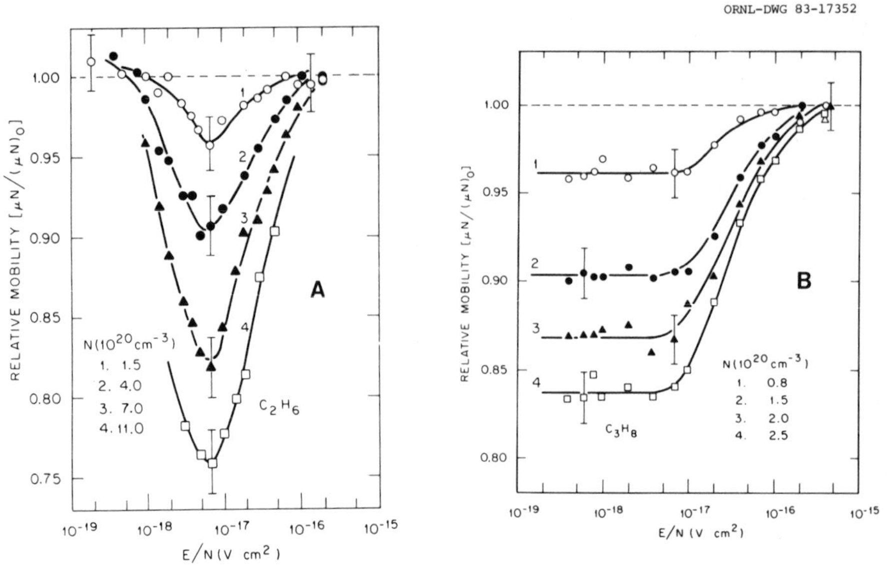

Fig. 2. Ratio $\mu N/(\mu N)_0$ of the electron mobility μN at gas number density N to the electron mobility at low gas number density $(\mu N)_0$ as a function of E/N and N for C_2H_6 (Fig. 2A) [14] and C_3H_8 (Fig. 2B) at T $\simeq$ 293 K [15]. The points are the experimental values; the lines have no significance.

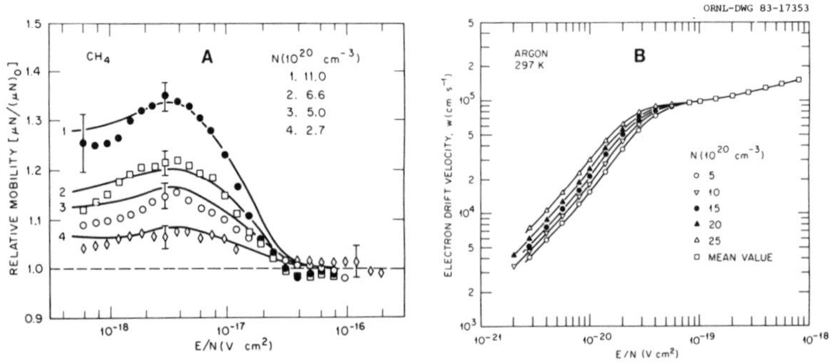

Fig. 3. Ratio $\mu N/(\mu N)_0$ of the electron mobility N at gas number density N to the electron mobility at low gas number density $(\mu N)_0$ as a function of E/N and N for CH_4 (T $\simeq$ 293 K) (Fig. 3A) and Ar (T = 297 K) (Fig. 3B). The points in A are the experimental data from [16], and the solid lines are the multiple-scattering calculations from [18, 19]. The experimental data in B are from [17].

ORNL-DWG 83-17354

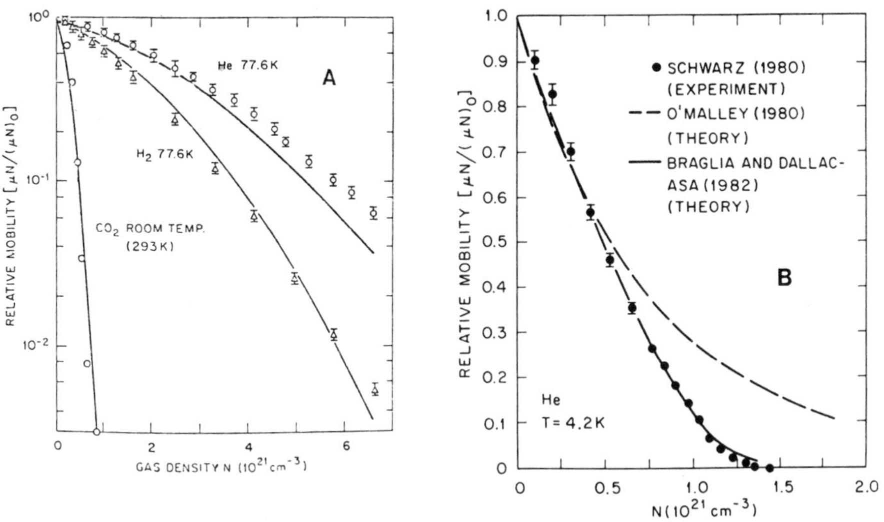

Fig. 4. (A) Relative electron mobility $\mu N/(\mu N)_0$ as a function of N for He, H_2 [13], and CO_2 [21]. The points are the experimental results, and the solid lines are the results of calculations by O'Malley using multiple-scattering theory [23]. (B) Relative electron mobility $\mu N/(\mu N)_0$ versus N in He at 4.2 K. The points are the measurements of Schwarz [24]. The multiple-scattering theory calculations of O'Malley [23] (---) and Braglia and Dallacasa [25](——) are shown in comparison with the experimental results [25].

ORNL-DWG 83-13737

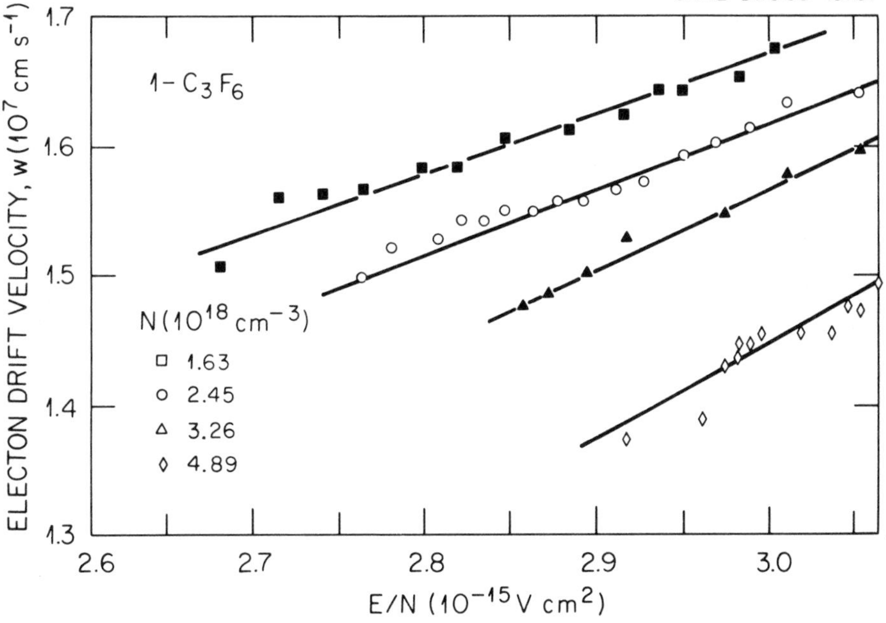

Fig. 5. w versus E/N and N for $1\text{-}C_3F_6$ ($T \simeq 293$ K) [26].

2.1.2. Polar Gases

The w measurements described above have all involved the motion of electrons in nonpolar or weakly polar gases. Attention has recently been focused on the influence that the electric dipole moment has on the electron motion in high-pressure polar gases. Measurements of w have been made in NH_3 [27, 28] and in H_2O [29] as a function of T and N. Examples of the measurements of w obtained by Christophorou et al. [28] at room temperature are shown in Fig. 6A; the thermal electron mobility measurements are plotted as a function of T and gas pressure in Fig. 6B. The considerably less accurate, but higher pressure, measurements of Krebs and coworkers [27, 29] are given in Figs. 7A and 7B for H_2O and NH_3, respectively. These measurements show a considerable decrease in electron mobility at $N > 2 \times 10^{19}$ molecules cm^{-3} and at lower gas temperatures, which has been interpreted by Christophorou et al. and Krebs and coworkers as a transition region from free electron motion in the gas (either binary *or* multiple electron scattering) to "trapped" electron motion (temporary and permanent electron capture by the high-density gas) at gas number densities near those of the liquid phase.

ORNL-DWG 83-18290

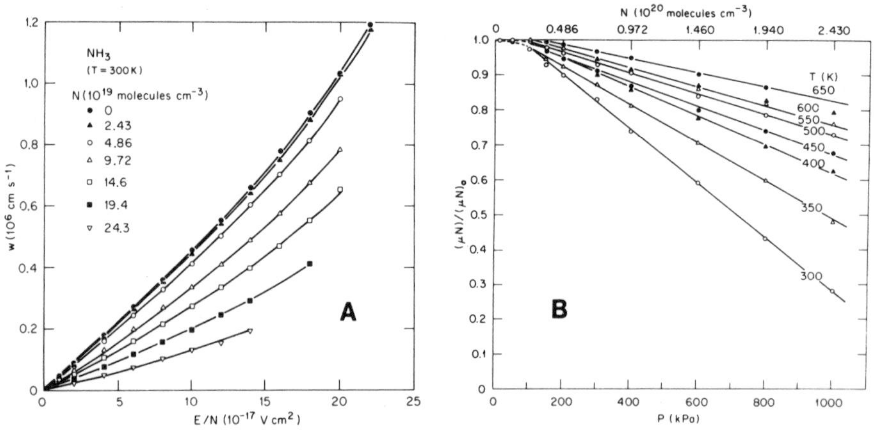

Fig. 6. (A) Electron drift velocity in NH_3 as a function of gas number density and E/N at T = 300 K [28]. (B) The ratio $\mu N/(\mu N)_0$ of the thermal electron mobility μN at gas density N to the electron mobility at low gas number density $(\mu N)_0$ as a function of gas pressure and temperature [28].

2.2. Theories of Electron Motion in High-Pressure Gases

2.2.1. Electron Drift Velocity

Following the observations by Grünberg [10, 11] of the large decrease in w with N in H_2 and N_2, Frommhold [30] suggested that this decrease originates from the fact that electrons spend a fraction of their mean free time between collisions as temporary negative ions formed by a Feshbach-type resonance

ORNL-DWG 82-18748

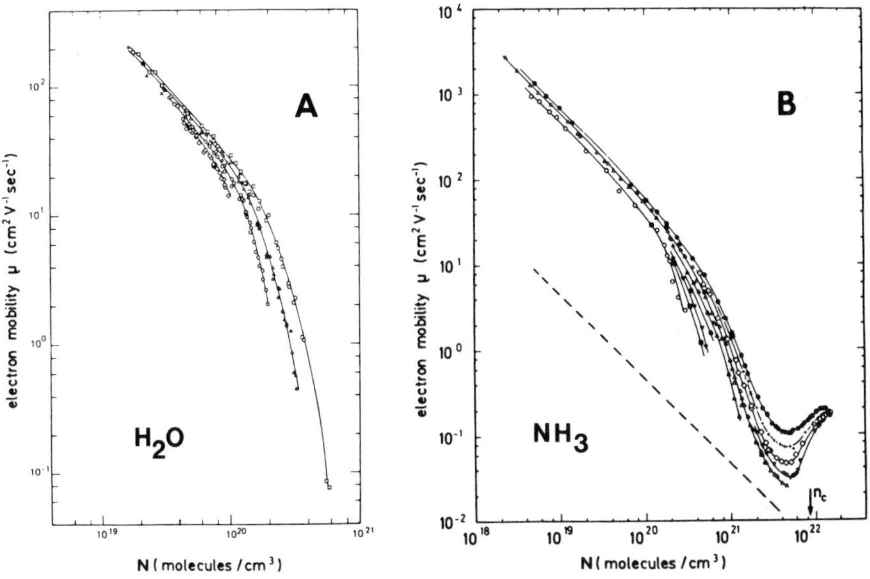

Fig. 7. (A) μ versus N in subcritical water vapor at 433 K (◇), 463 K (○), 493 K (△), and 523 K (□) [29]. (B) μ versus N in sub- and supercritical ammonia vapor at various temperatures: 300 K (○), 320 K (■), 340 K (▼), 360 K (◓), 380 K (▲), 400 K (△), 410 K (▽), 420 K (◇), 440 K (+), and 460 K (●). The arrow indicates the critical density of ammonia, and the dashed line represents the averaged mobility of unidentified impurity ions (T < 400 K) [27d].

associated with the rotational states of H_2 and N_2. Frommhold assumed that the correction to the low-pressure electron mobility $(\mu N)_0$ has the form (based on the analysis of trapped electron motion in a gas by Ritchie and Turner [31])

$$\mu N = (\mu N)_0 \, (1 + \nu_i \tau_i)^{-1} \simeq (\mu N)_0 \, (1 - \nu_i \tau_i) \ , \qquad (1)$$

where ν_i is the collision frequency for temporary negative ion formation, and τ_i is the mean autoionization lifetime of the negative ion. This hypothesis was proposed to account for the decreased mobility in several other molecular gases (e.g., CO_2 [Ref. 21], C_2H_6 [Ref. 14], and C_3H_8 [Ref. 15]). It, however, fails to explain the large decrease in the electron mobility as N increases at thermal energies in He [12, 13] (helium being an atomic gas cannot possess low-lying negative ion resonances associated with inelastic processes), and the large *increase* in mobility with increasing N in both CH_4 [16] and Ar [17] at room temperature (Fig. 3). The drastic decrease in the electron mobility in He with increasing N at low temperatures (Fig. 4B) is generally attributed to the formation of bubbles or pseudobubbles around the electrons in the gas that are known to exist in the liquid state [32]. Pseudobubbles are regions of reduced-gas density which surround and migrate with the electrons in the gas and are very sensitive to gas pressure and temperature. This hypothesis has been successful in accounting for the low-temperature high-pressure electron mobility measurements in He but is less successful at high temperatures and low pressures.

A new model for electron motion in a high-pressure gas was proposed by Legler [33] based on the observation that at high pressures the electron mean free path is approximately equal to the wavelength for thermal energy electrons. This led Legler to propose that multiple-scattering processes occur at these high pressures. Although Legler's theory was able to predict both the increase and decrease in the electron mobility with increasing N observed in several gases, the magnitude and number density dependence of the mobility corrections were considerably in error for a number of molecules. Atrazhev and Iakubov [18, 19] suggested that the increase in w with N can be accounted for by a density correction to the polarizability, which alters the scattering cross section as N increases. They were able to obtain good agreement with the observed increase in mobility with N in Ar and CH_4 (Fig. 3A) with the inclusion of a semi-empirical scattering parameter. The decreases in mobility with increasing N were explained in terms of a "quantum interference model" based on a multiple-scattering approach. They proposed a correction to the low-pressure electron mobility $(\mu N)_0$ as

$$\mu N = (\mu N)_0 \; (1 - N\langle\sigma\rangle \lambda\!\!\!^{-} F \sqrt{\pi}/2) \; , \tag{2}$$

where $\lambda\!\!\!^{-} = (\hbar/2\pi kT)^{\frac{1}{2}}$, $\langle\sigma\rangle$ is the average scattering cross section at thermal energies, and F is an empirical factor, which was assigned a value of 2 by Atrazhev and Iakubov [18] and a value of $2\pi/3 = 2.09$ by Iakubov and Polischuk [34, 35]. Expression (2) fits the observed pressure dependence of w in a number of cases.

Another theoretical attempt to rationalize the observed variations of the thermal electron mobilities with N was made by O'Malley [23]. In the situation where the electron mobility *increases* with increasing N, O'Malley's theory, based on a change in the kinetic energy of the electrons in collisions with the gas molecules at high N, gave a correction to the low-density mobility of the form

$$\frac{\mu N}{(\mu N)_0} \simeq \left[\frac{kT}{(kT + \Delta)}\right]^{\frac{1}{2}} \frac{\sigma_m(kT)}{\sigma_m(kT + \Delta)} \quad . \tag{3}$$

In Eq. (3), σ_m is the momentum transfer cross section at the electron energy equivalent to kT, and Δ is the real part in the shift in the electron kinetic energy at high N given by

$$\Delta \simeq \left(\frac{\hbar^2}{2m}\right) 4\pi c N R_e f_0 \ , \tag{4}$$

where $R_e f_0$ is the real part of the electron-atom (or molecule) forward scattering amplitude and c = 1 up to high N.

The decrease in w with increasing N was hypothesized by O'Malley to be due to the influence of the uncertainty principle on the shift in the kinetic energy at high N. O'Malley obtained a correction to the low density mobility given by

$$\mu N = (\mu N)_0 \exp\left[-\bar{\Gamma}/kT\right] \ , \tag{5}$$

where

$$\frac{\bar{\Gamma}}{kT} = g_0 \left(1 + \frac{g_0^2}{4}\right)^{\frac{1}{2}} + \frac{g_0^2}{2} \ , \tag{6}$$

$g_0 = 2N\langle\sigma\rangle c\lambda\!\!\!^{-}$, and c = 1 up to high N. In the low-pressure limit, Eq. (5) reduces to Eq. (2) with $F = 4/\sqrt{\pi} = 2.26$. O'Malley's treatment successfully predicted both the increase in mobility with increasing N in Ar at room temperature and high pressure, and the small decrease in mobility with increasing N in Ar at low temperature (90 K) and low pressure (<100 kPa). It also predicted well the decrease in mobility with increasing N up to moderate N values in a number of gases (Fig. 4A). However, at high N in He, the experimental measurements decrease more rapidly with increasing N than the theory predicts (Fig. 4B).

In another related effort, Braglia and Dallacasa [25] have developed a single theory to account for both the decrease and increase of the electron mobility with increasing N in terms of a quantum mechanical model based on multiple scattering effects. They obtained better agreement with the experimental measurements than O'Malley over a wider range of N in Ar and a number of gases where the mobility decreases with increasing N (Fig. 4B). As in the case of the O'Malley theory, however, their theoretical calculations deviate substantially from the experimental values at the higher number densities. Their theory also fails to correctly predict the density dependence of the mobility in CH_4 (and presumably also in C_2H_6 and C_3H_8). Gryko and Popielawski [36] have attempted to extend Braglia and Dallacasa's theory in Ar but have not obtained any substantial improvement in the agreement between theory and experiment.

The aforementioned theories fail to fully account for the effects of gas density on the electron mobility in the transition region from the high-pressure gas to the liquid. Schwarz [24] attempted to bridge the low- and high-temperature mobility measurements in He by incorporating a multiple scattering approach [18, 25b] at lower N with the gradual formation of bubbles at higher N in order to explain the large decrease in mobility that has been observed at gas number densities near the critical number density ($N \sim 10^{21}$ molecules cm^{-3}). Negative ion trapping has been proposed to account, at least in part, for the reduction in electron mobility with increasing N in CO_2 [1, 22] and 1-C_3F_6 [26] and in the polar gases NH_3 [27, 28] and H_2O [29]. It was observed by Christophorou et al. [28] that multiple scattering processes alone are unable to account for the electron mobility reduction in NH_3 at high gas number densities.

This brief discussion has shown that none of the theories proposed to account for the pressure dependence of the electron mobility can explain all of the observed phenomena. It seems reasonable to assume that, in most cases, two or more of these effects are responsible for the observed changes in the electron mobility in the density range from the low-pressure gas to the liquid phase. Further experimental and theoretical work is needed and, indeed, anticipated in this area.

2.2.2 Thermal Electron Diffusion

Recently, O'Malley [37] hypothesized that in certain instances the characteristic electron energy eD/μ at thermal energies should also be a function of N. He used the theoretical arguments he developed previously [23] on the changes in the electron mobility at high N to derive the relationship between eD/μ and N in the situations where μN both decreases and increases with increasing N. When μN decreases with increasing N, O'Malley found that the thermal diffusion coefficient ND depends on N as

$$ND(N) = ND_0 \exp[-\bar{\Gamma}/kT] , \qquad (7)$$

where $\bar{\Gamma}$ is given by Eq. (6), and D_0 is the classical low density diffusion coefficient. The characteristic energy in this case is given by

$$\frac{eND}{\mu N} = \frac{eD_0}{\mu_0} = kT , \qquad (8)$$

where μN is defined by Eq. (5). Thus, eD/μ is predicted to be independent of N when μN is a decreasing function of N.

Alternatively, when μN increases with N, O'Malley predicted that the characteristic energy will have the form

$$\frac{eD}{\mu} = kT + \frac{2}{3}\Delta_{av} , \qquad (9)$$

where Δ_{av} is the average of the shift in the electron kinetic energy as a function of N, and Δ is defined by Eq. (4). Consequently, the electron energy is predicted to increase with N when μN also increases with N. O'Malley contends that the very limited amount of eD/μ data as a function of N that are available in the literature tend to support his conclusions. Considerably more experimental data are required to test the validity of these relationships.

3. Electron Attachment to Molecules in Dense Gases

The gaseous medium in which electron attachment to a molecule occurs can have profound effects upon the mode of electron attachment and the magnitude of the corresponding rate constant. The interactions between the electrons, the attaching gas molecules, and the surrounding medium which cause these effects are still not understood in their entirety. In this section we briefly review the electron attachment mechanisms that occur, firstly in the low gas pressure regime where electron attachment mechanisms are generally thought to be well understood, and secondly in the high gas pressure regime where considerable work is still required to fully explore and understand the often subtle and complex effects of the medium in which electron attachment to molecules occurs.

3.1. Effect of the Nature and Density of the Gaseous Medium on Electron Attachment

It should, perhaps, be firstly indicated that the probing of the effects of the nature and density of the gaseous medium on electron attachment depends on the kind of the experimental method employed. Thus, in low-pressure ($P < 10^{-2}$ Pa) electron beam experiments only dissociative attachment and relatively long-lived (autodetachment lifetimes, $\tau_a > 10^{-6}$ s) parent (or fragment) negative ions are observable. Conversely, in electron swarm experiments long-lived parent anion formation and dissociative attachment processes result in electron attachment rate constants that are independent of gas pressure (assuming that collisional detachment from these negative ions is negligible), while moderately short-lived (τ_a in the range $\sim 10^{-7}$-10^{-12} s) parent negative ion formation appears as a pressure-dependent process. The rate constants for parent transient anion stabilization and destruction can be greatly influenced by the nature and density of the medium in which the electron-molecule attaching collisions take place.

Electron attachment in swarm experiments (where typically $P > 0.1$ kPa) can be characterized by the following reaction scheme:

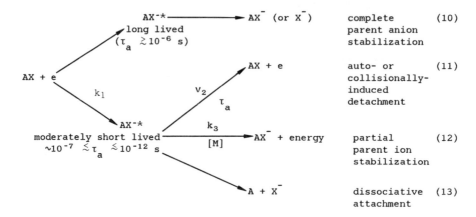

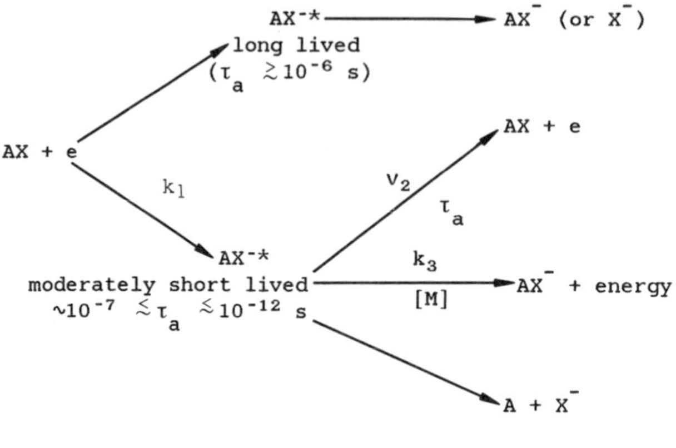

complete parent anion stabilization (10)

auto- or collisionally-induced detachment (11)

partial parent ion stabilization (12)

dissociative attachment (13)

In the absence of processes (10) and (13), competition between auto- or collisionally-induced detachment [process (11)] and collisional stabilization of the transient parent anion [process (12)] can result in a pressure-dependent attachment rate constant which is a function of the nature of the gaseous medium M and the total number density N_m. The apparent attachment rate constant k_a, for this reaction scheme is

$$k_a = k_1 \left[\frac{k_3 N_m}{\nu_2 + k_3 N_m} \right] . \qquad (14)$$

At low N_m or short τ_a, such that $\nu_2 = \tau_a^{-1} >> k_3 N_m$, then $k_a \propto N_m$ if dissociative attachment and other processes are absent. Alternatively, if the τ_a is long or N_m is large (i.e., $\tau_a^{-1} << k_3 N_m$), then $k_a = k_1$ and is independent of N_m or the nature of M. In typical swarm experiments (where $P \geq 0.1$ kPa), if $\tau_a > 10^{-6}$ s then virtully all transient parent anions are collisionally stabilized [i.e., process (10)]. However, when τ_a lies in the range 10^{-12} s $\lesssim \tau_a < 10^{-6}$ s, the observed attachment rate constant is highly dependent on the nature and density of the gaseous medium.

3.2. Normal Three-Body Attachment Processes--Perfluoroalkanes

At room temperature and relatively low gas pressures ($P_T \lesssim 20$ kPa), a number of gas molecules have been found to attach electrons by a three-body process which is well represented by processes (11) and (12) (e.g., O_2 [38, 39], SO_2 [40], and N_2O [41]). In a recent study by Hunter and Christophorou [42], the electron attachment rate constants for the perfluoroalkane series n-C_NF_{2N+2}

(N = 1-6) have been measured in a high-pressure swarm experiment (130 kPa $\leq P_T \leq$ 3.2 MPa) using N_2 and Ar as buffer gases. The normal three-body reaction scheme [processes (11) and (12)] was found to satisfactorily explain the observed pressure dependences in the attachment rate constants over the entire pressure range in both buffer gases. The observed attachment rate constants obtained as a function of mean electron energy at several gas pressures for C_3F_8 and n-C_4F_{10} are shown in Figs. 8A and 8B, respectively.

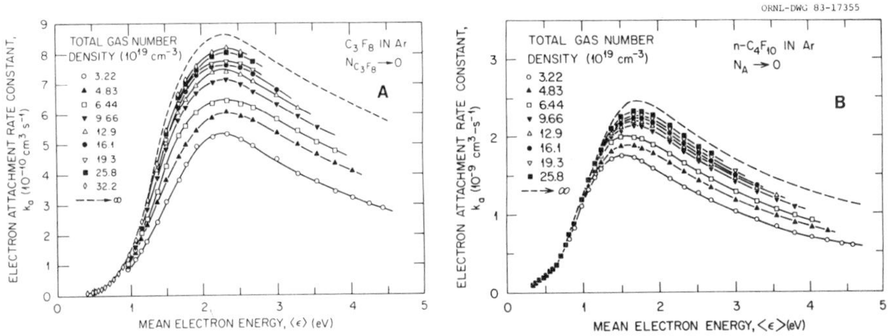

Fig. 8. Electron attachment rate constant, k_a, as a function of mean electron energy, ⟨ε⟩, and total gas pressure for C_3F_8 (Fig. 8A) and n-C_4F_{10} in Ar buffer gas (Fig. 8B) [42].

The electron attachment rate constant for the first two members of this series, namely, CF_4 and C_2F_6, has been found to be independent of gas pressure indicating that dissociative attachment [i.e., process (13)] is occurring for these two molecules. This is consistent with the results of electron beam studies [43]. For the larger members of this series, parent negative ion formation is the predominant mechanism, with the pressure dependence of k_a diminishing with increasing size of the molecule (Fig. 8). This behavior is due to the longer τ_a of the transient parent anions as the size of the molecule is increased thus requiring lower total gas pressures with which to saturate the anion stabilization process. Hunter and Christophorou [42] observed that k_a tended to saturate at $P_T \simeq$ 1.5 MPa for C_3F_8, $\simeq$ 1.0 MPa for n-C_4F_{10}, $\simeq$ 0.5 MPa for n-C_5F_{12}, and $\leq$ 0.1 MPa for n-C_6F_{14} and that increasing the total gas pressure beyond these values did not substantially increase k_a.

Analyses based on processes (11) and (12) often assume that each buffer gas molecule stabilizes the transient parent anion with unit efficiency. In general this is not the case, however, and the three-body electron attachment rate constant, k_{3m}, which can be found from Eq. (14) at low gas pressures, must be modified to

$$k_{3m} = \frac{k_a p}{N_m} , \qquad (15)$$

where p is the probability of collisional stabilization in each excited parent anion-buffer gas molecule collision. It was observed [42], for instance, that the stabilization efficiency of N_2 was different from that of Ar. Considerable variations in the ability of the buffer gas molecule to collisionally stabilize the transient parent anion have been observed for several other attaching gases [2, 44, 45]. The observation has been made [44, 45] that the rate of collisional stabilization at a given pressure increases (i.e., k_{3m} increases) with increasing complexity of the stabilizing molecule. Goans and Christophorou [46] have, for example, observed that C_2H_4 and C_2H_6 are considerably more efficient in stabilizing O_2^{-*} than is N_2 (Fig. 9A). Chanin et al. [38] have found that the greatest degree of stabilization is caused by O_2 itself, which appears to have a stabilization probability close to unity in this case. Similar observations were made by Rademacher et al. [47] for the three-body electron attachment process at low energies in SO_2, where C_2H_4 was again found to collisionally stabilize the transient parent anions at a considerably lower pressure than was the case for N_2 (Fig. 9B).

So far, this discussion has concentrated on electron attachment processes as a function of the total gas pressure which are well represented by processes (10) to (13). For a number of molecules, including O_2 and SO_2, the normal three-body process has been found to be inadequate in predicting the observed pressure dependences. Various explanations have been proposed to account for these anomalous pressure dependences. In essence, all these explanations are based on the interaction of the electron or transient parent anion with one or more of the surrounding gas molecules in the form of temporary or permanent dimers. Some of these types of processes are outlined below.

3.3. Dimer-Type Electron Attachment Processes

The dimer-type processes described in this section refer to molecular or ionic processes in which the molecules are bound together, either temporarily or permanently, by Van der Waals or other relatively weak forces and do not refer to permanent molecular complexes. These are described with reference to specific electron attaching gas molecules.

3.3.1. O_2

At low total gas pressures ($P_T \lesssim 10$ kPa) the low-energy, three-body attachment process in O_2 has been found by several researchers [38, 39] to obey the

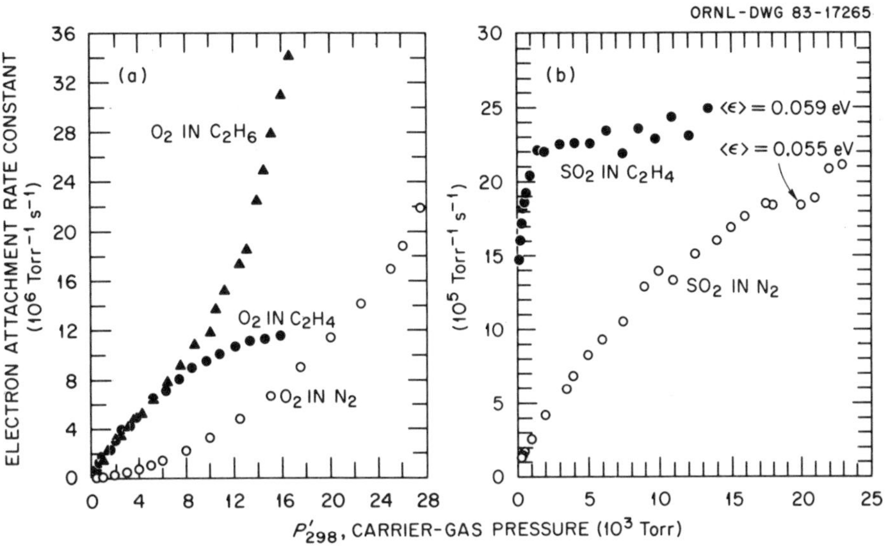

Fig. 9. Near-thermal energy (<ε> ≃ 0.05 eV) electron attachment rate constant as a function of carrier gas pressure for O_2 in N_2, C_2H_4, and C_2H_6 (Fig. 9A) [46]; and SO_2 in N_2 and C_2H_4 (Fig. 9B) [47]. Note that the units are in $Torr^{-1}$ s^{-1} (to change to cm^3 s^{-1} divide by the gas number density of 1 Torr at 298 K; i.e., by 3.24×10^{16} molecules cm^{-3}).

reaction scheme outlined by reactions (11) and (12). Subsequent studies have found that the rate of electron attachment does not saturate at higher gas pressures as predicted by Eq. (14). The measurements of k_a in O_2 by Goans and Christophorou [46] using N_2, C_2H_4, and C_2H_6 as buffer gases indicate that a normal three-body process is observable in the O_2/C_2H_4 mixture (Fig. 9A) over the entire pressure range of their measurements ($P_T \lesssim 2.2$ MPa), whereas the normal three-body processes can account for the observed attachment rate in the O_2/C_2H_6 mixtures only up to $P_T \lesssim 0.8$ MPa and that the rate of attachment actually increases instead of saturating at higher gas pressures. A similar observation was made in the O_2/N_2 gas mixture, with the nonsaturation effect occurring over the whole pressure range of the measurements (Fig. 9A). Goans and Christophorou [46] have interpreted the O_2/N_2 measurements as resulting from the influence of N_2 on the transient O_2^{-*} ion such as to perturb the O_2^{-*} potential energy curve at the higher gas pressures leading to enhanced electron attachment. The O_2/C_2H_6 measurements were interpreted as attachment to molecular clusters involving four or more C_2H_6 molecules.

Following the work of Christophorou and coworkers, Hatano, Shimamori, Fessenden, and their associates [48-50] made extensive measurements of thermal electron attachment to O_2 using many different buffer gases. Shimamori and Hatano [48] found that thermal electron attachment to O_2 followed a Bloch-Bradbury [processes (11) and (12)] mechanism in many binary gas mixtures at pressures below 20 kPa. Subsequent studies of thermal electron attachment to O_2 by Kokaku et al. [49] in several buffer gases have shown that the rate of electron attachment considerably exceeds that predicted by processes (11) and (12) at pressure $\lesssim 100$ kPa. The excess attachment was attributed to electron capture by pre-existing Van der Waals complexes ($O_2 \cdot M$) at higher pressures. Shimamori and Fessenden [50] have recently measured the temperature dependence of the three-body attachment rate constant in pure O_2 ($^{16}O_2$ and $^{18}O_2$), O_2/N_2, and O_2/CO mixtures at pressures $\lesssim 8$ kPa and have arrived at the conclusion that virtually all the electron attachment in O_2/N_2 mixtures over the temperature range $78 < T < 330$ K is by Van der Waals $O_2 \cdot N_2$ dimers. If the interpretation of those results proves to be correct, this work would suggest that Van der Waals molecules are important species in electron attachment processes not only at high but also at low pressures.

Finally, McMahon [51] has derived a generalized reaction scheme for the dependences of the rate constants of electron attachment to O_2 in various gaseous media. His reaction scheme includes collisional detachment and models the electron attachment processes in the O_2/buffer gases over a wide range of gas pressures using a single-step, two-step, and higher-order electron attachment processes.

3.3.2. N_2O, SO_2

Although electron attachment to O_2 has received the greatest attention both from an experimental and theoretical standpoint in recent years, several other molecular gases possess similar electron attaching processes. The molecules N_2O and SO_2, for example, have electron attachment mechanisms that are strikingly similar to those of O_2 (i.e., they possess a three-body electron attachment process at low energies, which decreases in magnitude with increasing energy and also dissociative electron attachment processes at high E/N values). The dissociative attachment to N_2O is complicated by competing pressure-dependent detachment and ion-molecule reaction processes. The initial negative ion formed at high E/N has been found by Parkes [52] to be O^- which quickly undergoes further ion-molecule reactions to form NO^- and other anions. The molecule NO has a very low electron affinity (EA ≤ 0.1 eV [53]), and thus NO^- can be easily collisionally detached or undergo further ion-molecule reactions to form more stable negative ion species [52, 54]. The addition of CO_2 and O_2 to N_2O has been found by Parkes [55] and Dutton et al. [56] to lead to a dramatic increase in stable negative ion production and pressure dependent high E/N electron attachment processes over a wide pressure range.

Electron attachment at thermal and epithermal electron energies has been found [41, 57, 58] to obey a three-body reaction scheme at low total gas pressures (<100 kPa). At higher gas pressures the results of more recent studies [59] on the electron attachment processes in mixtures of N_2O with several buffer gases led Shimamori and Fessenden [59] to suggest that thermal electron attachment to Van der Waals molecules significantly contributes to the electron attachment rate constants they measured, even at gas pressures of only a few tens of kilopascals. The nature of the buffer gas molecule has been found in these studies to have a great influence on the apparent three-body attachment rate constant for N_2O, similar to that observed in O_2/buffer gas mixtures [44, 45].

Fewer electron attachment studies have been performed in SO_2. Rademacher et al. [47] observed a three-body electron attachment process at thermal energies in agreement with more recent low-pressure studies [40]. In both of these studies, a dissociative attachment process was observed at higher E/N values. One of the dissociative attachment fragment anions, O^-, has been found [40, 60] to undergo associative detachment with SO_2 (i.e., $SO_2 + O^- \rightarrow SO_3 + e$). Consequently, pressure-dependent ion-molecule reaction processes similar to those observed for N_2O are expected to occur at higher gas pressures in SO_2 as well.

The electron attachment rate constant at energies just above thermal for SO_2 in N_2 and C_2H_4 are plotted in Fig. 9B as a function of total gas pressure. These measurements indicate that C_2H_4 is considerably more efficient than N_2 in stabilizing the transient SO_2^{-*} ion. They also indicate that while a normal three-body electron attachment reaction scheme is able to model the observed pressure dependence in the electron attachment rate over the whole pressure range, in SO_2/C_2H_4 gas mixtures and in SO_2/N_2 at lower total gas pressures (<0.5 MPa), additional electron attachment processes contribute to the observed rate constants in high pressure (P_T > 0.5 MPa) SO_2/N_2 gas mixtures. Mechanisms similar to those proposed by Goans and Christophorou [46] or Shimamori et al. [48, 59] in O_2 and N_2O may possibly account for the increased rate of electron attachment.

3.3.3. 1-C_3F_6

Electron attachment to 1-C_3F_6 has recently been observed to be greatly dependent on gas pressure and temperature, both in the pure gas and in gas mixtures. Hunter et al. [61] have observed that the electron attachment rate constant for 1-C_3F_6 in N_2 and Ar buffer gases (Fig. 10) increases dramatically with increasing total gas pressure and partial 1-C_3F_6 pressure. Measurements of the effective ionization coefficient in pure 1-C_3F_6 [62, 63] also exhibit a large pressure dependence. A very large variation in the rate of electron attachment with temperature was observed in 1-C_3F_6/N_2 and

1-C_3F_6/Ar gas mixtures [61, 64], the rate of electron attachment decreasing dramatically with increasing gas temperature above ~300 K (Fig. 11A). Recently, McCorkle et al. [64] have demonstrated that below ~300 K, the rate of electron attachment decreases with decreasing gas temperature (Fig. 11B), particularly at electron energies near thermal energy.

ORNL-DWG 83-17356

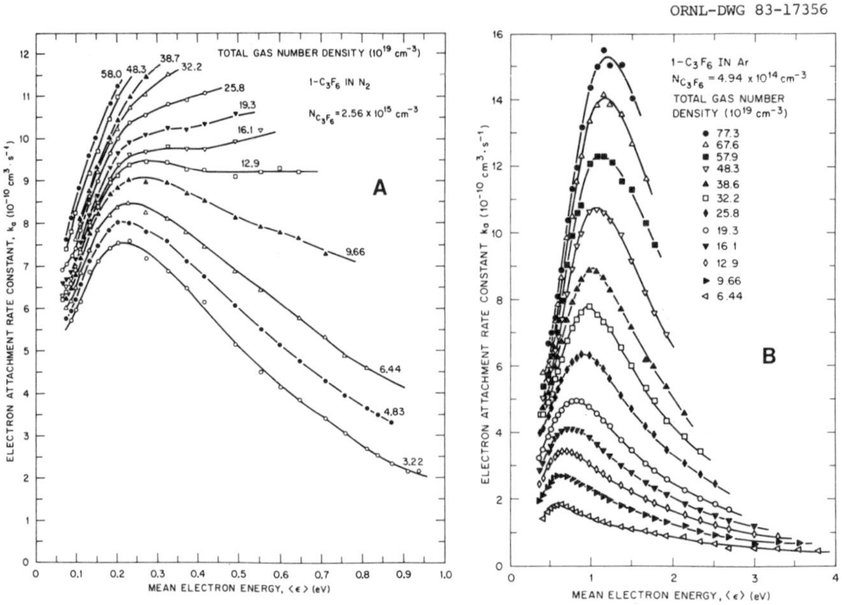

Fig. 10. Electron attachment rate constant, k_a, for 1-C_3F_6 at fixed concentrations of 1-C_3F_6 as a function of $\langle \varepsilon \rangle$ and the total gas pressure of N_2 (Fig. 10A) and Ar buffer gases (Fig. 10B) [61].

Mass spectrometric studies and temperature- and pressure-dependent breakdown measurements [61], along with the temperature- and pressure-dependent attachment studies, enabled Hunter et al. [61] to propose the following reaction scheme for electron attachment to 1-C_3F_6 below ~2 eV:

$$
\begin{array}{ccc}
1\text{-}C_3F_6 + e & & 2(1\text{-}C_3F_6) + e \\
\Big\uparrow [M]\ k_3 & & \Big\uparrow [M]\ k_6 \\
e + 1\text{-}C_3F_6 \xrightarrow{k_1} 1\text{-}C_3F_6^{-*} \xrightarrow[{[1\text{-}C_3F_6]}]{k_4 N_A} (1\text{-}C_3F_6)_2^{-*} \xrightarrow[{[M]}]{k_7 M} (C_3F_6)_2^{-} \\
\Big\downarrow v_2 & & \Big\downarrow v_5 \\
1\text{-}C_3F_6 + e & & 2(1\text{-}C_3F_6) + e \quad .
\end{array}
$$

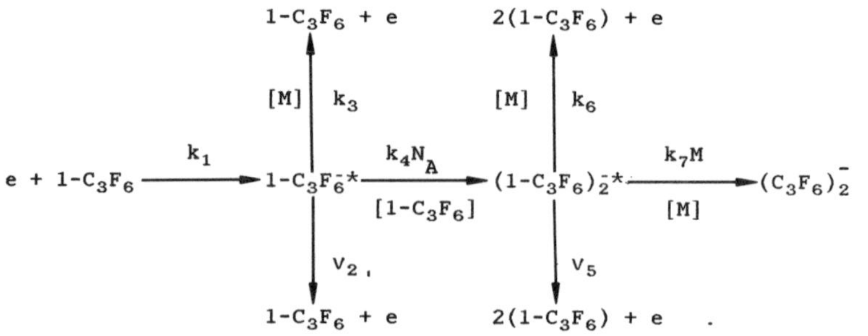

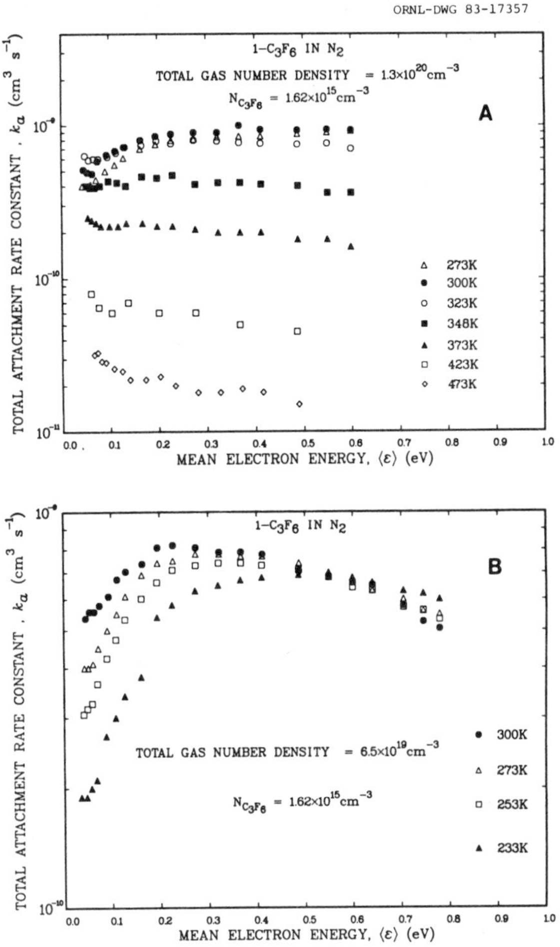

Fig. 11. Electron attachment rate constant, k_a, for 1-C_3F_6 as a function of $\langle\varepsilon\rangle$ at a fixed partial 1-C_3F_6 and total gas pressure for several gas temperatures in the range $273 \leq T \leq 473$ K (Fig. 11A) and $233 \leq T \leq 300$ K (Fig. 11B) [64].

In (16) an electron associatively attaches to the parent molecule with a rate constant k_1 to form a comparatively long-lived parent negative ion ($\tau_{a1} = 1/\nu_2 \lesssim 10^{-6}$ s) which may combine with another 1-C_3F_6 molecule with a large rate constant, k_4, to form a short-lived dimer ion ($\tau_{a2} = 1/\nu_5$ of the order 10^{-12} s) which can be stabilized by the buffer gas, M, with a rate constant k_7. This reaction scheme has been simplified by observing that at low electron energies ($\varepsilon < 2$ eV) dissociative electron attachment processes are negligible [65]. The mass spectrometric studies also indicated that stable parent anion formation via electron attachment to $1\text{-}C_3F_6$ is negligible. Thus, the observed electron attachment rate constant $k_a(N_A,M)$ for this reaction scheme at low electron energies can be expressed as

$$k_a(N_A,M) = \frac{k_1 k_4 k_7 N_A M}{(\nu_2 + k_3 M + k_4 N_A)(\nu_5 + k_6 M + k_7 M)} \quad . \tag{17}$$

The large temperature dependence of k_a has been attributed to the increase in ν_2 with increasing T [64]. McCorkle et al. [64] rewrote Eq. (17) as

$$k_a(N_a,M,T) = k_1 \frac{k_4 N_A}{(\nu_2 + k_4 N_A)} \times p \ , \tag{18}$$

where p is the probability of stable anion formation via the transient $(1\text{-}C_3F_6)^{-*}$. If it is assumed that p and k_4N_A are T independent and the temperature dependence of k_1 and ν_2 can be represented by an Arrhenius expression, then

$$k_a(N_A,M,T) = k_1' \frac{e^{-\left(\frac{\varepsilon_{max}}{kT + \langle\varepsilon\rangle'}\right)} \, k_4 N_A}{\nu_2' \, e^{-\left(\frac{E'}{kT}\right)} + k_4 N_A} \times p \ . \tag{19}$$

In Eq. (19), $\langle\varepsilon\rangle'$ is the contribution to the mean electron energy from the electric field, $kT + \langle\varepsilon\rangle'$ is the mean electron energy, ε_{max} is the *vertical attachment energy*, and E' is related to the EA of $1\text{-}C_3F_6$ which is expected to be small. From Eq. (19), $k_a(T)$ can be affected by changes in T in two opposite ways: (1) decrease with T because ν_2 increases with T thus destroying $1\text{-}C_3F_6^{-*}$ by autodetachment and (2) increase with T because k_1 may increase when increases in T make the transition to the negative ion state energetically more favorable; this, of course, is negligible when we consider values of k_a at $\langle\varepsilon\rangle \gg kT$. Thus, the T dependence of k_a at $\langle\varepsilon\rangle \gg kT$ is dominated by the T dependence of the autodetachment decay constant ν_2. In this case k_a increases with decreasing T as long as T is high enough for autodetachment to be significant (see Fig. 11). On the other hand, the T dependence of k_a at small values of $\langle\varepsilon\rangle$ may be influenced by the increases in k_1 with T, and in this case k_a may actually increase with T at low T where autodetachment is small (see Fig. 11 for $\langle\varepsilon\rangle < 0.1$ eV).

4. Concluding Remarks

While in this chapter we focused our discussion on the effect of the nature and density of the gaseous medium on the electron drift velocity and the electron attachment rate constant, similar effects on other quantities such as the electron diffusion coefficient and the electron scattering and ionization cross sections are being investigated [7]. Also, while our discussion on w(E/N) and $k_a(\langle\varepsilon\rangle)$ in this chapter has been mostly for gas number densities well below the critical number density, similar studies over the entire density range from a low-pressure gas to the liquid state have been made (e.g., see a recent review in [7]). Our understanding of the effects of gas density on the interactions of electrons in dense gases and, consequently, on the electron transport coefficients and reaction rate constants is still incomplete. Much experimental and theoretical work on the interactions of slow electrons in dense gases in particular--and on other medium-dependent processes [7] in general--is needed and indeed anticipated in the years ahead.

5. Acknowledgment

This research has been sponsored by the Office of Health and Environmental Research, U.S. Department of Energy, under contract W-7405-eng-26 and by the Office of Naval Research under contract DOE No. 40-1246-82 with the Union Carbide Corporation.

References

[1] Christophorou, L. G.: Rad. Phys. Chem. 7, 205 (1975).

[2] Christophorou, L. G.: Chem. Revs. 76, 409 (1976).

[3] Freeman, G. R.: In: Electron and Ion Swarms (Christophorou, L. G., Ed.), p. 93. New York: Pergamon Press, 1981.

[4] Iakubov, I. T. and Khrapak, A. G.: Electrons in Dense Gases and Plasmas. Moscow: Publishing House "Nauka", 1981.

[5] Christophorou, L. G. and Siomos, K.: In: Proceedings of the 3rd International Swarm Seminar (Lindinger, W., Villinger, H., and Federer, W., Eds.), p. 12. Austria: Innsbruck, 1983.

[6] Hunter, S. R. and Christophorou, L. G.: In: Electron-Molecule Interactions and Their Applications, Vol. 2, Chapter 3 (Christophorou, L. G., Ed.). New York: Academic Press (in press).

[7] Christophorou, L. G. and Siomos, K.: In: Electron-Molecule Interactions and Their Applications, Vol. 2, Chapter 4 (Christophorou, L. G., Ed.). New York: Academic Press (in press).

[8] Lowke, J. J.: Aust. J. Phys. 16, 115 (1963).

[9] Levine, J. L. and Sanders, T. M.: Phys. Rev. Lett. 8, 159 (1962).

[10] Grünberg, R.: Z. Phys. 204, 12 (1967).

[11] Grünberg, R.: Z. Naturforsch. 23a, 1994 (1968).

[12] Grünberg, R.: Z. Naturforsch. 24a, 1838 (1969).

[13] Bartels, A.: Appl. Phys. 8, 59 (1975).

[14] Huber, B.: Z. Naturforsch. 23a, 1228 (1968).

[15] Huber, B.: Z. Naturforsch. 24a, 578 (1969).

[16] Lehning, H.: Phys. Lett. 29A, 719 (1969).

[17] Bartels, A.: Phys. Lett. 44A, 403 (1973).

[18] Atrazhev, V. M. and Iakubov, I. T.: J. Phys. D 10, 2155 (1977).

[19] Atrazhev, V. M. and Iakubov, I. T.: High Temp. 18, 966 (1980).

[20] Robertson, A. G.: Aust. J. Phys. 30, 39 (1977).

[21] Lehning, H.: Phys. Lett. 28A, 103 (1968).

[22] Warman, J. M., Sowada, U., and Armstrong, D. A.: Chem. Phys. Lett. 82, 458 (1981).

[23] O'Malley, T. F.: J. Phys. B 13, 1491 (1980).

[24] Schwarz, K. W.: Phys. Rev. Lett. 41, 239(1978); Schwarz, K. W.: Phys. Rev. B 21, 5125 (1980).

[25] (a) Braglia, G. L. and Dallacasa, V.: Phys. Rev. A 26, 902 (1982).
(b) Braglia, G. L. and Dallacasa, V.: Phys. Rev. A 18, 711 (1978).
(c) Dallacasa, V.: J. Phys. B 12, 3125 (1979).

[26] Aschwanden, Th.: In: Gaseous Dielectrics III (Christophorou, L. G., Ed.), p. 32. New York: Pergamon Press, 1982.

[27] (a) Krebs, P. and Wantschik, M.: J. Phys. Chem. 84, 1155 (1980).
(b) Krebs, P.: Chem. Phys. Lett. 70, 465 (1980).
(c) Krebs, P., Giraud, V., and Wantschik, M.: Phys. Rev. Lett. 44, 211 (1980).
(d) Krebs, P. and Heintze, M.: J. Chem. Phys. 76, 5484 (1982).

[28] Christophorou, L. G., Carter, J. G., and Maxey, D. V.: J. Chem. Phys. 76, 2653 (1982).

[29] Giraud, V. and Krebs, P.: Chem. Phys. Lett. 86, 85 (1982).

[30] Frommhold, L.: Phys. Rev. 172, 118 (1968).

[31] Ritchie, R. H. and Turner, J. E.: Z. Phys. 200, 259 (1967).

[32] Levine, J. L. and Sanders, T. M.: Phys. Rev. 154, 138 (1967); Harrison, H. R., Sander, L. M. and Springett, B. E.: J. Phys. B 6, 908 (1973).

[33] Legler, W.: Phys. Lett. 31A, 129 (1970).

[34] Iakubov, I. T. and Polischuk, A. Y.: Phys. Lett. 91A, 67 (1982).

[35] Iakubov, I. T. and Polischuk, A. Y.: J. Phys. B 15, 4029 (1982).

[36] Gryko, J. and Popielawski, J.: Phys. Rev. A 24, 1129 (1981).

[37] O'Malley, T. F.: Phys. Lett. 95A, 32 (1983).

[38] Chanin, L. M., Phelps, A. V., and Biondi, M. A.: Phys. Rev. 128, 219 (1962).

[39] Grünberg, R.: Z. Naturforsch. 24a, 1039 (1969).

[40] Lakdawala, V. K. and Moruzzi, J. L.: J. Phys. D 14, 2015 (1981).

[41] Phelps, A. V. and Voshall, R. E.: J. Chem. Phys. 49, 3246 (1968).

[42] Hunter, S. R. and Christophorou, L. G.: Journal of Chemical Physics (submitted).

[43] Spyrou, S. M., Sauers, I., and Christophorou, L. G.: J. Chem. Phys. 78, 7200 (1983).

[44] Caledonia, G. F.: Chem. Revs. 75, 333 (1975).

[45] Christophorou, L. G.: Rad. Phys. Chem. 12, 19 (1978).

[46] Goans, R. E. and Christophorou, L. G.: J. Chem. Phys. 60, 1036 (1974).

[47] Rademacher, J., Christophorou, L. G., and Blaunstein, R. P.: J. Chem. Soc., Faraday Trans. II 71, 1212 (1975).

[48] Shimamori, H. and Hatano, Y.: Chem. Phys. 12, 439 (1976); Shimamori, H. and Hatano, Y.: Chem. Phys. Lett. 38, 242 (1976); Shimamori, H. and Hatano, Y.: Chem. Phys. 21, 187 (1977).

[49] Kokaku, Y., Hatano, Y., Shimamori, H., and Fessenden, R. W.: J. Chem. Phys. 71, 4883 (1979); Kokaku, Y, Toriumi, M., and Hatano, Y.: J. Chem. Phys. 73, 6167 (1980).

[50] Shimamori, H. and Fessenden, R. W.: J. Chem. Phys. 74, 453 (1981); Shimamori, H. and Hotta, H.: J. Chem. Phys. 78, 1318 (1983).

[51] McMahon, D.R.A.: In: Electron and Ion Swarms (Christophorou, L. G., Ed.), p. 117. New York: Pergamon Press, 1981; McMahon, D.R.A.: Chem. Phys. 63, 95 (1981); McMahon, D.R.A.: Chem. Phys. 66, 67 (1982).

[52] Parkes, D. A.: J. Chem. Soc., Faraday Trans. I 68, 2103 (1972).

[53] Hughes, B. M., Lifshitz, C., and Tiernan, T. O.: J. Chem. Phys. 59, 3162 (1973); Chen, E.C.M. and Wentworth, W. E.: J. Phys. Chem. 87, 45 (1983).

[54] McFarland, M., Dunkin, D. B., Fehsenfeld, F. C., Schmeltekopf, A. L., and Ferguson, E. E.: J. Chem. Phys. 56, 2358 (1972).

[55] Parkes, D.A.: J. Chem. Soc., Faraday Trans. I 68, 2121 (1972).

[56] Dutton, J., Harris, F. M., and Hughes, D. B.: J. Phys. D 8, 313 (1975); Dutton, J., Harris, F. M., and Hughes, D. B.: J. Phys. D 8, 1640 (1975).

[57] Chaney, E. L. and Christophorou, L. G.: J. Chem. Phys. 51, 883 (1969).

[58] Warman, J. M., Fessenden, R. W., and Bakale, G.: J. Chem. Phys. 57, 2702 (1972); Warman, J. M. and Fessenden, R. W.: J. Chem. Phys. 49, 4718 (1968).

[59] Shimamori, H. and Fessenden, R. W.: J. Chem. Phys. 68, 2757 (1978); Shimamori, H. and Fessenden, R. W.: J. Chem. Phys. 69, 4732 (1978); Shimamori, H. and Fessenden, R. W.: J. Chem. Phys. 70, 1137 (1979); Shimamori, H. and Fessenden, R. W.: J. Chem. Phys. 71, 3009 (1979).

[60] Doussot, C., Bastien, F., Marode, E., and Moruzzi, J. L.: J. Phys. D 16, 2451 (1982).

[61] Hunter, S. R., Christophorou, L. G., McCorkle, D. L., Sauers, I., Ellis, H. W., and James, D. R.: J. Phys. D 16, 573 (1982).

[62] Aschwanden, Th., Böttcher, H., Hansen, D., Jungblut, H., and Schmidt, W. F.: In: Gaseous Dielectrics III (Christophorou, L. G., Ed.), p. 23. New York: Pergamon Press, 1982.

[63] Verhaardt, M.F.A. and van der Laan, P.C.T.: Paper 33.12 in Proceedings of the Fourth International Symposium on High Voltage Engineering, Athens, Greece, September 5-9, 1983.

[64] McCorkle, D. L., Christophorou, L. G., and Hunter, S. R.: In: Proceedings of the 3rd International Swarm Seminar (Lindinger, W., Villinger, H., and Federer, W., Eds.), p. 37. Austria: Innsbruck, 1983.

[65] Harland, P. W. and Thynne, J. C.: Int. J. Mass Spectrom. Ion Phys. 9, 253 (1972); Lifshitz, C. and Grajower, R.: Int. J. Mass Spectrom. Ion Phys. 10, 25 (1972/73).

Electrons in Inert Gases

I. Ogawa

Department of Physics, Rikkyo University, Toshima-Ku, Tokyo, 171, Japan

Studies of the behaviour of an electron swarm in various inert gases are not only important in practical applications such as the electrical discharge engineering and the development of radiation detectors or gaseous lasers, but also of great value, or even indispensable, in obtaining precise information about collision processes, especially the elastic scattering, of low-energy electrons with gas atoms.

Since inert gases are monatomic and their atoms are of closed-shell structure, the analysis to be used for deriving cross sections from swarm data naturally becomes the simplest and hence the best established, and may be performed quite accurately since only elastic collisions are essentially relevant.

Theoretically, also, the collision of an electron with an inert gas atom is the most typical case of electron-atom collision processes and has long been investigated in detail by a number of authors. The presence of remarkable Ramsauer-Townsend effect in Ar, Kr, and Xe adds a particular interest to these gases.

In spite of these significances and interest, however, agreement among the reported values of various swarm parameters and derived cross sections for electrons in inert gases has not been very satisfactory in many cases, owing to several experimental difficulties related, for example, to the extreme sensitivity of various swarm properties for these gases to the gaseous impurity. A substantial disagreement remains to be seen, in particular, with regard to the position and the shape (the depth, in particular) of the Ramsauer minimum in the momentum transfer cross section for electrons in Ar, Kr, and Xe as a function of electron energy. Also, no modernized measurement had ever been reported until very recently of the characteristic energy, $\varepsilon_k \equiv eD/\mu$, for electrons in Ne, Kr and Xe, one of the most important swarm parameters which may be ranked with the drift velocity, w. In the definition made above, e denotes the elementary charge, D the "transverse" diffusion coefficient for electrons in a direction perpendicular to the applied electric field, and μ the electron mobility.

Recently, however, these experimental difficulties have been gradually overcome technically and some long-lacking swarm data became first available with a high accuracy, including, for example, the characteristic energy data for Ne and Xe. This will certainly contribute greatly to the resolution of the existing discrepancy among the reported swarm data and derived cross sections, and also to the evaluation of different theoretical estimates.

In the present article, the present status of such a recent progress will be reviewed briefly, after summarizing the general features of an electron swarm in inert gases. Attempts will be made on the way to point out the tasks still remaining to be pursued. Because of the limited available space, our discussion will be concentrated on the case of electron swarms in complete drift-equilibrium under lower reduced electric fields, E/N, where any inelastic process may safely be neglected. Here, E denotes the strength of the applied electric field and N the number density of gas atoms.

1. Feature of an Electron Swarm in Inert Gases

1.1 General Features

In the absence of any inelastic interaction between an electron and a gas atom, the approximate solution of the Boltzmann transport equation becomes greatly facilitated and the distribution function $f(\varepsilon)$ of electron energy ε proves to be given by a simple closed formula (known as the Davydov distribution /1/):

$$f(\varepsilon) = A \exp \{ - \frac{6m}{M} \int_0^{\varepsilon} \frac{\varepsilon}{\varepsilon_1^2 + \frac{2m}{M} kT\varepsilon} d\varepsilon \}, \tag{1}$$

where m and M denotes, respectively, the mass of an electron and an atom, k the Boltzmann constant, T the gas temperature, $\varepsilon_1 \equiv eE/Nq_m$, and q_m the momentum transfer cross section for electrons as defined below, while A is the normalizing constant to be determined from the condition $\int f(\varepsilon)d\gamma = 1$. Here, $d\gamma$ denotes the infinitesimal volume in the velocity space.

As a result, the procedure to determine the momentum transfer cross section

$$q_m \equiv \int q(\theta)\ (1 - \cos\theta)\ d\omega = q_s\ (1 - \overline{\cos\theta}) \tag{2}$$

as a function of electron energy from the observed swarm data (e. g., the drift velocity and/or the characteristic energy as a function of E/N and T) becomes remarkably simplified. Here, $q(\theta)$ denotes the differential scattering cross section, θ the scattering angle, $q_s = \int q(\theta)\, d\omega$ the total scattering cross section, and $d\omega = 2\pi \sin\theta\, d\theta$ the infinitesimal solid angle.

This is the great merit of inert gases in investigating the relevant collision processes from the observed swarm behaviours as compared with molecular gases. At the same time, however, the following facts make electron swarm experiments with inert gases considerably difficult in a peculiar way:

(1) The absence of inelastic processes like rotational or vibrational excitation of a molecule makes the achievement of drift equilibrium much slower than in molecular gases. Such an effect manifests itself most strongly in Ramsauer gases like Ar, Kr and Xe.

(2) The presence of an extremely small amount of molecular (not necessarily electron-attaching) impurities, sometimes as low as a few ppm, may well affect strongly the electron energy distribution and hence exert a serious influence upon the magnitude of various swarm parameters, deteriorating the accuracy of estimated cross sections, especially in the case of Ramsauer gases.

(3) By the same reason as in (1), the average random energy, and hence the average random velocity of electrons in inert gases, is much greater than in molecular gases, resulting in a strong diffusivity of swarm electrons. This often hinders the passage of electrons through a narrow slit, a small aperture hole, or a dense grid, the electrons being readily trapped by the periphery or grid wires, reduces the swarm current, and eventually lower the accuracy in measuring swarm parameters particularly at lower gas pressures. The only practical solution of this problem would be to use higher gas pressures, which in itself involves various experimental troubles.

(4) Again by the same reason, the drift velocity of electrons in inert gases is generally much slower than in molecular gases, and the diffusion (longitudinal, in particular) often takes place rapidly during their drift, thus limiting sometimes the accuracy of the time of flight (exactly,drift) method widely employed to measure the drift velocity, or impairing some basic performances of applicational instruments like the spatial or temporal resolution of a radiation detector.

Fortunately, however, these difficulties have been gradually surmounted and the lacking important data are being steadily acquired in recent years.

1.2 Comparison of Swarm Parameters for Individual Species of Inert Gases

Let us consider semiquantitatively how the two major swarm parameters, *i.e.*

the drift velocity w and the characteristic energy ε_k, depend upon the mass M of a gas atom and the rough magnitude of the momentum transfer cross section $\bar{q}_m$ for electrons in individual inert gases, by neglecting for simplicity the gas temperature T and by following an elementary treatment similar to the one attempted early by Alfven /2/.

Denoting by m the electron mass, by v the average random velocity of electrons, and by $\nu_m \equiv N\bar{q}_m v$ the momentum transfer collision frequency, the two parameters may be readily shown (See the Appendix) to be approximately given by

$$w \cong c_1\ eE/(m\nu_m) = c_2\ (m/M)^{1/4}\ [(2/m)(eE/N\bar{q}_m)]^{1/2}$$
$$= c_3\ [M(amu)]^{-1/4}[\bar{q}_m(A^2)]^{-1/2}[(E/N)(Td)]^{1/2} \quad cm/\mu s \tag{3}$$

and

$$\varepsilon_k \cong c_4(M/m)^{1/2}(eE/N\bar{q}_m)$$
$$= c_5\ [M(amu)]^{1/2}[\bar{q}_m(A^2)]^{-1}[(E/N)(Td)] \qquad eV, \tag{4}$$

respectively. Here, c_i's (i = 1,2,...,5) are numerical constants of the order of unity and $\bar{q}_m$ represents the momentum transfer cross section averaged over an energy region where the substantial part of electron energy distribution spreads depending on the magnitude of E/N.

According to these results, the drift velocity should be roughly proportional to the square root of E/N with a factor inversely proportional to the product of the fourth root of the atomic mass and the square root of the averaged momentum transfer cross section $\bar{q}_m$, provided that E/N (and hence ε_k) is large enough to neglect the thermal motion of gas atoms.

For smaller E/N's, $v \cong \sqrt{3kT/m}$ approximately, so that w is given as

$$w \cong c_6(e/\sqrt{3mkT})(\bar{q}_m)^{-1}(E/N), \tag{5}$$

where c_6 is a numerical constant close to unity, i. e., w becomes proportional to E/N with a factor independent of M, inversely proportional to $\bar{q}_m$, and also inversely proportional to the square root of gas temperature.

Thus, $w \propto E/N$ for lower E/N's and $\propto \sqrt{E/N}$ for higher E/N's if $\bar{q}_m$ varies little with electron energy. This is indeed approximately the case with He and Ne, but no longer with Ramsauer gases like Ar, Kr and Xe, for which the observed w vs. E/N curve shows a peculiar shoulder near a particular value of E/N where $\bar{q}_m$ becomes minimum.

As regards the characteristic energy, the above result shows that it is roughly proportional to the squre root of the atomic mass divided by the averaged momentum transfer cross section $\bar{q}_m$ for sufficiently large E/N's. Therefore,

ε_k is generally more sensitive to $\bar{q}_m$ when compared with w which is inversely proportional to the square root of $\bar{q}_m$. It should be noted here that the quantity $\varepsilon_k/(E/N) = (eD/\mu)/(E/N)$, which is often referred to in order to indicate the "diffusivity" of drifting electrons in a gas, may be taken as a measure of microscopic quantity $\sqrt{M} / \bar{q}_m$ for larger E/N's for which $\varepsilon_k >> kT$.

In He and Ne, experimental data show indeed that ε_k is roughly proportional to E/N for sufficiently large E/N's (For Ne, see Fig. 1). Furthermore, the ratio of the observed values of ε_k for Ne and He at E/N = 0.2 Td, for example, i. e. $\varepsilon_k(Ne)/ \varepsilon_k(He) = 0.83/0.11 \cong 7.6$, is in fact in good agreement with the calculated ratio of $(\sqrt{M}/\bar{q}_m)(Ne)/(\sqrt{M}/\bar{q}_m)(He) = \sqrt{20/4}\ (6.08/1.69) \cong 8.0$ from eq.(4).

In Ar, Kr and Xe, however, ε_k is no longer proportional to E/N as a result of Ramsauer effect and increases first rapidly at a certain value of E/N as is seen in Figs. 2 and 3 and then continues to rise slowly. In these gases, the magnitude of the averaged cross section $\bar{q}_m$ for each value of E/N cannot readily be estimated definitely by inspection because of the violent variation in $q_m(\varepsilon)$ and the resulting ambiguity in the shape of energy distribution for electrons. In spite of these unfavourable circumstances, the preceding semiquantitative result is still capable of explaining fairly well the general trend of the observed ε_k vs. E/N characteristics for individual inert gases including even those with Ramsauer effect.

For instance, the values of E/N at which ε_k is expected to take on a common value of 0.3 eV, as calculated from eq.(4) assuming tentatively as $c_5 = 1.5$, are in fact fairly close to the observed values as shown in Table 1. Thus, they reproduce fairly well the observed order followed by $(E/N)_{obs}(\varepsilon_k = 0.3$ eV)'s as well as their relative ratios for individual gas species, except for a minor disorder (reversal in order) for Ne and Xe.

Table 1. Comparison of E/N's for various inert gases giving the same value of $\varepsilon_k = 0.3$ eV as calculated from eq.(4) with $c_5 = 1.5$.

	He	Ne	Ar	Kr	Xe
M (amu)	4	20	40	84	131
$\bar{q}_m (A^2)$	6.4	1.1	0.16	0.75	3.5
$(E/N)_{calc}$ (Td)	0.64	0.05	0.005	0.016	0.06
$(E/N)_{obs}$ (Td)	0.66	0.050	0.0053	0.012	0.032

2. Present Status of the Measurement of Major Swarm Parameters

The most important transport parameters for an electron swarm in a gas in drift equilibrium are the drift velocity w and the characteristic energy ε_k as defined earlier. Among other useful parameters are the "longitudinal" diffusion coefficient D_L in the direction parallel to the electric field, the "longitudinal" characteristic energy $\varepsilon_L \equiv eD_L/\mu$, the anisotropy ratio $S \equiv D_L/D$ for diffusion, and the "magnetic" drift velocity w_M to be obtained under a transverse magnetic field. In the present subsection, a brief review is given of the present status of measurement only of the three major parameters w, ε_k and D_L in high purity inert gases.

Table 2 lists the names of author of some significant papers worthy of particular attention for each item. The references are given at the end of the present article. A more comprehensive list of references before about 1979 has been published by Beaty, Dutton and Pitchford /3/. A compilation of swarm data prior to 1972 has been made earlier by Dutton /4/. Reference may be also made to the critical surveys given in the textbooks written by Gilardini /5/ and by Huxley and Crompton /6/, although the data and references collected there are limited to those earlier than about 1971 - 2.

As is seen in the Table, the drift velocity has already been measured most accurately for almost all the inert gases, usually by means of electrical shutter method. Phelps and his colleagues /7/,/8/ are among the first who gave the best w data available at that time for all of He, Ne, Ar, Kr and Xe at room and lower temperatures as early as 1961 - 2. For He, Ne and Ar, however, Crompton et al. /9/,/10/ and Robertson /11/,/12/ have later attempted improved measurements and their results are now believed to be most accurate.

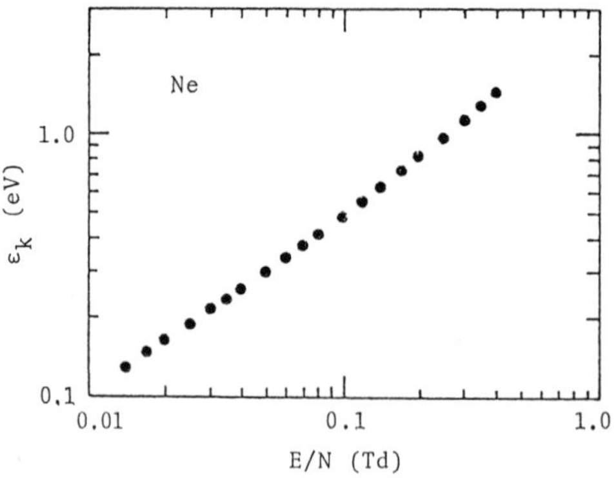

Fig. 1. The characteristic energy of electrons in Ne as observed by Ogawa, Koizumi, Murakoshi, Yamamoto and Shirakawa /15/ at 292 ± 1 K.

Table 2. Present status of the measurement of basic swarm parameters and the determination of the momentum transfer cross section for electrons in inert gases. References are given at the end of the present article.

Gas	Drift velocity, w	Characteristic energy $\varepsilon_k \equiv eD/\mu$	Longitudinal diffusion coefficient, D_L	Momentum transfer cross section, q_m experimental	Momentum transfer cross section, q_m theoretical
He	Pack and Phelps('61) /7/ Crompton, Elford and Jory ('67; 293K) /9/ Crompton, Elford & Robertson('70;77K)/10/	Warren and Parker ('62) /13/ Crompton, Elford and Jory ('67; 293K) /9/	Wagner, Davis and Hurst ('67) /18/ Elford ('74) /19/	Crompton, Elford and Jory ('67) /9/ Crompton, Elford and Robertson ('70) /10/ Milloy and Crompton ('77) /22/	Yau et al.('78) /27/ Nesbet ('79) /28/ O'Malley, Burke and Berrington ('79) /29/ McEachran and Stauffer ('83) /30/
Ne	Pack and Phelps('61) /7/ Robertson('72;77K & 293K) /11/	Ogawa, Koizumi, Murakoshi, Yamamoto and Shirakawa ('83) /15/	No data available	Robertson ('72) /11/ O'Malley and Crompton ('80) /23/ Sol et al.('75) /24/ Golovanivsky('81)/25/ Ogawa et al.('83)/15/	McDowell ('71) /31/ Thompson ('71) /32/ Garbaty et al.('71)/33/ Yau et al. ('78) /27/ McEachran and Stauffer ('83) /30/
Ar	Pack and Phelps('61) /7/ Robertson ('77) /12/	Warren and Parker ('62) /13/ Milloy and Crompton ('77) /14/	Wagner, Davis and Parker ('67) /18/ Robertson and Rees ('72) /20/	Frost and Phelps('64) /17/ Milloy, Crompton,Rees and Robertson ('77) /26/	Thompson ('71) /32/ Garbaty et al.('71)/33/ Yau, McEachran et al. ('80) /34/ McEachran and Stauffer ('83) /35/
Kr	Pack, Voshall and Phelps ('62) /8/	Ogawa, Koizumi and Shirakawa ('83) /16/	No data available	Frost and Phelps('64) /17/	Yau, McEachran and Stauffer ('80) /34/ Sin Fai Lam ('82) /36/ McEachran and Stauffer ('83) /37/
Xe	Pack, Voshall and Phelps ('62) /8/	Ogawa, Koizumi and Shirakawa ('83) /16/	No data available	Frost and Phelps('64) /17/	Yau, McEachran and Stauffer ('80) /34/ Sin Fai Lam ('82) /36/ McEachran and Stauffer ('83) /37/

Meanwhile, the characteristic energy ε_k had not been measured at all until very recently for high purity inert gases except for He /9/,/13/ and Ar /13/, /14/, in spite of the fact that this quantity is much more sensitive to the magnitude of q_m than the drift velocity, particularly for relatively higher energies from about 0.1 to a few eV as was pointed out earlier, and therefore the more useful in deriving q_m's for these energies.

This situation prompted the author and his colleagues to attempt a measurement of ε_k in Ne, Xe and Kr by the Townsend method /6/ at room temperature. The result for Ne is plotted in Fig. 1. Full data are given in Ref./15/. The momentum transfer cross section $q_m(\varepsilon)$ derived from the data as a function of energy ε proved to be in excellent agreement with previous estimates by other methods as will be described in the next subsection.

As for Xe and Kr, the author and his colleagues are still carrying on the measurement, but some preliminary data are shown in Fig. 2 (Xe) and Fig. 3 (Kr) /16/. Although there still remain a considerable scatter of data (presumably due to the insufficiency of collected current, particularly in Kr) and a certain pressure dependence of unknown origin, the observed characteristic energy clearly exceeds the early estimates from the drift velocity data by Frost and Phelps /17/, being twice or thrice as large as the latter, both in Xe and in Kr, except for lower E/N's.

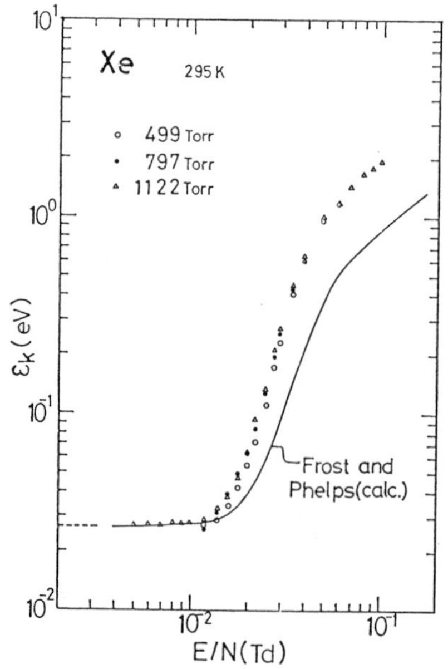

Fig. 2. The characteristic energy of electrons in Xe as observed by Ogawa _et al_. /16/.

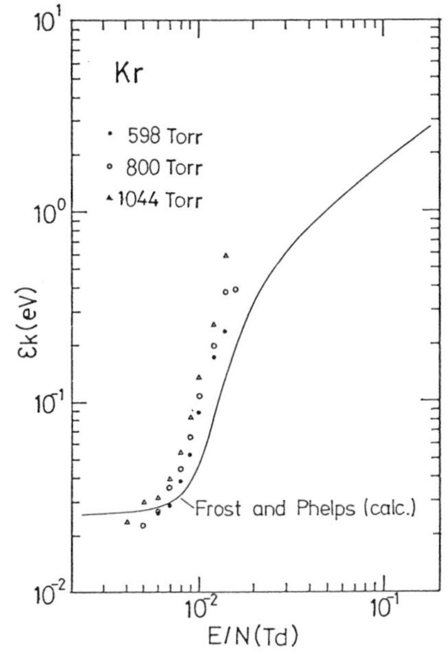

Fig. 3. The characteristic energy of electrons in Kr as observed by Ogawa et al. /16/ (preliminary data)

As concerns the "longitudinal" diffusion coefficient D_L or its related quantities such as $\varepsilon_L \equiv eD_L/\mu$ and $S \equiv D_L/D$, experimental data are even scantier, especially for lower E/N's for which the Ramsauer effect, if any, becomes most important. Apart from the pioneer work by Wagner, Davis and Hurst /18/ (only on He and Ar as for inert gases) over a rather limited range of E/N, the only available data at present appear to be those on D_L in He by Elford /19/ and in Ar by Robertson and Rees /20/ extended to lower E/N's to observe the peculiar peak due to the Ramsauer effect.

Since the parameters like D_L, ε_L, and S are all particularly sensitive, even more than ε_k, to rapid variations with energy ε of the momentum transfer cross section $q_m(\varepsilon)$ for the electrons, as has early been shown theoretically by Lowke and Parker /21/, they are expected to be extremely useful in determining the fine structure of $q_m(\varepsilon)$ such as the exact position and the shape (depth, width, etc.) of a Ramsauer minimum. It would be highly desirable, therefore, to measure these parameters as accurately as possible especially for Xe and Kr, in order to investigate in detail the Ramsauer minimum for these gases.

Measurements of basic swarm parameters for mixtures of inert gases appear to have hardly been attempted so far, despite their possible utility in examining closely the cross sections already proposed for individual inert gases. The author et al. have recently measured ε_k for He-Ne mixtures /15/.

3. Momentum Transfer Cross Section for Low Energy Electrons in Inert Gases

3.1 General Remarks

The momentum transfer cross section (MTCS) q_m for low energy electrons in most of the inert gases has been derived since 1960's usually from drift velocity data, though sometimes also from some other swarm data like the characteristic energy data, the results of microwave afterglow experiments or the observations of electron cyclotron resonances (ECR). For higher energies (above a few or several eV, say), the MTCS has also been derived, at least for He, Ne and Ar, from the differential scattering cross section obtained by beam experiments. In He and Ar, the MTCS's derived independently from recent swarm and beam data are in good agreement with each other, indicating a high reliability of the both results.

Of the two major swarm parameters, the drift velocity w depends more strongly upon the MTCS's at lower energies, while the characteristic energy ε_k is expected to be more sensitive to the MTCS values for higher energies, as is readily seen from the basic relationships $w \propto \langle 1/(q_m v) \rangle$ and $\varepsilon_k \propto \langle v/q_m \rangle / \langle 1/(q_m v) \rangle$, where $\langle\rangle$ denotes the average over electron energy (or velocity) distribution and v the random velocity of an electron.

Moreover, w decreases approximately linearly (as E/N) with decreasing E/N. Hence, even under very weak fields, the mobility $\mu \equiv w/E$ may well yield information about the MTCS for thermal energies. In contrast, ε_k approaches the common thermal (Einstein) limit of kT when E/N tends to zero, irrespective of the gas species. Therefore, it is essentially difficult to obtain information about q_m for very low energies from ε_k data. Meanwhile, ε_k values for slightly higher E/N's are extremely sensitive to the magnitude and the shape of $q_m(\varepsilon)$ in a rather higher energy region such as the Ramsauer minimum in Ar, Kr and Xe as was already emphasized repeatedly. This is due to the fact that even a slight difference in $q_m(\varepsilon)$ may well affect drastically the shape of electron energy distribution, in the higher energy side in particular, and consequently the values of ε_k. This makes the measurement of ε_k particularly suited to the detailed study of the MTCS in Ramsauer gases.

Quantum mechanical calculation of the MTCS in various inert gases has been attempted by many authors for long years as is seen in Table 2 shown before. For He, recent results by Nesbet /28/ and also by some other authors /29/, /30/ are in excellent agreement with experimentally derived cross sections over an energy range from about 0.01 to 12 eV. For other inert gases, theoretical results are not so satisfactory in general as in He, although considerable improvement has been made in the latest years.

3.2 Present Status of Cross Section Determination for Each Gas Species

A brief but broad critical review has been presented by Phelps in 1979 on the determination of cross sections for various gases, atomic and molecular, from swarm data at the first International Swarm Seminar held in Tokyo /39/. Besides, Itikawa has attempted in 1974 and 1978 a relative evaluation of various proposed values of $q_m(\varepsilon)$ for several familiar gases including five inert gases /40/. Similar attempts are also made for some common gases in the monographs written by Gilardini /5/ as well as by Huxley and Crompton /6/. Meanwhile, Hayashi /40/ has recently proposed a series of recommended cross sections (including $q_m(\varepsilon)$) in five inert gases as well as in some molecular gases for practical purposes by compromising among a few conflicting estimates, some based on swarm data while others on beam data.

3.2.1 Helium

Helium is the inert gas for which the electron swarm parameters have long been investigated most thoroughly. The values of the MTCS, which are widely approved to be most reliable and accurate at the present time for low energies below about 12 eV, are those derived by Crompton, Elford and Robertson /10/ and also by Milloy and Crompton /22/ from their own drift velocity data together with their characteristic energy data /9/. They are in satisfactory agreement with the cross section derived from beam experiments performed by Andrick and Bitsch /42/. Also they are reproduced very well within about 2 % or less by a theoretical calculation made by Nesbet /28/ as was mentioned earlier. At higher energies, the swarm-based MTCS is much less certain /41/.

3.2.2 Neon

The values of the MTCS for electrons in neon that have been most widely accepted are those which were derived by Robertson /11/ from his drift velocity data using the modified effective range theory (MERT) for energies ranging from 0.03 to 7.00 eV. Later, however, O'Malley and Crompton /23/ made an attempt to apply an improved MERT approximation (the extended MERT, EMERT) to the same drift velocity data and derived a slightly different set of q_m values, together with the estimated s-wave scattering length of 0.214 ± 0.005 a. u.

The author and his colleagues /15/ have recently measured the characteristic energy for the first time and have derived the MTCS for energies ranging from 0.01 to 1.00 eV from the experimental result with an estimated error limit of about ± 4 %, as was mentioned earlier in Sec. 2. As is shown in Fig. 4, the obtained q_m values proved to be in very good agreement not only with those derived by Robertson /11/ as well as those by O'Malley and Crompton /23/, but also with those estimated bv Sol, Devos and Gauthier /24/ from

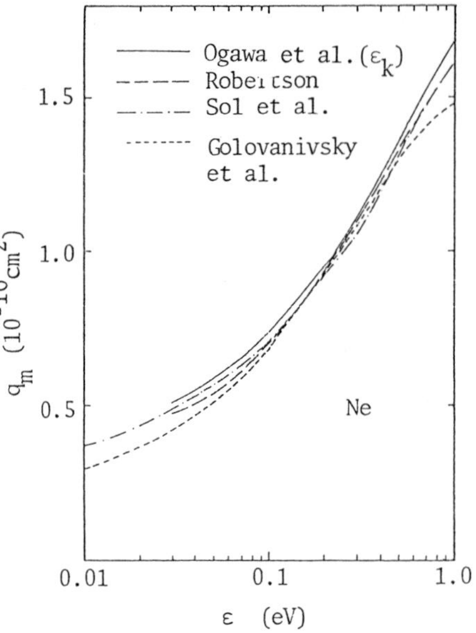

Fig. 4. The momentum transfer cross section for electrons in Ne derived from characteristic energy data /15/, as compared with some previous estimates.

microwave afterglow experiments and those by Golovanivsky and Kabilan /25/ from electron cyclotron resonance, each with the relevant error limits. This means that the values of the MTCS have been almost established with an accuracy of about ± 4 - 5 % at least in the energy range from 0.1 to 1.0 eV. As regards the s-wave scattering length a_0, however, the author et al. /15/ have estimated it to be about 0.24 a. u., in considerable disagreement with O'Malley et al.'s result of 0.214 a.u./23/.

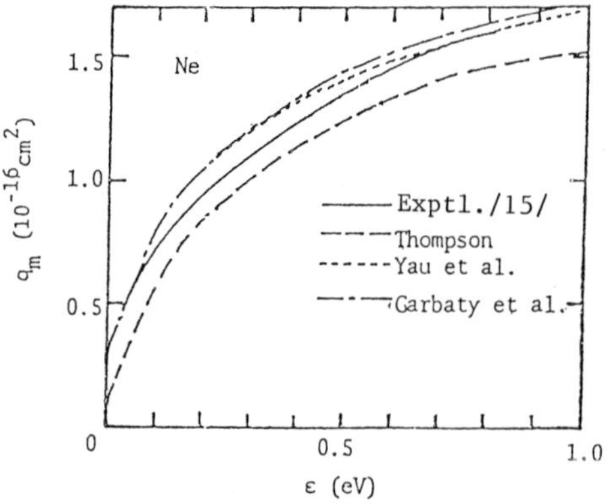

Fig. 5. Comparison of the momentum transfer cross section for electrons in Ne derived experimentally from characteristic energy data /15/ with some theoretical results /27/,/32/, /33/.

As is seen in Fig. 5, these experimental results, including the one obtained by the author et al., are all in fair agreement with any of the three existing theoretical results /27/,/32/,/33/ reported so far for energies ranging from 0.01 to 1.0 eV. On finer inspection, however, it is seen in Fig. 5 that the experimental results are about 10 - 20 % greater than Thompson's theoretical results /32/ and about 0 - 10 % smaller than those reported by Garbaty and La Bahn /33/ and also by Yau, McEachran and Stauffer /27/. Also, the experimental value of $a_0 \cong 0.24$ a.u. is somewhat larger than expected from any of the three theories.

3.2.3 Argon

With respect to this familiar inert gas which is also well-known as a typical Ramsauer gas, the momentum transfer cross section for 0 - 4 eV as derived by Milloy, Crompton, Rees and Robertson /26/ from the drift velocity data obtained by Robertson /12/ and the characteristic energy data by Milloy and Crompton /14/ is considered to be most accurate at the present time.

In their paper, Milloy *et al.*/26/ have emphasized that the characteristic energy ε_k is much more sensitive to the depth of the Ramsauer minimum in the MTCS than the drift velocity and demonstrated clearly that none of the MTCS's derived previously from the data other than ε_k, *e.g.* the one by Frost and Phelps from the drift velocity data /17/,the one by Golden from beam experiments /43/, and the one by McPherson from microwave experiments /44/, were compatible with the ε_k data used by Milloy *et al.* to derive their cross section.

It should also be mentioned that the same authors /26/ have calculated from their derived MTCS the "longitudinal" diffusion coefficient D_L, a quantity more sensitive to the Ramsauer minimum, as a function of E/N and compared it with the experimental results observed by Robertson and Rees /20/.

3.2.4 Xenon and Krypton

Figures 6 and 7 show some of the estimated momentum transfer cross sections $q_m(\varepsilon)$ for electrons in Xe and Kr, respectively, in a low energy region including the Ramsauer minimum, that have been either derived experimentally or calculated theoretically.

Of these, the one shown by a solid line was derived as early as 1964 by Frost and Phelps /17/ from the drift velocity measured by Pack, Voshall and Phelps /8/ and has long been widely accepted as the almost single reliable experimental estimates in this energy region where beam experiments are extremely difficult to carry out.

In Figs. 8 and 9 are shown with a solid line as a function of E/N the characteristic energy of electrons in Xe and Kr, respectively, as calculated by a Boltzmann analysis employing the MTCS just mentioned. In the same Figures

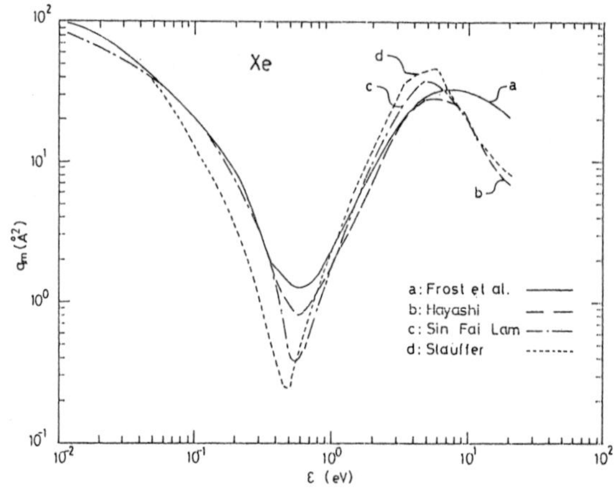

Fig. 6. Momentum transfer cross sections for electrons in Xe. a: experimental, Frost and Phelps /17/; b: experimental, Hayashi /38/; c: theoretical, Sin Fai Lam /36/; d: theoretical, McEachran and Stauffer /37/.

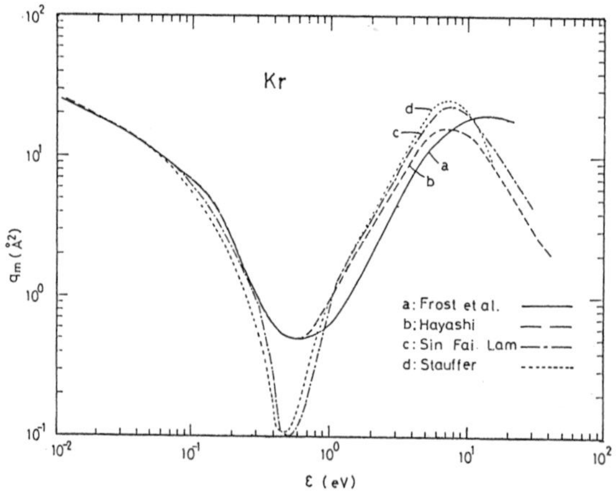

Fig. 7. Momentum transfer cross sections for electrons in Kr. a: experimental, Frost and Phelps /17/; b: experimental, Hayashi /38/; c: theoretical, Sin Fai Lam /36/; d: theoretical, McEachran and Stauffer /37/.

(Figs. 8 and 9) are also plotted with a broken line the calculated ε_k values based on slightly revised values of q_m that were proposed by Hayashi /38/ referring to the recent results of beam experiments in the higher energy region.

Meanwhile, Sin Fai Lam /36/ and McEachran and Stauffer /37/ have recently carried out a theoretical calculation of the MTCS for electrons in Xe and Kr quite independently. The results are plotted with a dot-&-dash line in Figs. 6 and 7. The ε_k values calculated by using these theoretical cross

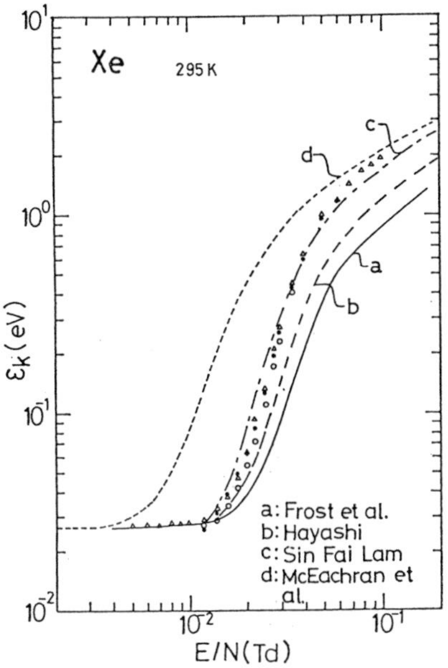

Fig. 8. The characteristic energy of electrons in Xe as a function of E/N. Comparison with the calculated values based on the various proposed momentum transfer cross sections shown in Fig. 6.

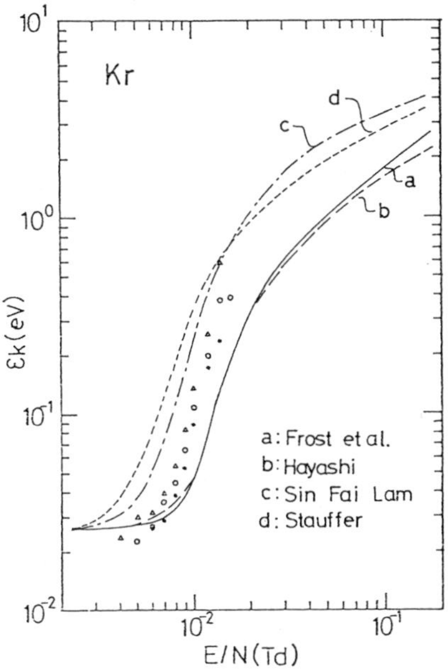

Fig. 9. The characteristic energy of electrons in Kr as a function of E/N. Comparison with the calculated values based on the various proposed momentum transfer cross sections shown in Fig. 7.

sections are shown with a dot-&-dash line for Sin Fai Lam's case and with a dotted line for McEachran _et al_'s case in Figs. 8 and 9, respectively.

As is seen in Figs. 8 and 9, the observed ε_k values for Xe are closest to the calculated values obtained with Sin Fai Lam's theoretical cross section, whereas those observed for Kr are definitely smaller than either of the two sets of calculated values based on theoretical cross sections, one by Sin Fai Lam /36/ and the other by McEachran and Stauffer /37/. Also, the observed ε_k's for Kr are considerably greater than the calculated values based on the cross section derived experimentally by Frost et al. and also the one recommended by Hayashi.

Numerical calculations are in progress in the author's laboratory to derive the MTCS capable of giving best fit to the observed characteristic energy data. It should be noted in passing that, both in Xe and Kr, the MERT or the EMERT approximation is not applicable unfortunately to the derivation of the MTCS near and above the Ramsauer minimum which lies in an energy region as high as 0.5 - 0.7 eV in both gases.

3.2.5 Radon

Because of its peculiar nature as the "daughter" element of the radioactive decay of Ra and also of its own alpha-radioactivity with a half-life of about 3.8 days, it is extremely difficult to make an electron swarm experiment with this gas and no swarm parameters for Rn have ever been measured so far. It is of interest to note, however, that Sin Fai Lam predicts theoretically /36/ that the Rn gas will exhibit the Ramsauer-Townsend effect also in the vicinity of 1 eV.

Appendix: Derivation of Equations (3) and (4) in the Text

During a free flight between two successive momentum transfer collisions with gas atoms, an "average" electron under an electric field E moves (on the average) toward the anode by a distance $s = c_1 (eE/m)\, \tau^2$ ($c_1 \cong \frac{1}{2}$) approximately. Here, c_1 denotes a numerical constant with the magnitude of the order of unity depending on the degree of approximation, m the electron mass, and τ the mean free time for electrons from momentum transfer collision.

Let us denote by $\nu_m \equiv 1/\tau$ the momentum transfer collision frequency, by N the number density of gas atoms, by v the average random velocity of electrons, and by $\bar{q}_m$ the momentum transfer cross section averaged over an energy region in which the substantial part of electron energy distribution is contained. Then $\nu_m = N\bar{q}_m v$ and the drift velocity w is given as

$$w = s/\tau = c_1 (eE/m)\tau = c_1 (eE/(m\nu_m)) = c_1 (e/\bar{q}_m)(E/N)(1/(mv)). \qquad \text{(A1)}$$

Meanwhile, the diffusion constant D for electrons (assumed as isotropic, for simplicity) is well known to be approximately given by $D \cong \lambda v/3 = v/(3N\bar{q}_m)$ from the conventional kinetic theory of gases, where λ denotes the momentum transfer mean free path for an electron, and the mobility μ is derived immediately from Eq.(A1) as $\mu \equiv w/E = c_1(e/N)/(m\overline{vq_m})$. Combining these two expressions, the characteristic energy ε_k is expressed simply as

$$\varepsilon_k \equiv eD/\mu = (2/3)mv^2. \tag{A2}$$

On the other hand, the energy given to an "average" electron from the electric field per unit time is eEw. In a steady state, this must be equal to $\overline{\Delta\varepsilon}\,\nu_m$, the energy lost by the electron through momentum transfer (elastic) collisions per unit time, so that

$$eE\,w = \overline{\Delta\varepsilon}\,N\overline{q_m v}, \tag{A3}$$

where $\overline{\Delta\varepsilon}$ denotes the mean energy loss per collision for an electron.

For elastic collisions of an electron of mass m with a gas atom of mass M (>> m) at rest, $\overline{\Delta\varepsilon}$ is again well known to be given in good approximation by

$$\overline{\Delta\varepsilon} = 2(m/M)\varepsilon, \quad \text{with} \quad \varepsilon \equiv (1/2)mv^2. \tag{A4}$$

Substituting (A4) in (A3) and solving simultaneous equations (A1), (A2) and (A3) for w, ε_k and v, we readily obtain the formulae (3) and (4) in the text.

References

1. Davydov, P.: Phys. Z. Sowjetunion 8, 59-70 (1935).
2. Alfven, H.: Cosmical Electrodynamics, 1st ed., p.43-47. Oxford: At the Clarendon Press. 1950.
3. Beaty, E. C., Dutton, J., Pitchford, L. C.: A Bibliography of Electron Swarm Data. (JILA Information Center Report, No. 20).University of Colorado, Dec. 1, 1979.
4. Dutton, J.: A Survey of Electron Swarm Data. J. Phys. Chem. Ref. Data 4, 577-856 (1975).
5. Gilardini, A. L.: Low Energy Electron Collisions in Gases. (Wiley Series in Plasma Physics). New York-London-Sydney-Toronto: John Wiley & Sons. 1972.
6. Huxley, L. G. H., Crompton, R. W.: The Diffusion and Drift of Electrons in Gases. (Wiley Series in Plasma Physics). New York-London-Sydney-Toronto: John Wiley & Sons. 1974.
7. Pack, J. L., Phelps, A. V.: Phys. Rev. 121, 798-806 (1961).
8. Pack, J. L., Voshall, R. E., Phelps, A. V.: Phys. Rev. 127, 2084-2089 (1962).
9. Crompton, R. W., Elford, M. T., Jory, R. L.: Aust. J. Phys. 20, 369-400 (1967).
10. Crompton, R. W., Elford, M. T., Robertson, A. G.: Aust. J. Phys. 23,

667-681 (1970).

11. Robertson, A. G.: J. Phys. B: Atom. Molec. Phys. 5, 648-664 (1972).
12. Robertson, A. G.: Aust. J. Phys. 30, 39-49 (1977).
13. Warren, R. W,, Parker, J. H., Jr.: Phys. Rev. 128, 2661-2671 (1962).
14. Milloy, H. B., Crompton, R. W.: Aust. J. Phys. 30, 51-60 (1977).
15. Ogawa, I., Koizumi, T., Murakoshi, H., Yamamoto, S., Shirakawa, E.: Proceedings of the 3rd International Swarm Seminar (Lindinger, W., Villinger, H., Federer, W., eds.), p. 46-54. Innsbruck, Austria: August, 1983.
16. Ogawa, I., Koizumi, T., Shirakawa, E.: unpublished.
17. Frost, L. S., Phelps, A. V.: Phys. Rev. 136, A 1538-1545 (1964).
18. Wagner, E. B., Davis, F. J., Hurst, G. S.: J. Chem. Phys. 47, 3138-3147 (1967).
19. Elford, M. T.: Aust. J. Phys. 27, 193-209 (1974).
20. Robertson, A. G., Rees, J. A.: Aust. J. Phys. 25, 637-639 (1972).
21. Lowke, J. J., Parker, J. H.: Phys. Rev. 181, 302-311 (1969).
22. Milloy, M. B., Crompton, R. W.: Phys. Rev. A15, 1847-1850 (1977).
23. O'Malley, T. F., Crompton, R. W.: J. Phys. B: Atom. Molec. Phys. 13, 3451-3464 (1980).
24. Sol, C., Devos, F., Gauthier, J-C.: Phys. Rev. A 12, 502-507 (1975).
25. Golovanivsky K. S., Kabilan, A. P.: Abstracts of Contributed Papers. The Twelfth International Conference on the Physics of Electronic and Atomic Collisions (Datz, S., ed.), p. 106-107. Gatlinburg, Tennessee, USA: 1981.
26. Milloy, H. B., Crompton, R. W., Rees, J. A., Robertson, A. G.: Aust. J. Phys. 30, 61-72 (1977).
27. Yau, A. W., McEachran, R. P., Stauffer, A. D.: J. Phys. B: Atom. Molec. Phys. 11, 2907-2921 (1978).
28. Nesbet, R. K.: Phys. Rev. A 20, 58-70 (1979).
29. O'Malley, T. F., Burke, P. G., Berrington, J.: J. Phys. B: Atom. Molec. Phys. 12, 953-965 (1979).
30. McEachran, R. P., Stauffer, A. D.: J. Phys. B: Atom. Molec Phys. 16, 255-274 (1983).
31. McDowell, M. R. C.: J. Phys. B: Atom. Molec. Phys. 4, 1649-1660 (1971).
32. Thompson, D. G.: J. Phys. B: Atom. Molec. Phys. 4, 468-482 (1971).
33. Garbaty, E. A., La Bahn, R. W.: Phys. Rev. A 4, 1425-1431 (1971).
34. Yau, A. W., McEachran, R. P., Stauffer, A. D.: J. Phys. B: Atom. Molec. Phys. 13, 377-384 (1980).
35. McEachran, R. P., Stauffer, A. D.: J. Phys. B: Atom. Molec. Phys. 16, 255-274 (1983) and its succeeding paper (to be published).
36. Sin Fai Lam, L. T.: J. Phys. B: Atom. Molec. Phys. 15, 119-142 (1982).
37. McEachran, R. P., Stauffer, A. D.: Abstracts of Contributed Papers. The

Thirteenth International Conference on the Physics of Electronic and Atomic Collisions (Eichler, J., Fritsch, W., Hertel, I. V., Stolterfoht, N., Wille, U., eds.), p.89. Berlin, FRG: 1983; Stauffer, A. D.: private communication.

38. Hayashi, M.: Recommended Values of Transport Cross Sections for Elastic Collision and Total Collision Cross Section for Electrons in Atomic and Molecular Gases. IPPJ-AM-19 (Institute of Plasma Physics, Nagoya University, Nagoya, Japan), November, 1982.
39. Phelps, A. V.: Proceedings of the International Seminar on Swarm Experiments in Atomic Collision Research (Ogawa, I., ed.), p.23-32. Tokyo, Japan: September 6-7, 1979.
40. Itikawa, Y.: Atomic Data and Nucl. Data Tables 14, 1-10 (1974); ibid. 21, 69-75 (1978).
41. Phelps, A. V., Pack, J. L., Frost, L. S.: Phys. Rev. 117, 470-474 (1960).
42. Andrick, D., Bitsch, A.: J. Phys. B: Atom. Molec. Phys. 8, 393-410(1975).
43. Golden, D. E.: Phys. Rev. 151, 48-51 (1966).
44. McPherson, D. A., Feeney, R. K., Hooper, J. W.: Phys. Rev. A 13, 167-179 (1976).

Temperature Dependences of Positive-Ion Molecule Reactions

N. G. Adams and D. Smith

Department of Space Research, University of Birmingham, Birmingham B 15 2TT, England

1. Introduction

The thermalised afterglow plasma, in association with appropriate diagnostic techniques, is a suitable medium for investigating a wide variety of ionic processes under well-defined conditions. The pioneering work of S.C.Brown and M.A.Biondi and their co-workers using the pulsed (stationary) afterglow/ microwave cavity diagnostic technique has provided a wealth of data,notably that relating to electron-ion dissociative recombination. Pulsed afterglows have also been used successfully by H.J.Oskam, D.Smith and W.C.Lineberger and their colleagues to study a wide variety of processes. Brief summaries of these studies, including descriptions of the experimental techniques, are given in the book by McDaniel and Mason [1]. Later the flowing afterglow was conceived,developed and exploited by E.E.Ferguson, F.C.Fehsenfeld and A.L.Schmeltekopf at Boulder to study ion-molecule reactions at thermal energies. The flowing afterglow has subsequently been exploited to great effect by the Boulder group and by several other groups, and this work has laid the foundations for an understanding of ion-molecule reactions at thermal energies. The flowing afterglow technique has been discussed in detail by Ferguson et al. [2], and it has been compared and contrasted with the more recently developed Selected Ion Flow Tube (SIFT) technique by Smith and Adams [3].

The standard diagnostic used in most flowing afterglow apparatuses is a downstream mass spectrometer; indeed, this is the only diagnostic required for ion-molecule reaction studies. For the study of most other plasma reaction processes, it is necessary to be able to determine the charged particle number densities within the body of the plasma, that is along the afterglow plasma column. This is the unique feature of the Flowing After-

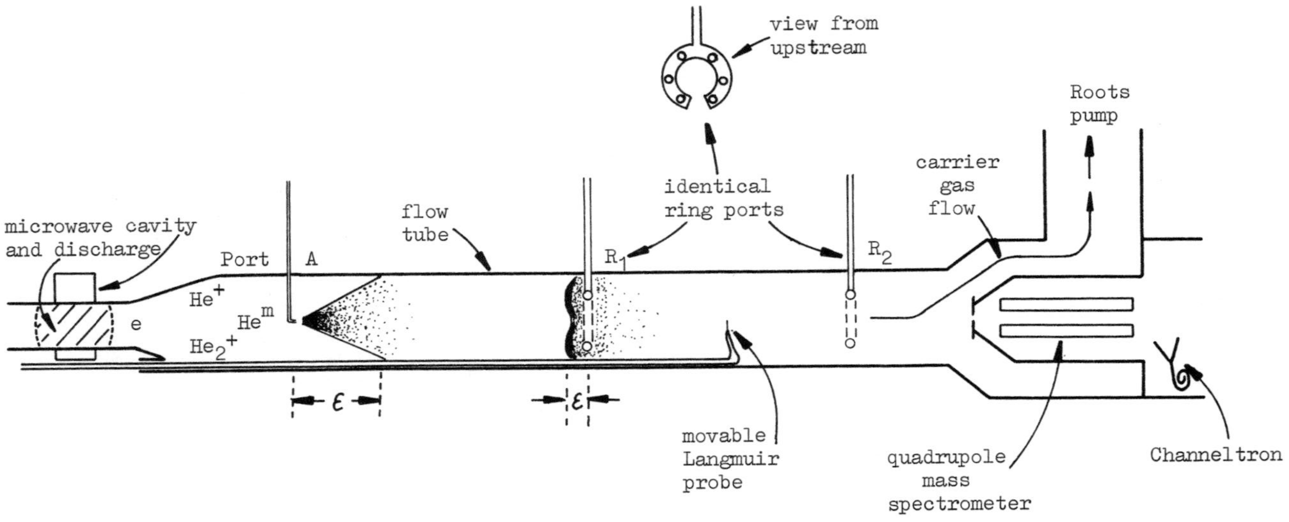

Fig.1 Schematic representation of the Flowing Afterglow/Langmuir Probe (FALP) experiment indicating the flow tube, the microwave discharge, the density distribution of ion source and/or reactant gases entering the axial inlet port A and the ring ports R_1 and R_2, the Langmuir probe and the quadrupole mass spectrometer and Channeltron detector. The "end corrections" for each type of port are labelled as ε. The flow tube is surrounded by a vacuum vessel to facilitate its temperature variation but this has been omitted for clarity.

glow/Langmuir Probe (FALP) apparatus in which a cylindrical Langmuir probe is used to determine electron temperature (T_e), electron density (n_e), positive ion density (n_+) and negative ion density (n_-) along the axis of the afterglow column with a spatial resolution of about one millimetre. A brief description of the FALP technique is given in the next section. To date, it has been exploited to study ambipolar diffusion, electron temperature relaxation, electron-ion dissociative recombination, electron attachment and ion-ion recombination, in many cases over appreciable temperature ranges. The results from these studies will be summarised in this paper.

2. The FALP Technique

In essence, a flowing afterglow consists of a flow tube along which a carrier gas (usually pure helium) is constrained to flow by the action of a large Roots-type pump, and in which ionization is created in the upstream region by a gas discharge or some other ion source. An afterglow plasma is thus distributed along the flow tube and, under favourable conditions, the charged particle energies will be relaxed in the afterglow to those appropriate to the carrier gas temperature. Distance along the flow tube and the residence (or reaction) time of the individual ionized or neutral species in the plasma are coupled via their respective flow velocities which can readily be determined if the flow dynamics are understood (a detailed appraisal is given in the paper by Adams et al. [4]). The various reaction processes can then be studied by adjusting n_e, n_- and n_+ to suitable values (appropriate to the particular process to be studied, e.g. recombination, attachment, etc.) and by adding controlled amounts of appropriate reactant gases into the thermalised afterglow plasma via one or more inlet ports situated at fixed positions along the flow tube. (In some conventional flowing afterglow apparatuses, a single inlet port is used which can be moved along the flow tube axis [5]). Different reactant gases are often added simultaneously via separate inlet ports in order to generate suitable plasma media or, for example, to remove helium metastable atoms from the afterglow (specific examples are given in the following sections). A major objective is to create afterglow plasmas in which the loss of a particular charged species occurs solely by the process which it is desired to study.

The original FALP apparatus [6], like the original flowing afterglow [2], was constructed from Pyrex glass but a change to stainless steel was inevitable with the desire to study processes over a wide temperature range. A schematic of the latest FALP is shown in Figure 1. The stainless steel flow tube is approximately one metre long and eight centimetres in diameter. The ionization source is a microwave discharge through the carrier gas at pressures of the order of one Torr. In helium carrier gas, the maximum n_e in the upstream region of the afterglow is about $10^{11}cm^{-3}$ as determined with the Langmuir probe (discussed below). The axial inlet ports (such as port A in Fig.1), which are used in most flowing afterglow experiments, have large mixing lengths (or end corrections, $\mathcal{E} \sim 10cm$). In contrast, the ring ports, R_1 and R_2, have a small $\mathcal{E}$ which is essential for the study of fast processes such as electron-ion recombination and electron attachment. With these ring ports the reactant gas is introduced into the afterglow against the flow of carrier gas (a contraflow technique) and this reduces $\mathcal{E}$ to about 1 or 2 cm [7]. A differentially-pumped mass spectrometer is located at the downstream end of the flow tube so that the positive and negative ion species in the plasmas can be identified.

The Langmuir probe technique has been used for decades to study steady state discharge plasmas (see, for example, the book by Swift and Schwar [8])but, prior to the work in our laboratory, it had not been satisfactorily applied to the study of decaying (afterglow) plasmas. Our probe technique was first developed for use in pulsed (stationary) afterglows [9] and subsequently we combined it with the flowing afterglow to create the FALP apparatus [6]. The details of the probe technique are given in several papers [6, 9-14] and only an outline is necessary here. A small cylindrical wire (the probe) is swept in potential,relative to the local plasma potential and a current/voltage characteristic is recorded and analysed to provide values for T_e,n_e, n_+ and n_- as appropriate. In fact, it is the gradients of these parameters along the axis(defined here as the z coordinate)of the flowing plasma which provides values for attachment and recombination coefficients etc. under truly thermalised conditions. The complete flow tube is enclosed in a vacuum vessel which facilitates the temperature variation of the flow tube over the range 80-600 K and minimises temperature gradients. The Langmuir probe data also provide values for the mass ratios m_+/m_e in positive ion/electron plasmas and m_+/m_- in positive ion/negative ion plasmas (or mean values of these ratios in plasmas containing more than one species of positive or negative ion [14, 15]). This is a valuable supplement to the mass spectrometric data.

3. Ambipolar Diffusion and Electron Temperature Relaxation Studies

Ambipolar diffusion is always a finite loss process for ions and electrons in flowing afterglow plasmas. Sufficiently far away from the disturbed upstream region of the afterglow, fundamental mode diffusive loss prevails and n_e and n_+ in an electron-ion plasma decrease exponentially with z. Thus it is a straightforward matter to determine $\partial n_e/\partial z$ and hence to derive values for the ambipolar diffusion coefficients, D_a, for plasmas of various ionic compositions

[6]. Theory predicts [16] that, at a given temperature, $D_a p$ = const. (where p is the gas pressure), a prediction amply supported by experiment [1]. Since D_a is inversely proportional to p then, when ambipolar diffusion needs to be inhibited, i.e. when studying other reaction processes, the FALP must be operated at suitably high pressure. Of course, ambipolar diffusive loss also occurs in ion-ion plasmas (but at a slower rate than in electron-ion plasmas, since $D_- << D_e$ (D_- and D_e are the free diffusion coefficients of negative ions and electrons respectively). In plasmas in which electrons and negative ions co-exist, the situation is more complicated with respect to diffusion (see Section 5).

In the region near to the discharge, which generates the flowing afterglow, T_e is much greater than T_+ or T_g (the ion and gas temperatures respectively) and there the diffusive loss is most rapid (since $D_a = D_+(1 + T_e/T_+)$[1]). The 'hot' electrons which exist in this region are 'cooled' in collisions with positive icns and carrier gas atoms during their passage down the flow tube. It is important to estimate this rate of cooling, dT_e/dt, so that it can be ascertained where downstream thermalisation has been reached (i.e. where $T_e = T_+ = T_g$). The Langmuir probe can be used to determine T_e and hence dT_e/dt (knowing the plasma flow velocity which can also be readily determined, again using the probe [4]). Such T_e relaxation has been studied in both stationary afterglows [17] and in the FALP [18,19]. The rate of electron temperature relaxation is described by:

$$\frac{dT_e}{dt} = - \frac{1}{\tau} (T_e - T_{+,g}) \qquad (T_+ \sim T_g \text{ in these plasmas}) \qquad (1)$$

The time constant, τ, describes the net effect due to electron-ion ($\tau_{e,+}$) and electron-neutral ($\tau_{e,n}$) collisions such that $\tau^{-1} = \tau_{e,+}^{-1} + \tau_{e,n}^{-1}$. The small fractional ionization in these afterglow plasmas ($\sim 10^{-6}$) ensures that electron-neutral collisions control the rate of T_e relaxation. This is especially so when the neutral gas is molecular since electron cooling can occur via excitation of rotational and vibrational states of the molecules.

τ_{en} has been determined in inert gas stationary afterglows and thus cross sections for momentum transfer in electron-inert gas atom collisions have been deduced [17]. Using the FALP, the fractions of energy transferred in electron collisions with the molecular gases N_2 and O_2 have been determined as a function of T_e from τ_{en} measurements [18,19]. Such measurements can also provide other fundamental data such as quadrupole moments.

It should be noted that T_e relaxation is inhibited when helium metastable atoms are present in an afterglow due to heating of the electrons in superelastic collisions with these excited atoms. To avoid this undesirable situation, a small amount of argon can be added to the upstream region of the afterglow to quench the metastable atoms (via Penning reactions [6]). The addition of argon also reactively removes any He_2^+ ions which are produced in the three-body association of He^+ with He [20]. This does of course mean that, in addition to He^+ ions, Ar^+ ions will be present in the afterglow, however this does not result in severe complications in practice when studying other processes.

The quantity of data so far obtained which relates to T_e relaxation is relatively small, but it very well illustrates the potential of the FALP for such studies.

4. Electron-Ion Dissociative Recombination Studies

Dissociative recombination reactions of molecular positive ions with electrons have been studied for many years because they are an important loss process of ionization both in laboratory plasmas such as gas lasers [21] and in naturally occurring plasmas such as the ionosphere and inter-stellar gas clouds (for further discussion of these subjects see the reviews by Smith and Adams [22,23]). The primary objective of most studies has been to determine the recombination coefficients, α, for particular positive ion species and how these coefficients vary (i) with T under truly thermal equilibrium conditions such that $T_e = T_+ = T_g$ when α_t is obtained, and (ii) with T_e for $T_e > T_+, T_g$ when α_e is obtained. Notable amongst the many techniques used to determine α_t and α_e is the stationary afterglow [24,25]. The variation of the dissociative recombination cross section with electron energy, σ_E, has been studied using ion traps [26,27] and merged beams [28]. Very recently, the FALP technique has been used for studies of α_t [7].

As previously mentioned, it is crucial in plasma experiments such as the FALP to establish conditions such that the process under investigation is the dominant loss process. In the FALP for dissociative recombination studies this is achieved in the following manner. The active species upstream in the afterglow plasma are electrons, He^+ and He_2^+ ions and He metastable atoms, He^m. Argon is introduced upstream (via port A, Fig.1) to destroy He_2^+ and He^m. The plasma then contains only electrons and atomic ions (He^+ and Ar^+). Recombination of atomic ions with electrons in these plasmas is negligibly slow and so ambipolar diffusion followed by wall recombination is the only loss process for ionization. A molecular reactant gas is then added to this diffusing plasma via one of the ring ports (R_1, R_2) in sufficient quantity to convert the atomic ions to appropriate molecular positive ions (e.g. O_2 and NH_3 to produce O_2^+ and NH_4^+ etc.). Knowledge of the vast literature on ion-molecule reactions and careful use of the downstream mass spectrometer are essential in creating the desired positive ion species as the only ionic species in the plasma. At low temperatures and high pressures a potential problem is the creation of ion clusters (e.g. $O_2^+.O_2$, $NH_4^+.NH_3$, etc.) and then careful use of the mass spectrometer is especially important.

When the desired plasma conditions have been established, the electron density, n_e, is determined along the flow tube axis using the Langmuir probe and an obvious increase in $\partial n_e/\partial z$ is observed downstream of the molecular gas inlet port (see Fig.2a as an example). Under conditions of large n_e and sufficiently high helium pressure, recombination loss dominates diffusion loss and then we have :

$$\frac{1}{n_e(z_1)} - \frac{1}{n_e(z_2)} = \alpha_t \frac{(z_2 - z_1)}{v_p} \qquad (2)$$

where $n_e(z_1)$, $n_e(z_2)$ are the values of n_e at positions z_1 and z_2, and v_p is the plasma flow velocity. Hence α_t can readily be obtained from a plot of $1/n_e$ versus z. Typical plots are shown in Fig.2b to illustrate the quality of the data obtained. Experiments can be carried out within the temperature range 80-600 K, although the low temperature limit is higher than 80 K when it is necessary to use ion source gases which are condensible (e.g. H_2O, NH_3 etc.).

The measurement of $\alpha_t(O_2^+)$ has become a 'test' for recombination studies, since its value has been very well established using a variety of techniques. It has been measured in the FALP using O_2 as the reactant gas at 95,205,295, 420,530 and 590 K and the data are shown in the log-log plot in Fig.3. The linearity of the graph is good and and the power law: $\alpha_t(O_2^+) = 1.95 \times 10^{-7} (300/T)^{0.7} cm^3 s^{-1}$ closely describes $\alpha_t(O_2^+)$ over the complete temperature range. Good agreement is found with the values obtained previously at 300 K using a variety of techniques (see e.g. Ref.26) and with pulsed stationary afterglow data obtained above 300 K [29]. The small discrepancy at low temperatures between the present values and the pulsed afterglow data [29] at low temperatures is thought to be due to the presence of cluster ions in the higher pressure stationary afterglow experiment. The results are also in good agreement with pulsed afterglow data for $\alpha_e(O_2^+)$[30] and with results for $\alpha_e(O_2^+)$ derived from cross sections measured with the ion trap technique [26]. All these experiments essentially indicate a common $T^{-0.7}$ dependence of $\alpha_t(O_2^+)$ and $\alpha_e(O_2^+)$. A somewhat weaker $T^{-0.5}$ temperature dependence was derived from cross section measurements made using the merged beam technique [31].

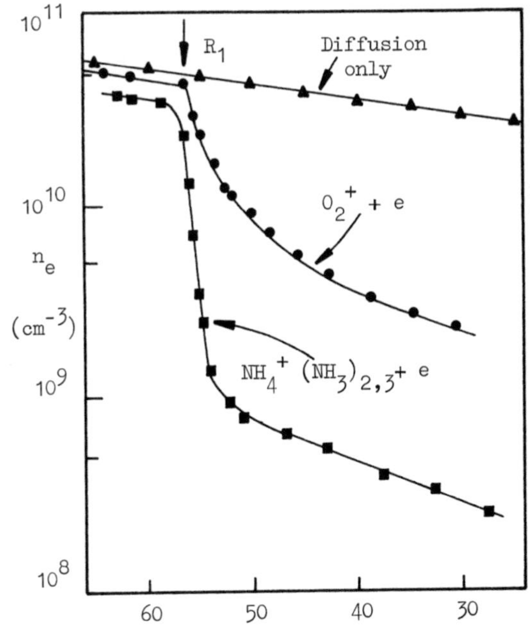

Fig.2(a) n_e versus z data (z is measured from the mass spectrometer sampling orifice) obtained using the FALP at a helium pressure of 0.6 Torr and a temperature of 300 K for: ▲, diffusion controlled plasma (He^+, Ar^+, electrons) with the addition of Ar through port A (see Fig.1) to destroy He^m and He_2^+ but without the addition of ion source gases: ● with addition of O_2 via port R_1(see Fig.1) which results in an increased loss rate of electrons due dissociative recombination of O_2^+ ions: ■ with addition of NH_3; the rapid decrease of n_e is due to dissociative recombination of $NH_4^+ (NH_3)_{2,3}$ ions.

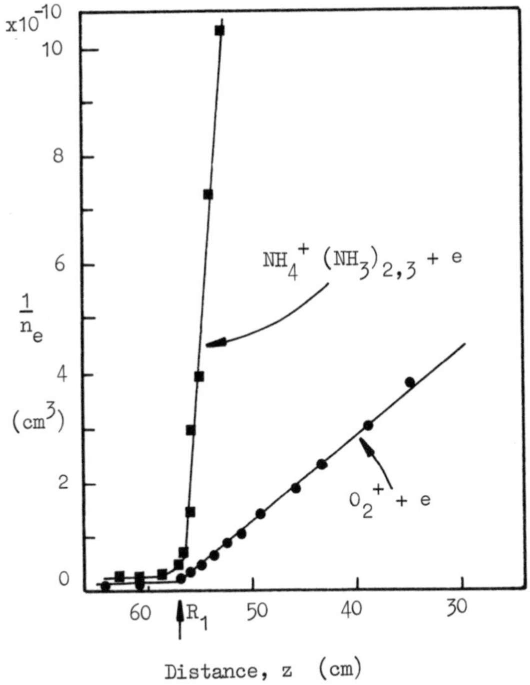

Fig.2(b) $1/n_e$ versus z plots of the data for O_2^+ and $NH_4^+(NH_3)_{2,3}$ shown in (a). The slopes of these plots provide values for the recombination coefficients, α_t. The very different slopes indicate that $\alpha_t(NH^+_4(NH_3)_{2,3})$ is much larger than α_t (O_2^+) at 300 K. Note that these plots are linear over a factor of more than 20 which is quite sufficient to define the α_t accurately. At small z (and low n_e) upcurving of these plots occurs (not shown) as ambipolar diffusion rather than dissociative recombination becomes the dominant loss process for ionization.

An important point to make, which is relevant to all the dissociative recombination reactions we have studied to date, is that the molecular ions, for which α_t have been measured, are confidently expected to be in their ground electronic and equilibrium vibrational/rotational states. This is so because the ions, although on production may well be internally excited, will undergo resonant charge or proton transfer with their parent molecules (which are present in the afterglow). Such reactions are known to quench both electronic and vibrational excitation very efficiently [32,33]. Rotational relaxation is ensured by the high collision frequency of the molecular ions with the ambient carrier gas atoms.

The temperature dependence of $\alpha(NO^+)$ has been in dispute for several years. The dispute centred around the difference between the stationary afterglow data of Biondi and co-workers [25,34] and the ion trap data of Walls and Dunn [26]. The stationary afterglow data indicate that α_t (for $200 < T < 450K$) has a markedly different temperature dependence than α_e (for $500 < T_e < 2000K$), whereas no indication of such a deviation was noticable in the α_e derived from σ_E data obtained in the ion trap experiment. Thus, FALP studies of α_t were undertaken in an attempt to clarify this situation.

NO^+ ions are generated in the FALP by introducing NO (from which HNO_3 impurities had been removed to avoid negative ion production) into the He^+/Ar^+ afterglow. $\alpha_t(NO^+)$, which was measured at 205,295,465 and 590 K, is very well described by the power law $\alpha_t(NO^+) = 4.0 \times 10^{-7}(300/T)^{0.9} cm^3 s^{-1}$, see Fig.3. This is in good agreement with the pulsed afterglow data for $\alpha_t(NO^+)$ and with the $\alpha_e(NO^+)$ ion trap data. Results obtained from satellite measurements in the ionosphere [35] indicate that $\alpha_e(NO^+) \sim T_e^{-0.85}$ and are in remarkable agreement in magnitude with the FALP data at the temperature common to both data sets. However, the $\alpha_e(NO^+)$ stationary afterglow results cannot be reconciled with the FALP data for $\alpha_t(NO^+)$. It does seem that on balance, a common $T^{-0.9}$ dependence for $\alpha_t(NO^+)$ and $\alpha_e(NO^+)$ is appropriate below about 10^3K. Values of $\alpha_e(NO^+)$ derived from merged beam data [31] are not reconcilable with the other data referred to above. A more detailed discussion of all these data is given by Alge et al. [7].

$\alpha_t(NH_4^+)$ was measured at 295,415,460,540 and 600K in the FALP. The NH_4^+ ions were generated by adding small amounts of NH_3 to the afterglows (large NH_3 flows initiated ion clustering, see below). The data obtained (see Fig.3) are reasonably well described by the power law $\alpha_t(NH_4^+) = 1.35 \times 10^{-6}(300/T)^{0.6} cm^3 s^{-1}$. This is in good agreement with stationary afterglow data in the overlapping temperature range [36], although not in good agreement with ion

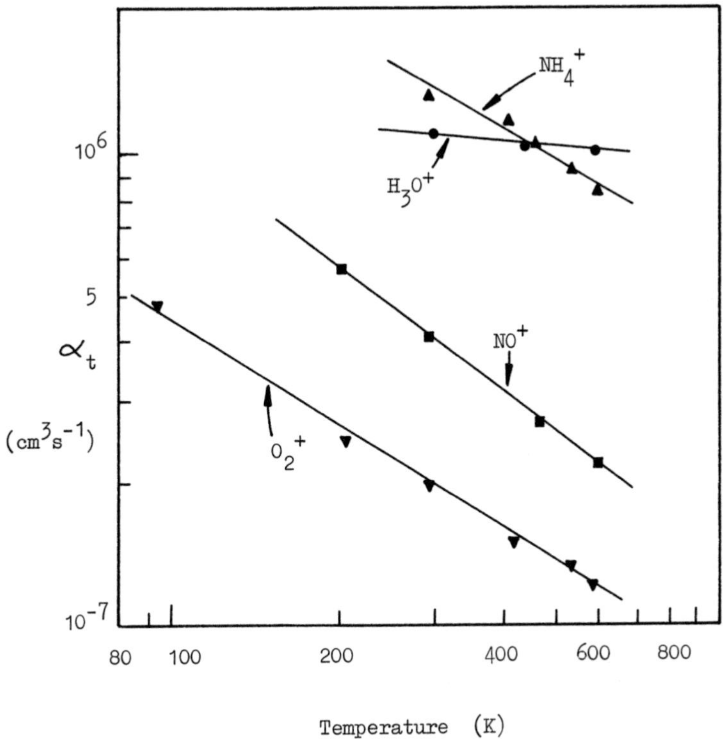

Fig.3 Electron-ion dissociative recombination coefficients, α_t (i.e. for $T_e = T_+ = T_g$), for O_2^+, NO^+, NH_4^+ and H_3O^+ determined at several temperatures using the FALP technique. These data indicate that $\alpha_t\ (O_2^+) \sim T^{-0.7}$, $\alpha_t\ (NO^+) \sim T^{-0.9}$, $\alpha_t\ (NH_4^+) \sim T^{-0.5}$ and that $\alpha_t\ (H_3O^+)$ does not vary significantly with temperature over the available range.

trap data (see Ref. 7). By adding excess NH_3 to the afterglow at 300 K, on equilibrium ratio of $NH_4^+(NH_3)_2$ and $NH_4^+(NH_3)_3$ cluster ions could be established. A composite α_t of $2.8 \times 10^{-6} cm^3 s^{-1}$ was obtained for these ions which is in excellent agreement with the stationary afterglow value for $NH_4^+(NH_3)_2$ cluster ions at 300 K [36].

$\alpha_t(H_3O^+)$ was measured at 295,450 and 600K and, somewhat surprisingly, no significant variation was observed with temperature, a constant value of $1.0 \times 10^{-6} cm^3 s^{-1}$ being obtained. An identical value was obtained in a stationary afterglow experiment at 540 K [37]. Recent values for $\alpha_e(H_3O^+)$ derived from merged beam data [38] are significantly smaller than the FALP values and also show a marked 'temperature' (energy) dependence. A similar marked 'temperature' dependence for $\alpha_e(H_3O^+)$ was also indicated by ion trap data [27]. Collectively, the data may be indicating that $\alpha_t(H_3O^+)$ and $\alpha_e(H_3O^+)$ have different T and T_e dependences.

The most recent FALP studies of dissociative recombination have been directed towards the measurement of α_t at suitably low temperatures for ions which are known (or expected) to exist in interstellar gas clouds [23,39,40]. Thus the α_t for electron recombination of the ions HCO^+, DCO^+, N_2H^+, N_2D^+,CH_5^+, H_3^+ and D_3^+ have been determined at a temperature of 95 K [41]. At this temperature,it was not possible to use argon to destroy helium metastables in the afterglow; rather, an excess of the gases from which the ions were generated (ion source gases) was introduced to rapidly destroy the metastables. The very slow Penning reactions of He^m with most gases expected at low temperatures [42] required that relatively large amounts of ion source gases were needed, especially in the case of H_2 and D_2. The major ion in the helium afterglow at 95 K, prior to addition of the ion source gases, was He_2^+ which is known to react rapidly with most gases, including H_2 [43]. Thus, to study $\alpha_t(H_3^+)$, H_2 was added in excess and Penning reactions and ion molecule reactions established H_3^+ as the major ionic species. At excessively high H_2 pressures, the three-body association reaction $H_3^+ + H_2 + He \rightarrow H_5^+ + He$ led to appreciable concentrations of H_5^+ in the afterglow. Whilst the data for $\alpha_t(H_3^+)$ are as yet preliminary, the startling indication is that $\alpha_t(H_3^+)$ at 95 K is very small ($< 10^{-7} cm^3 s^{-1}$) contrary to expectations and to pulsed afterglow data [44,45]. If confirmed, this will have a profound effect on current thinking on interstellar chemistry and physics! Similar experiments indicate that $\alpha_t(D_3^+)$ is also very small. However, $\alpha_t(H_5^+)$ and $\alpha_t(D_5^+)$ are much larger ($\sim 10^{-6} cm^3 s^{-1}$ at 95 K) in keeping with previous data [44,45] and with general expectations for polyatomic ions [46].

$\alpha_t(HCO^+)$ and $\alpha_t(N_2H^+)$ and their deuterated analogues were determined by adding small amounts of CO and N_2 respectively into the H_3^+(or D_3^+) afterglows whence rapid proton transfer reactions occurred generating HCO^+ and N_2H^+ respectively as the only ion species in the afterglow plasmas. Rapid recombination was immediately observed with the introduction of the CO or N_2 indicating that $\alpha_t(HCO^+)$ and $\alpha_t(N_2H^+)$ are much greater than $\alpha_t(H_3^+)$ (and similarly for the deuterated ions). The actual values obtained at 95 K are $\alpha_t(HCO^+) = 2.9 \times 10^{-7} cm^3 s^{-1}$, $\alpha_t(N_2H^+) = 4.9 \times 10^{-7} cm^3 s^{-1}$. $\alpha_t(DCO^+)$ and $\alpha_t(N_2D^+)$ were insignificantly different from the values for their hydrogenated analogues. $\alpha_t(HCO^+)$ has been measured at 200 and 300 K by other workers [47]; these low temperature FALP data are not inconsistent with expectations from these earlier data. Addition of CH_4 to the H_3^+ afterglow rapidly generated CH_5^+ ions and hence $\alpha_t(CH_5^+)$ was readily determined to be $1.5 \times 10^{-6} cm^3 s^{-1}$ at 95 K, which is quite typical of α_t for polyatomic ions previously measured, albeit at higher temperatures.

These FALP data are the first data obtained for α_t at such low temperatures. Much more still remains to be done. The FALP technique also offers the possibility of determining the products of dissociative recombination - a very worthwhile pursuit for the future.

5. Electron Attachment Studies

Electron attachment reactions have been studied for decades using a variety of techniques. Much of the early work has been reviewed in the books by Christophorou [48] and by Massey [49]. Since the publication of these books, many research papers have reported measurements of attachment coefficients (designated here as β in units of $cm^3 s^{-1}$; see the review by Christophorou et al.[50] and the papers by Smith et al.[51] and Alge et al.[52] for comprehensive references). Negative ion formation in electron-molecule interactions can occur via a two-body process in which fragmentation (dissociation) of the molecule occurs e.g.

$$CCl_4 + e \longrightarrow Cl^- + CCl_3 \qquad (3)$$

or via a three-body interaction generating the negative ion of the parent molecule e.g.

$$O_2 + O_2 + e \longrightarrow O_2^- + O_2 \qquad (4)$$

Reaction such as (4) usually occur at appreciable rates only at relatively high pressures, although there are exceptions to this (e.g. SF_6 attachment). Some two-body reactions such as (3) are exoergic (in which case they are loosely called thermal attachment reactions), although this does not mean that they are necessarily rapid at thermal energies because often activation energy

Table 1. Electron attachment coefficients, β(300 K), determined using the FALP at 300 K.

Molecule	CCl_4	SF_6*	CCl_3F	CH_3I	C_6F_6*	CH_2Br_2
β(300K)	3.9(-7)	3.1(-7)	2.6(-7)	1.2(-7)	1.1(-7)	9.3(-8)
Product Ion	Cl^-	SF_6^-	Cl^-	I^-	$C_6F_6^-$	Br^-
E_a (meV)	-	-	~20	-	-	~48

Molecule	C_7F_{14}	CF_3Br	$CHCl_3$	CCl_2F_2	Cl_2	CH_3Br
β(300 K)	8.1(-8)	1.6(-8)	4.4(-9)	3.2(-9)	2.0(-9)	6(-12)
Product Ion	$C_7F_{14}^-$	Br^-	Cl^-	Cl^-	Cl^-	Br^-
E_a(meV)	42	86	120	150	50	307

The measurements were made in helium carrier gas at a pressure of 0.6 Torr. The units of β are cm^3s^{-1} and, for example, 3.9(-7) = 3.9×10^{-7}. Reactions with the molecules marked with an asterisk are presumably saturated three-body attachment reactions since the parent ions are the observed products and so the β are the equivalent two-body attachment coefficients. All of the other reactions proceed via dissociative attachment giving the product ions indicated. The β values for several of the reactions increase with increasing temperature (see Fig.4) and then the activation energies, E_a (given in milli-electronvolts (meV)) have also been obtained.

barriers exist (see below). Endoergic (non-thermal) two-body attachment reactions are those which have threshold energies for dissociative electron attachment such as the $e + O_2 \rightarrow O^- + O$ reaction.

The variable-temperature FALP is very suitable for the study of thermal attachment reactions, that is for the determination of β coefficients and how they vary with temperature under truly thermal conditions. The product negative ions are also readily identified in the experiments. The principle of the method is straightforward; controlled amounts of the attaching gas are introduced into the thermalised He^+/Ar^+ afterglow (see previous section) and the density gradient of electrons ($\partial n_e/\partial z$) is monitored in the usual manner. Two practical points which must be recognised if accurate values of β are to be obtained are the following. Firstly, n_e must be sufficiently small so that any molecular positive ions formed in reactions of He^+ or Ar^+ with the attaching gas cannot recombine with electrons at a significant rate (and therefore contribute to $\partial n_e/\partial z$). Secondly, n_e must be much smaller than n_m, the number density of attaching molecules in the afterglow, so that n_m is invariant with z (i.e. $\partial n_m/\partial z = 0$), otherwise data interpretation would be prohibitively difficult. These constraints place a practical lower limit of $\sim 10^{-11}\,cm^3 s^{-1}$ on the value of β which can be accurately measured. When these potential difficulties are avoided, the loss of electrons is only via ambipolar diffusion and attachment, and then :

$$v_p \frac{\partial n_e}{\partial z} = D_a \nabla^2 n_e - \beta n_e n_m \qquad (5)$$

This equation cannot be solved analytically to give $n_e(z)$ because D_a is a function of n_-/n_e. However, Oskam [53] and Biondi [54] have shown how $n_e(z)$ can be obtained when negative ion formation is occurring. Thus :

$$n_e(z) = \frac{n_e(0)}{1 - \upsilon_D/\upsilon_a} \left[\exp\left(-\frac{\upsilon_a z}{v_p}\right) - \frac{\upsilon_D}{\upsilon_a} \exp\left(-\frac{\upsilon_D z}{v_p}\right)\right] \qquad (6)$$

where $\upsilon_D = D_{ae}/\Lambda^2$ (D_{ae} is the diffusion coefficient in the absence of negative ions, Λ is the characteristic diffusion length of the flow tube), $\upsilon_a = \beta n_m$, v_p is the plasma flow velocity and $n_e(0)$ is the electron density at the position of entry of the attaching gas into the plasma. When $\upsilon_D << \upsilon_a$, equation (6) reduces to a single exponential solution and the derivation of β from the $n_e(z)$ data is easy. In practice, however, this condition cannot be realised and so $n_e(z)$ is first determined in the absence of attaching gas to derive υ_D, $n_e(z)$ then determined for given values of n_m and then υ_a (and hence β) is obtained by computer fitting $n_e(z)$ according to equation (6).

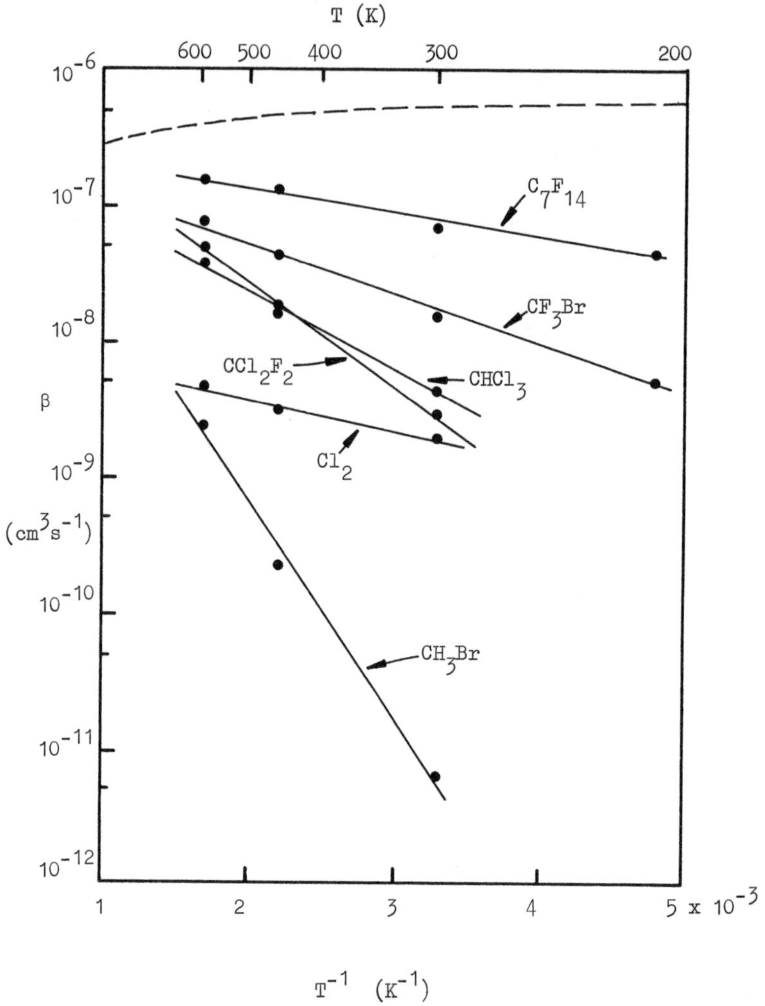

Fig.4 Arrhenius plots (ln β versus 1/T) derived from FALP data for the attachment reactions of electrons with C_7F_{14}, CF_3Br, $CHCl_3$, CCl_2F_2, Cl_2 and CH_3Br. The slopes of the lines provide values for the activation energies, E_a, for the reactions and these values are given in Table 1. The dashed line describes $\beta_{max} = 5 \times 10^{-7}(300/T)^{\frac{1}{2}} cm^3 s^{-1}$ which (following Warman and Sauer [55] gives the approximate upper limit values of β for any reaction (a more sophisticated treatment has been given by Klots [56], see text).

The twelve molecular gases included to date in FALP studies of electron attachment are given in Table 1 together with the measured β values at 300 K arranged in order of their magnitudes. They range from the very large $\beta(CCl_4) = 3.9 \times 10^{-7} cm^3 s^{-1}$ to that for $\beta(CH_3Br)$ which is some five orders-of-magnitude smaller. The measured $\beta(CCl_4)$ is near to the theoretical maximum for electron attachment which is given by Warman and Sauer [55] as $\beta_{max} \sim 5 \times 10^{-7}(300/T)^{\frac{1}{2}} cm^3 s^{-1}$, a result obtained from a consideration of the electron de Broglie wavelength. A more sophisticated theoretical treatment by Klots [56] yields a value for $\beta(CCl_4)$ of $3.29 \times 10^{-7} cm^3 s^{-1}$ at room temperature. $\beta(CCl_4)$ exhibits a weak inverse temperature dependence within the range 200 to 600 K(the temperature range over which most of these FALP attachment studies were carried out). However, the β values for the reactions involving several of the other molecular gases increased markedly with increasing temperature. This is attributed to energy barriers in these reactions and the associated activation energies, E_a, have been estimated from the slopes of the Arrhenius plots shown in Fig.4. The E_a values range from about 20 meV for the CCl_3F reaction to 307 meV for the CH_3Br reaction (see Table 1). The manifestation of the large value for $E_a(CH_3Br)$ is the rapid increase in $\beta(CH_3Br)$ with increasing temperature. The origin of activation energy barriers in these attachment reactions has been discussed by Wentworth et al.[57] and Christodoulides et al.[58].

To date, all except three of the reactions included in the FALP studies proceed via dissociative attachment with the production of atomic negative ions (see Table 1). The three exceptions are the SF_6, C_6F_6 and C_7F_{14}, reactions in which the parent negative ions SF_6^-, $C_6F_6^-$ and $C_7F_{14}^-$ respectively are the products. Presumably, therefore, these are three-body reactions involving first the binary attachment of an electron to the molecule followed by a collision of the excited negative ion with a helium carrier gas atom:

$$\text{e.g.} \quad SF_6 + e \longrightarrow (SF_6^-)^* \xrightarrow{+He} SF_6^- + He^* \qquad (7)$$

Actually, above 300 K, SF_5^- becomes an increasingly important product of the SF_6 reaction. Detailed studies of Fehsenfeld [59] have shown that E_a=0.43 eV for SF_5^- production.

The values of β derived from these studies are considered to accurate to ±15% within the range 200 to 400 K and ±20% at the higher temperatures (up to 600 K). Further details of the result of these studies are given elsewhere [51,52] together with comparisons of these FALP data with corresponding β and E_a values obtained from other studies. Very rarely have reliable β and E_a values been obtained previously in the same experiment. This is the great advantage of the FALP.

6. Ion-Ion Recombination Studies

Ion-Ion (or ionic) recombination describes reactions in which a positive ion and a negative ion neutralize each other. This can occur (i) in a two-body collision (binary mutual neutralization) such as

$$NO^+ + NO_2^- \longrightarrow NO + NO_2 \qquad (8)$$

which is usually considered to involve only electron transfer at a pseudo-crossing of potential curves [60] or (ii) in an interaction in which, prior to neutralization, collisions occur between the reactant ions and neutral third bodies (i.e. the bath gas atoms or molecules in which the reaction is occurring) resulting in the dissipation of energy as the ions accelerate in their mutual Coulombic field. This process is usually referred to as three-body ionic recombination [61,62] and is written as, for example,

$$NO^+ + NO_2^- + He \longrightarrow \text{products} + He \qquad (9)$$

This process (9) is less well defined than mutual neutralization and can involve ion-ion coalescence and chemical bond disruption and formation. A further process known as collision-enhanced mutual neutralization, which combines the essential features of (i) and (ii), has also been identified [63].

The FALP is essentially a low pressure experiment (in relation to ionic recombination) and so the most detailed FALP studies have been of the two-body process. The major motivation for such studies has been to provide critical data for tropospheric and stratospheric ion chemical models and to this end most of the initial FALP work was directed towards ionic recombination reactions involving atmospheric ions. Prior to the FALP work, no reliable data at thermal energies were available relating to this important class of ionic interactions, although data were available at higher energies from merged beam studies [60].

For FALP studies of ionic recombination it is necessary to create afterglow plasmas devoid of electrons and only containing the desired species of positive and negative ions. Then either $n_+(z)$ or $n_-(z)$ (or preferably both) are measured from which the ionic recombination coefficients, α_i, are determined in the same way as the electronic recombination coefficients, α_e, are determined (see equation (2) et seq.) i.e. from reciprocal n_- (or n_+) versus z/v_p plots. Sample plots for two reactions having very different α_i are shown in Fig.5. It is also essential (as in the electronic recombination studies) to ensure that n_+ and n_- are sufficiently large that ionic recombination dominates over ambipolar diffusion in the ion-ion plasmas. These ion-ion plasmas are created by adding sufficient quantities of appropriate electron attaching gases to the thermalised afterglows.

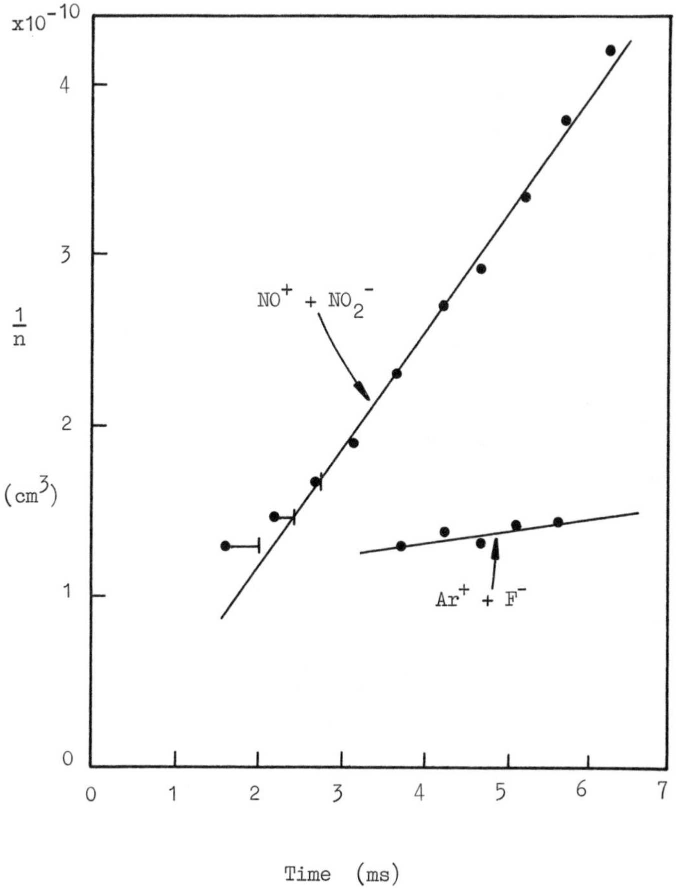

Fig. 5 Reciprocal ion density versus time (= z/v_p) plots obtained at 300 K using the FALP technique for $NO^+ + NO_2^-$ and $Ar^+ + F^-$ ion-ion plasmas. The data were obtained in argon carrier gas (rather than the usual helium carrier gas) in order to minimise loss of ions by ambipolar diffusion. The steeper slopes of the line relating to the molecular ion plasma results from ion-ion recombination and provides a value for the recombination coefficient, α_i for this reaction. This contrasts with the line of shallower slope for the atomic ion plasma from which only an upper limit to α_i for the $Ar^+ + F^-$ reaction can be obtained since the loss of ions is predominantly via ambipolar diffusion (see text and Table 2).

Familiarity with ion-molecule reactions and electron attachment reactions is invaluable in creating such plasmas, but also the mass spectrometer is essential to identify the positive and negative ions present in the plasmas. In order to determine n_+ or n_- from Langmuir probe data, it is necessary to know m_+ and m_- (the ionic masses as, if course, it is necessary to know m_e for n_e determinations). The essential details of these aspects of the technique are given in an early paper [64] and in a recent review [65].

Many ionic recombination reactions have been studied involving positive and negative ions of varying complexity, including reactions involving 'simple' molecular ions (e.g. NO^+, O_2^+, NO_2^-, NO_3^-)[66-69] 'cluster' ions (e.g. $H_3O^+.(H_2O)_n$, $NO_3^-.(HNO_3)_m$) [68,70-72] and atomic ions (e.g. Ar^+, Cl^-)[69]. Much attention has been given to reaction (8). The α_i for this reaction was first measured at 300 K [66] and then as a function of temperature which showed that it varied as $T^{-\frac{1}{2}}$ [71] in accordance with theoretical predictions [73]. Emission spectroscopy was then used to establish that the energy released in the reaction was largely converted into electronic excitation of the product NO [74].

A few representative α_i for reactions involving both 'simple' and 'cluster' ions are given in Table 2 where it can be seen that there is remarkably little variation of α_i with the complexity of the reaction molecular ions (see the recent review [65] for a detailed list) ranging only from(3-10) x $10^{-8}cm^3s^{-1}$ at 300 K. The most recent measurements of α_i have been for the relatively simple reactions $O_2^+ + Cl^-$ and $NO^+ + Cl^-$ (unpublished data). For these reactions, the α_i are the smallest measured to date for any reaction involving molecular ions. The magnitudes of α_i for reactions involving only atomic ions are, in general, much smaller because of the sparcity of available energy levels in the neutral atomic products into which the majority of the reaction energy must be deposited [75]. Indeed, the binary α_i expected for most of these reactions are too small to be measured using the FALP and, to date, only upper limits to α_i for several such reactions have been determined (see Fig.5 and the examples in Table 2).

The large majority of the ionic recombination reactions studied in the FALP have been carried out in helium carrier gas at pressures below one Torr. That the α_i are independent of pressure in this régime identifies the recombination as a two-body process. Two reactions (i.e. $NO^+ + NO_2^- \longrightarrow$ products and $SF_3^+ + SF_5^- \longrightarrow$ products) have been studied in helium up to pressures of eight Torr and the α_i are observed to increase with pressure more rapidly than expected on the basis of much higher pressure data [65,76]. This has been interpreted as a manifestation of the phenomenon of collision-

Table 2. Some examples of two-body ion-ion recombination coefficients, α_i, determined using the FALP at 300 K.

	Reaction	$\alpha_i \times 10^8 cm^3 s^{-1}$	$\overline{\sigma_i} \times 10^{12} cm^2$
(a)	$NO^+ + NO_2^-$	6.4	1.1
	$SF_5^+ + SF_6^-$	3.9	1.3
	$Cl_2^+ + Cl^-$	5.0	1.0
	$O_2^+ + CO_3^-$	9.5	1.7
(b)	$H_3O^+(H_2O)_3 + NO_3^-$	5.5	1.3
	$NH_4^+(NH_3)_2 + Cl^-$	7.9	1.4
	$H_3O^+(H_2O)_3 + NO_3^-.HNO_3$	5.7	1.5
(c)	$Ar^+ + F^-$	<0.2	<0.04
	$Xe^+ + Cl^-$	<0.5	<0.07

(a) one (or both) of the reactant ions is a 'simple' molecular ion, (b) one (or both) of the reactant ions is a 'cluster' ion, (c) both reactant ions are atomic. Comprehensive lists of α_i, values are given in Refs. 65 and 69. Note the very large mean thermal cross sections, $\overline{\sigma_i}$, for these reactions which have been determined by dividing α_i by the mean relative velocity between the reactants at 300 K. Note, also, the relative insensitivity of α_i (and especially $\overline{\sigma_i}$) to the complexity of the reactant molecular ions, and the relating small values of α_i and $\overline{\sigma_i}$ when both reactant ions are atomic.

enhanced binary recombination referred to earlier in this section, although much more work is required to substantiate this.

7. Summary

The exploitation of the variable temperature FALP apparatus is only just beginning. Data acquisition using a microcomputer is now fast and accurate and optical viewports have been included at several positions along the flow tube. Thus unique opportunities exist to study in detail electronic and ionic recombination, electron attachment, Penning processes etc. over a wide temperature range, including the determination of the rate coefficients for these processes and - via radiation emission studies- the states of excitation of the product ions or neutrals of some of these reactions. Also, a SIFT-type ion injector has now been added to the FALP thus providing the opportunity to inject mass-analysed ion beams into afterglow plasmas. If this can be achieved,it will enormously expand the range of ionic reaction which can be studied and will result in a further growth in the understanding of thermal energy reaction processes.

Acknowledgments

We gratefully acknowledge the major contribution made by Dr. Erich Alge to this work during the last three years.

References

1. McDaniel,E.W.,Mason,E.A.: The Mobility and Diffusion of Ions in Gases, New York: Wiley. 1973.
2. Ferguson,E.E.,Fehsenfeld,F.C.,Schmeltekopf,A.L.: Adv.Atom.Mol.Phys.5,1 (1969).
3. Smith,D.,Adams,N.G. In : Gas Phas Ion Chemistry, Vol.1 (Bowers,M.T.,Ed.), p.1 New York: Academic Press. 1979.
4. Adams,N.G., Church,M.J., Smith,D. : J.Phys.D.8, 1409 (1975).
5. Farragher, A.L. : Trans.Faraday Soc. 66, 1411 (1970).
6. Smith, D., Adams, N.G., Dean, A.G., Church, M.J. : J.Phys.D.8, 141 (1975).
7. Alge, E., Adams, N.G., Smith, D. : J.Phys. B. 16, 1433 (1983).
8. Swift, J.D., Schwar, M.J.R. : Electrical Probes for Plasma Diagnostics, London : Iliffe. 1970.
9. Goodall, C.V., Smith, D. : Plasma Phys. 10, 249 (1968).
10. Smith, D.,Goodall, C.V., Copsey, M.J. : J.Phys. B. Ser. 2,1, 660 (1968).
11. Smith, D., Plumb, I.C. : J. Phys. D. 5, 1226 (1972).
12. Smith, D. : Planet.Space Sci. 20, 1717 (1972).
13. Smith, D., Plumb, I.C. : J. Phys. D. 6, 196 (1973).
14. Smith, D., Church, M.J. : Int. J.Mass Spectrom. Ion Phys. 19, 185 (1976).
15. Smith, D., Church, M.J., Miller, T.M. : J.Chem. Phys. 68, 1224 (1978).
16. McDaniel, E.W. : Collision Phenomena in Ionized Gases, New York : Wiley. 1964.
17. Dean, A.G., Smith, D., Adams, N.G. : J.Phys. B. 7, 644 (1974).
18. Smith, D., Dean, A.G. : J.Phys. B. 8, 997 (1975).
19. Dean, A.G., Smith, D. : J.Atmos.Terres. Phys. 37, 1419 (1975).
20. Smith, D., Copsey, M.J. : J.Phys. B. Ser.2, 1, 650 (1968).

21. Biondi, M.A.In : Applied Atomic Collision Physics, Vol.3 (Massey,H.S.W., Bederson, B., McDaniel, E.W. Eds.) p.173, New York : Academic Press. 1982.
22. Smith, D., Adams, N.G.In : Topics in Current Chemistry, Vol.89 (Vepřek, S., Venugopalan, M.Eds.) p.1, Berlin : Springer-Verlag. 1980.
23. Smith, D., Adams, N.G.: Int. Revs. Phys.Chem. 1, 271 (1981)
24. Frommhold, L., Biondi, M.A., Mehr, F.J. : Phys.Rev. 165, 44 (1968).
25. Weller, C.S., Biondi, M.A. : Phys. Rev. 172, 198 (1968).
26. Walls, F.L., Dunn, G.H. : J.Geophys. Res. 79, 1911 (1974).
27. Heppner, R.A., Walls, F.L., Armstrong, W.T., Dunn, G.H. : Phys.Rev.A13, 1000 (1976).
28. Auerbach, D.,Cacak, R., Caudano, R., Gaily, T.D., Keyser, C.J., McGowan, J.Wm., Mitchell, J.B.A., Wilk, S.F.J. : J.Phys. B. 10, 3797 (1977).
29. Kasner, W.H., Biondi, M.A. : Phys. Rev. 174, 139 (1978).
30. Mehr, F.J., Biondi, M.A. : Phys.Rev. 181, 264 (1969).
31. Mul, P.M., McGowan, J.Wm. : J.Phys. B. 12, 1591 (1979).
32. Albritton, D.L. In : Kinetics of Ion-Molecule Reactions, (Ausloos,P.Ed.) p 119, New York : Plenum Press. 1979.
33. Lindinger, W., Howorka, F., Lukac, P., Kuhn, S., Villinger, H., Alge,E., Ramler, H. : Phys.Rev. A23, 2319 (1981).
34. Huang, C-M., Biondi, M.A., Johnsen, R. : Phys.Rev. A11, 901 (1975).
35. Torr, M.R., St.Maurice, J.P., Torr, D.G. : J.Geophys. Res.82, 3287(1977).
36. Huang, C-M., Biondi, M.A., Johnsen, R. : Phys. Rev. A14, 984 (1976).
37. Leu, M.T., Biondi, M.A., Johnsen, R. : A7, 292 (1973).
38. Mul, P.M., McGowan, J.Wm., Defrance, P., Mitchell, J.B.A. : J.Phys. B. 16, 3099 (1983).
39. Herbst, E., Klemperer, W. : Ap.J. 185, 505 (1973).
40. Dalgarno, A., Black, J.H. : Rept.Prog. Phys. 39, 573 (1976).
41. Adams, N.G., Smith, D., Alge, E. : in preparation.
42. Lindinger, W., Schmeltekopf, A.L., Fehsenfeld, F.C. : J.Chem.Phys. 61, 2890 (1974).
43. Albritton, D.L. : Atom.Data Nucl.Data Tables 22, 1 (1978).
44. Leu, M.T., Biondi, M.A., Johnsen, R. : Phys. Rev. A8, 413 (1973).
45. MacDonald, J.A., Biondi, M.A., Johnsen, R. : American Gaseous Electronics Conf. Dallas, Texas, 19-22 Oct. 1982.
46. Biondi, M.A. : Comments Atom.Molec. Phys. 4, 85 (1973).
47. Leu, M.T., Biondi, M.A., Johnsen, R. : Phys. Rev. A8, 420 (1973).
48. Christophorou, L.G. : Atomic and Molecular Radiation Physics, London : Wiley - Interscience. 1971.
49. Massey, H.S.W. : Negative Ions, 3rd Edition, Cambridge : Cambridge University Press. 1976.
50. Christophorou, L.G., James, D.R., Pai, R.Y. In : Applied Atomic Collision Physics, Vol.5 (Massey, H.S.W., McDaniel, E.W., Bederson, B. Eds.)p.87, New York : Academic Press. 1982.
51. Smith, D., Adams, N.G., Alge, E. : J.Phys. B. (1984) in press.
52. Alge, E., Adams, N.G., Smith, D. : in preparation.
53. Oskam, H.J. : Philips Res.Rept. 13, 335 (1958).
54. Biondi, M.A. : Phys. Rev. 109, 2005 (1958).
55. Warman, J.M., Sauer, M.C.Jr. : Int.J.Radiat.Phys.Chem. 3, 273 (1971).
56. Klots, C.E. : Chem. Phys. Lett. 38, 61 (1976).
57. Wentworth, W.E., George, R., Keith, H. : J.Chem.Phys. 51, 1791 (1969).
58. Christodoulides,A.A., Schumacher, R., Schindler, R.N. : J.Phys.Chem. 79, 1904 (1975).
59. Fehsenfeld, F.C. : J.Chem. Phys. 53, 2000 (1970).
60. Moseley, J.T., Olson, R.E., Peterson, J.R. In : Case Studies in Atomic Physics, Vol.5 (McDowell, M.R.C., McDaniel, E.W. Eds.) p.1, Amsterdam : North Holland Publ. Co. 1975.
61. Mahan, B.H. In : Advances in Chemical Physics, Vol.23 (Prigogine,I., Rice, S.A. Eds.) p.1, New York : Wiley. 1973.
62. Flannery, M.R. In : Applied Atomic Collision Physics Vol.3 (Massey,H.S.W., Bederson, B., McDaniel, E.W.Eds.) p.141, New York : Academic Press.1982.

63. Bates, D.R. : J.Phys. B. 14, 4207 (1981).
64. Smith, D., Dean, A.G., Adams, N.G. : J.Phys. D. 7, 1944. (1974).
65. Smith, D., Adams, N.G. In : Physics of Ion-Ion and Electron-Ion Collisions (Brouillard, F., McGowan, J.Wm.Eds.) p.501, New York :Plenum. 1983.
66. Smith, D., Church, M.J. : Int.J.Mass Spectrom. Ion Phys. 19, 185 (1976).
67. Church, M.J., Smith, D. : Int.J.Mass Spectrom.Ion Phys. 23, 137 (1977).
68. Smith, D., Church, M.J., Miller, T.M. : J.Chem.Phys. 68, 1224 (1978).
69. Church, M.J., Smith, D. : J.Phys. D. 11, 2199 (1978).
70. Smith, D., Adams, N.G., Church, M.J. : Planet.Space Sci.24, 697 (1976).
71. Smith, D., Church, M.J. : Planet.Space Sci. 25, 433 (1977).
72. Smith, D., Adams, N.G., Alge, E. : Planet.Space Sci. 29, 449 (1981).
73. Olson, R.E. : J.Chem. Phys. 56, 2979 (1972).
74. Smith, D., Adams, N.G., Church, M.J. : J.Phys. B. 11, 4041 (1978).
75. Olson, R.E. : Combustion and Flames 30, 243 (1977).
76. Smith, D., Adams, N.G. : Geophys. Res. Lett. 9, 1085 (1982).

Subject Index

Druck: Novographic, Ing. Wolfgang Schmid, A-1230 Wien.